Zoophysiology   Volume 35

Editors:
S.D. Bradshaw W. Burggren
H.C. Heller S. Ishii H. Langer
G. Neuweiler D.J. Randall

Springer
Berlin
Heidelberg
New York
Barcelona
Budapest
Hong Kong
London
Milan
Paris
Santa Clara
Singapore
Tokyo

Zoophysiology

Volumes already published in the series:

H. Kobayashi Y. Takei

The Renin-Angiotensin System

Comparative Aspects

With 40 Figures and 13 Tables

Springer

Professor Dr. HIDESHI KOBAYASHI
Professor Emeritus
University of Tokyo
Zenyaku Kogyo Co., Ltd.
2-33-7 Ohizumi-machi
Nerima-ku
Tokyo, Japan

Professor Dr. Yoshio Takei
Ocean Research Institute
University of Tokyo
1-15-1 Minamidai
Nakano-ku
Tokyo, Japan

ISBN-13: 978-3-642-64725-3   e-ISBN-13: 978-3-642-61164-3
DOI:10.1007/978-3-642-61164-3

Library of Congress Cataloging-in-Publication Data. Kobayashi, Hideshi. The renin-angiotensin system: comparative aspects/H. Kobayashi, Y. Takei. □ p. □ cm. – (Zoophysiology: v. 35) Includes bibliographical references and index. ISBN 978-3-642-64725-3   □ 1. Angiotensin – Physiology. 2. Renin – Physiology. 3. Physiology, Comparative. □ I. Takei, Y. (Yoshio), 1951- □. II. Title. III. Series. QP572.A54K63 1996 □ 596'.011 – dc20 96-18478

Cover design: Design & Production GmbH, Heidelberg

Typesetting: Scientific Publishing Services (P) Ltd, Madras

SPIN: 10063092   31/3137/SPS – 5 4 3 2 1 0 – Printed on acid-free paper

Dedicated to Mr. Hiroshi Hashimoto, Chairman of the Board, to Mr. Kazuhiro Hashimoto, President, to Ms. Ayako Sakai, Managing Director, of the Zenyaku Kogyo Company, with respect and gratitude, and to the memories of the late Professors Donald S. Farner and Kiyoshi Takewaki for their inspiration.

Preface

The renin-angiotensin system and the mechanisms regulating this system developed during the adaptive evolution of vertebrates, along with many other systems involved in the integrated survival of the organism. Because animal species have evolved from common ancestral populations, a basis for the comparison of body structures and physiological processes exists among animal groups belonging to different classifications. The comparative approach provides a better understanding of the structure and function of adaptive systems and facilitates the development of general principles governing these systems among animal groups; further, this approach reveals significant characteristics specific to certain animal groups. As the evolution of adaptation of animals to environmental conditions is explored, directions for future research are suggested. In this book, advances in research on the renin-angiotensin system are described with emphasis on the comparative aspects. However, since studies on the renin-angiotensin system of birds, reptiles, amphibians and fishes are limited compared with those conducted in mammals, in some chapters descriptions are concerned primarily with mammals. It has taken a long time to write this volume, and the topic is a broad one, with new data always emerging; therefore, certain aspects, and sometimes the most recent information, may not be included. Chapters 1-3 and sections 8.1-8.4, 8.6, 8.7 were written by H.K. ; Chapters 4-7 and Section 8.5 by Y.T. ; Chapter 9 was written by both authors.

We are indebted to Drs. H. Uemura, M. Nozaki, Y. Okawara, T. Karakida, Y. Tezuka, Messsrs. T. Tsukawara, N. Itazu, K. Hagiwara, K. Ichinohe, A. Komine, Y. Chiba and Mrs. M. Kurihara, who cooperated in conducting experiments on drinking behavior during the period when H.K. was Professor at Toho University, and to Misses J. Okubo and S. Nishida when Y.T. was at Kitasato University. We owe a special debt to Dr. K. Yamaguchi, Department of Physiology, Niigata University, who helped with the radioimmunoassay of angiotensin (ANG) II and supplied ANG II antibody, and to Professor K. Tanaka, Department of Agriculture, Gifu University, who cooperated in the radioimmunoassay of steroids.

We are grateful to the late Professor Donald S. Farner, then Coordinating Editor of *Zoophysiology*, for inviting us to write this volume and to Professor Susumu Ishii, recently appointed an Editorial Board member and Professor Howard A. Bern, University of California, for encouraging us to finish this book. Dr. Judith Turiel and Mrs. Barbara M. Licht were very helpful in editing the chapters written by H.K. The chapters written by Y.T. were critically read by Dr. Neil Hazon, Gatty Marine Laboratory, University of St. Andrews. We are grateful to them. Several people kindly lent original figures and tables, which are acknowledged in the appropriate place. We thank all these people and the publishers who gave permission to reproduce figures or tables from journals or books. Thanks are also due to Dr. M. Ohtomi, Misses Y. Osada and C. Yamada, and Messrs. K. Fujii, N. Ohta and K. Hasegawa for assistance in preparing the manuscript, figures and tables.

Finally, we are indebted to Mr. K. Hashimoto, the President, Zenyaku Kogyo Co., Ltd., for his unwavering encouragement and support, without which this work could not have been completed.

Tokyo, Summer 1996   *Hideshi Kobayashi*
                                                      Yoshio Takei

Contents

Chapter 1

Introduction

1.1 Evolutionary Aspects of the Renin-Angiotensin System (RAS)

In invertebrates, distribution of an ANG II-like substance has been studied immunocytochemically or chemically in four species: three annelids and one protochordate. In annelids, an ANG II-like substance was found immunocytochemically in the nervous system of the leech, *Theromyzon tessulatum*, (Verger-Bocquet et al. 1992; Salzet et al. 1992, 1993), and the earthworm, *Eisenia foetida*, (Ohta, Haruta, Kobayashi, unpubl. data; Fig. 1.1A,B). Recently, Salzet (1995) found, using a chemical method, ANG II amide in the nervous system of the leech, *Erpodella octoculata*. This peptide was diuretic in the leech. Laurent et al. (1995) identified chemically an ANG I-like substance in the nervous system of *Theromyzon tessulatum*, and suggested the presence of a complete RAS in the leech. It is worth noting that ANG II injection induces water absorption through the foot (contact-rehydration) in the terrestrial slug, *Limax maximus* (Makra and Prior 1985). In protochordates, this peptide was demonstrated immunocytochemically in the spinal cord of the amphioxus, *Branchiostoma belcheri* (Uemura et al. 1994; Fig. 1.1C). These findings suggest that biochemical processes generating ANG I- and ANG II-like substances are widely distributed among invertebrates, especially in the nervous system.

ANG II has also been demonstrated in the various regions of the vertebrate brain (see Ganten et al. 1982 for a book and Bunnemann et al. 1991, 1993 for reviews; see also Chap. 5). It seems, therefore, that an ANG II-like substance originally appeared in the nervous system of invertebrates, and this peptide was preserved during the evolutionary process from invertebrates to vertebrates. It is assumed, further, that an ANG II-like substance was primarily involved in nervous function and/or neurosecretion. The presence and function of ANG II in the vertebrate brain are considered evolutionarily as original features compared to those of circulating ANG II, although brain ANG II was discovered more recently than circulating ANG II.

In invertebrates, a biochemical process generating an ANG II-like substance has not been studied, except for recent preliminary studies on the leech (Salzet et al. 1995; Laurent et al. 1995). However, in vertebrates, this process has been investigated extensively with circulating ANG II and is now designated as the renin–angiotensin system (RAS) (Fig. 1.2). The RAS consists of renin produced by the juxtaglomerular cells, angiotensinogen produced by the liver and angiotensin converting enzyme present mostly in the lung endothelium. The renin and angiotensinogen are released into the general circulation, and ANG I

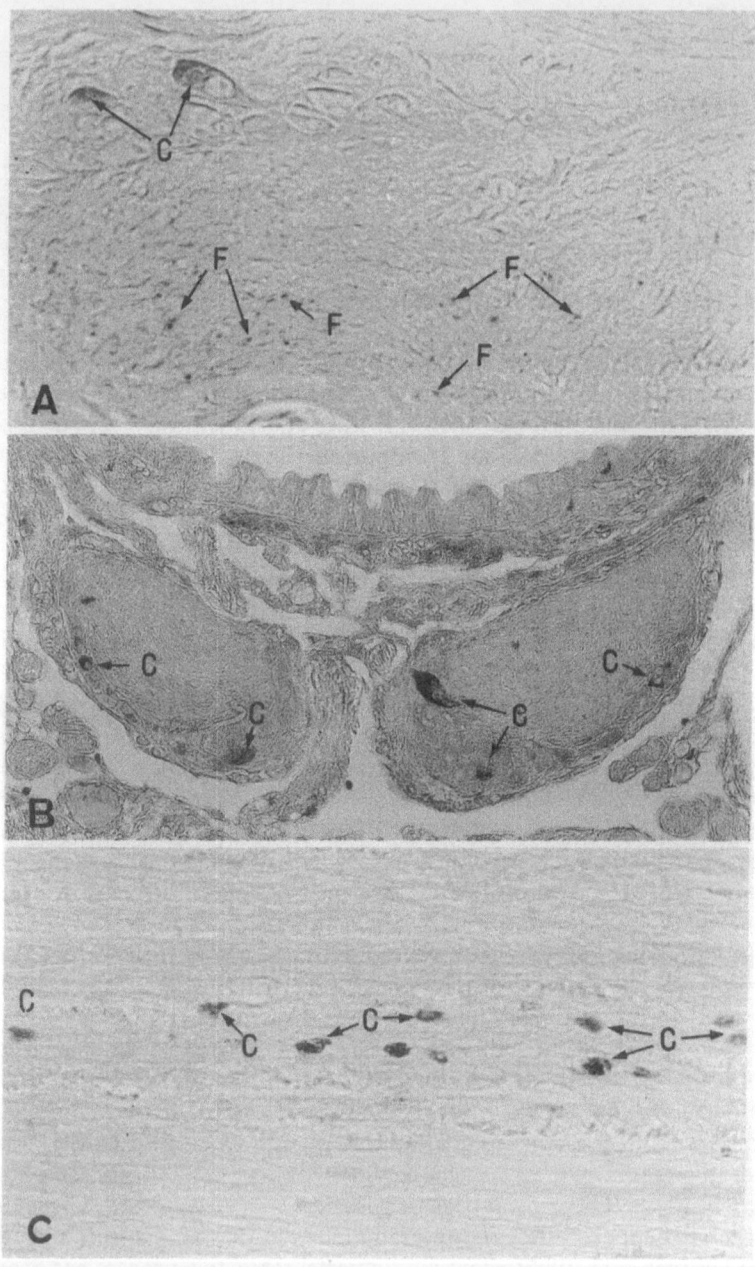

Fig. 1.1. A Horizontal section of the brain of the earthworm, *Eisenia foetida*; × 260. B Frontal section of the anterior portion of the subesophageal ganglion of *Eisenia foetida*; × 210. C Parasagittal section of the posterior portion of the spinal cord of the amphioxus, *Brachiostoma belcheri*; × 350. *C* Cells containing immunoreactive ANG II-like substance; *F* fibers containing immunoreactive ANG II-like substance

2

ANGIOTENSINOGEN

Asp-Arg-Val-Tyr-Ile-His-Pro-Phe-His-Leu⌐Val-Ile-His-R

    Renin substrate, a glycoprotein (m.w. about 58,000)
    produced by the liver and contained in the plasma.

    ←——————— Renin, an enzyme produced by the
                  juxtaglomerular cells and released
                  into the general circulation.

ANGIOTENSIN I (ANG I)

Asp-Arg-Val-Tyr-Ile-His-Pro-Phe⌐His-Leu

    ←——————— Converting enzyme. Conversion from
                  ANG I to ANG II takes place mostly
                  in the vascular epithelium of the
                  lung when ANG I passes through the
                  pulmonary circulation. The enzyme is
                  present in the lung, kidney, brain
                  and other organs.

ANGIOTENSIN II (ANG II)

Asp⌐Arg-Val-Tyr-Ile-His-Pro-Phe

    ←——————— Aminopeptidase

ANGIOTENSIN III (ANG III)

    Arg-Val-Tyr-Ile-His-Pro-Phe

    ←——————— Angiotensinases

Peptide fragments

Fig. 1.2. The human circulating RAS. Not illustrated are the pathway from ANG I to des-Asp[1]-ANG I by aminopeptidase and then to des-Asp[1]-ANG II (ANG III) by converting enzyme, nor the pathway from angiotensinogen directly to ANG II by tonin (Schiffrin and Genest 1983). *Arrowheads* show the peptide bonds cleaved by renin, converting enzyme or aminopeptidase (see Fig. 8.9). Other biologically active ANG II fragments such as ANG-(1–7) are discussed in Sections 4.4 and 8.3.3).

is produced in the blood. ANG I is converted into ANG II, a main final active substance, in the pulmonary capillary bed (Figs. 1.2, 2.1). Thus, ANG II circulates systemically to reach the various target tissues or organs, resulting in induction of a wide spectrum of physiological functions.

    As mentioned above, not only ANG II but also the RAS was identified in the vertebrate brain. Further, the RAS itself was recently found in the heart, kidney, adrenal, pituitary, ovary, testis and other organs of vertebrates (see

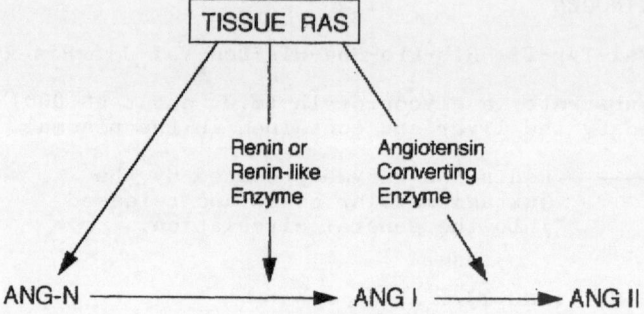

Fig. 1.3. The tissue RAS. ANG II produced by tissue RAS may act locally in a paracrine or autocrine fashion. ANG III is produced by aminopeptidase A from ANG II. *ANG-N* Angiotensinogen.

Campbell 1987b; Phillips et al. 1993 for reviews; Chap. 4, Table 4.1). The RAS present in tissues is called tissue RAS in contrast to circulating RAS. In these tissues or organs, products of the tissue RAS, ANG II or its derivatives (Fig. 1.3), seem to act locally through an autocrine or paracrine interaction. Thus, there are two ways for the RAS to function: one systemically and the other locally.

Since an ANG II-like substance is present mainly in the nervous tissue in invertebrates, it is assumed that the primary action of an ANG II-like substance is associated with nervous functions and/or neurosecretion in the brain. Therefore, the classical circulating RAS and the tissue RAS in the various tissues, except for the brain, are considered to be features that were acquired secondarily, during the evolutionary process from invertebrates to vertebrates, as body organization and adaptive physiological functions became more complex.

In invertebrates, as mentioned above, ANG II amide found in the nervous system is diuretic in the leech, and injection of ANG II induces water absorption from the foot in the terrestrial slug. Therefore, it seems that ANG II produced in the nervous system is involved in body fluid regulation as neurosecretory products. In vertebrates, ANG II produced by the brain RAS stimulates vasopressin release (Sect. 8.6.1) and drinking, except in amphbians (Sect. 8.2), and elevates blood pressure (Sect. 8.8). Circulating ANG II has similar functions. It seems that the primary function of ANG II is the regulation of body fluid homeostasis. Thus, involvement of ANG II in water balance of the body has a phylogenetic significance. Among actions of ANG II for body fluid homeotasis, the participation of ANG II in cardiovascular regulation has greatly increased during the evolutionary process to enhance locomotor activity by delivering fresh blood and nourishment to body tissues for two indispensable phenomena: hunting other animals for food and escaping from enemies. The distribution of ANG-like substances, their generating process and physiological actions should be studied further in invertebrates.

1.2 Historical Background of RAS

In the latter half of 1800s, Brown-Séquard presented an idea that various organs produce a substance (or substances) which might significantly contribute to the general activity of the body. Brown-Séquard and d'Arsonval (1892) studied whether the kidney influenced some phenomena in the body through "inner secretion" in addition to its excretory function. They injected, subcutaneously or intravenously, kidney extracts which were prepared in the same way as "liquide orchitique" into nephrectomized rabbits and guinea pigs, and found that the recipients maintained a remarkably better condition and lived longer than those without injections. These experiments and similar clinical observations by Brown-Séquard stimulated Tigerstedt to start studying the effects of kidney extracts on circulatory organs. Until then, only one study had been done (Oliver 1897), showing no consistent effects with kidney extracts upon the peripheral vessels. Tigerstedt and Bergman (1898) administered, intravenously, saline extracts from fresh rabbit kidneys into other rabbits and found a prolonged rise in arterial blood pressure. They named this pressor substance "renin", and regarded it as an internal secretion of the kidney. Further, they observed a moderate rise in blood pressure after injection of blood from the renal vein into the vessels of rabbits which had undergone bilateral nephrectomy.

These experiments were repeated by Lewandowsky (1899). He observed a transient pressor effect, but concluded that the effect was due to some pressor substance(s) in defibrinated blood and not to an internal secretion of the kidney. Lewandowsky's conclusions appeared to contradict that of Tigerstedt. Negative observations were obtained by Pearce (1909) who reported that injection of saline extracts of rabbit, dog and cat kidneys did not result in significant pressor responses. Hartwich and Hessel (1932) obtained a pressor response from autolyzed renal press juice, but they concluded that the pressor substance(s) were probably amines produced by decomposition. On the other hand, a number of reports were consistent with the findings of Tigerstedt and Bergman (Livon 1898; Vincent and Sheen 1903; Shaw et al. 1906; Bingel and Strauss 1909; Bingel and Claus 1910; Cash 1924). Thus, until around the early 1930s, there were confusing data and considerable doubt regarding the presence of the "renin" proposed by Tigerstedt and Bergman (1898).

Meanwhile, Goldblatt et al. (1934) rekindled interest in the role of "renin" in the physiological control of blood pressure. They found arterio- and arteriolosclerosis in the kidneys of hypertensive patients, and hypothesized that the stenosing effect of the sclerosis might reduce blood flow through the kidneys, causing hypertension. Goldblatt duplicated this condition experimentally by clamping renal arteries in dogs: renal artery constriction produced elevated blood pressure. Further, he found that complete sympathectomy and destruction of the spinal cord could not prevent the elevation of blood pressure following renal artery constriction; ligation of the renal veins, however, did prevent hypertension. Thus, Goldblatt showed that hypertension possessed a humoral basis and that humoral substance(s) were released from the kidney (see Goldblatt 1947, 1948). His important studies stimulated other investigators to search for the humoral substance. Prinzmetal and Friedman (1936) and Harrison et al. (1936–1937) observed that saline extracts of the ischaemic kidney of dogs produced greater rises in blood pressure than did saline extracts of normal kidney. Pickering and Prinzmetal (1938) confirmed the results of Tigerstedt and Bergman (1898) in rabbits, and showed that the active substance "renin" was present in the cortex but not in the medulla of the kidney.

Helmer and Page (1939) prepared a "renin" solution from the kidney cortex of pigs and found that their purified "renin" preparation needed to be mixed with a protein-like substance in plasma to reveal its vasoconstrictor activity. They suggested that "renin" was probably an enzyme, not a direct pressor substance. The factor which reacted with "renin" was initially designated "renin activator"; later they renamed it "renin substrate". Page and Helmer (1940) extracted a potent vasoconstrictor and pressor substance from plasma resulting from the action in vitro of renin. They named this substance "angiotonin".

In 1923, Houssay in Buenos Aires became interested in hypertension, motivated by the death of his most brilliant student who died of malignant hypertension at the age of 33 years (cited in Helmer 1962). Houssay's group observed that extirpation of the ischaemic kidney or discontinuation of arterial compression resulted in a return of blood pressure to normal levels (Fasciolo et al. 1938). They transplanted the ischaemic kidneys of dogs with chronic hypertension into the neck of normal of nephrectomized dogs and observed an immediate increase of blood pressure, whereas no change was observed after grafting normal kidneys. They thus demonstrated that a humoral mechanism was involved in the type of experimental hypertension produced by Goldblatt and his coworkers. Houssay and Taquini (1938) and Fasciolo et al. (1938) found that plasma from the venous blood of ischaemic kidneys contained an active substance inducing vasoconstriction. Houssay asked Braun-Menéndez, Fasciolo and Leloir to isolate this substance. Braun-Menéndez et al. (1940a) found "renin" to be an enzyme which produced a pressor substance in vitro when incubated with blood plasma. They named this substance "hypertensine" (Braun-Menéndez et al. 1940b). For the substrate in the plasma, they used the name "hypertensinogen". As mentioned earlier, Page and Helmer (1940) in America had named the vasoconstrictor substance "angiotonin" while Muñoz et al. (1939) and Braun-Menéndez et al. (1940b) in Argentina called it "hypertensin". In 1958, Braun-Menéndez and Page agreed to the name "angiotensin" for the new substance, "angiotensinogen" for the substrate and "angiotensinase" for enzymes which hydrolyze angiotensins.

In the 1950s, isolation, purification and analysis of the chemical structure of angiotensin were carried out in several laboratories. Synthesis of angiotensin was finally accomplished by Bumpus et al. (1957) and Rittel et al. (1957). Since then, the chemical structures of renin, angiotensinogen and the angiotensin II-generating system have been studied extensively (Fig. 1.2), and the receptors and angiotensinases have been characterized (see Skeggs et al. 1980; Haber and Carlson 1983; see Robertson and Nicholls 1993 for reviews).

6

Recently, tissue RAS was found in the brain, ovary, pituitary, adrenal, vascular tissues and other organs in mammals (see Campbell 1987a,b; Phillips et al. 1993 for reviews; see also Chap. 4). The autocrine or paracrine actions of ANG II generated in these tissues and organs are being studied. Investigations of the renin–angiotensin system were initiated with a predominantly medical emphasis using mammalian species; not until the late 1960s did comparative studies on this system begin in lower vertebrates. It is very recent that studies of ANG II-like substances have been started in invertebrates (Verger-Bocquet et al. 1992; Salzet et al. 1992, 1993, 1995; Uemura et al. 1994; Laurent et al. 1995).

1.3 Survey of Books and Reviews

An excellent review of the structure of the juxtaglomerular apparatus, written by Barajas and Müller, appears in the book, *The Renin–Angiotensin System*, edited by Johnson and Anderson 1980). The functional anatomy of the juxtaglomerular apparatus is described in a recent monograph, *The Juxtaglomerular Apparatus*, by Taugner and Hackenthal (1989). Observations in this monograph, based on light and electron microscopy, are mostly on mammals. There is one small chapter on phylogeny and ontogeny. Persson and Boberg (1988) edited a book, *The Juxtaglomerular Apparatus*, concerned primarily with the structure of the juxtaglomerular apparatus and the tubuloglomerular feedback control mechanism in mammals. Reviews of comparative studies of the juxtaglomerular apparatus are limited, but some older reviews are still useful, including those by Sokabe and Ogawa (1974), Nishimura (1980a,b) and Wilson (1984a). Recently, Henderson and Deacon (1993) described concisely the phylogeny and comparative physiology of the RAS with beautiful illustrations. Dantzler (1989) has written about comparative renal anatomy and function in nonmammalian vertebrates in his book, *Comparative Physiology of the Vertebrate Kidney*.

There are several books concerning thirst, including Peters et al. (1975), Fitzsimons (1979), Rolls and Rolls (1982), deCaro et al. (1986), Grossman (1990) and Ramsay and Booth (1991). Chapters dealing with thirst in connection with the RAS appear in books by Ganten et al. (1982), Harding et al. (1988) and Fitzsimons (1993). The organization of ANG II immunoreactive cells and fibres is documented in detail for the rat brain (Lind et al. 1985a,b). The role of the RAS in sodium appetite was reviewed in books written by Fitzsimons (1979) and Denton (1982) and more recently reviewed by Epstein and his colleagues (Yang and Epstein 1991) and Fitzsimons (1993). Other biological actions, such as effects on hormone release, are comprehensively reviewed by Saavedra (1992) and Wright and Harding (1992) and mentioned in several chapters in a book edited by Robertson and Nicholls (1993). The local action of the tissue RAS is a recent topic of study and has been reviewed by Campbell (1987a,b), Dzau (1988), Phillip et al. (1993) and by others in Robertson and Nicholls (1993).

The biochemistry of the RAS reviewed by Peach in 1977 is still the most comprehensive in this field. Recent biochemical data on the RAS and ANG II receptors are reviewed by several experts in the book (Vol. 1) edited by Robertson and Nicholls (1993). They have edited two books entitled *The Renin-Angiotensin System*: Vol. 1 covers biochemistry and physiology and Vol. 2, pathophysiology and therapeutics.

The mechanism regulating renin release has been extensively described in reviews by Davis and Freeman (1976), Keeton and Campbell (1981) and Gibbons et al. (1984). The distribution of ANG II receptors is detailed in reviews by Mendelsohn (1985) and Saavedra et al. (1986b), and the subcellular mechanism of ANG II actions is reviewed by Smith (1986). A recent review by Smith et al. (1992) on specific receptor antagonists is a useful source of information on the biochemical characterization of ANG II receptors.

Survey of Books and Reviews

Buckley J, Ferrario CM (1977) Central actions of angiotensin and related hormones. Pergamon Press, New York

Campbell DJ (1987) Tissue renin-angiotensin system: sites of angiotensin formation. J Cardiovasc Pharmacol 10: S1–S8

Dantzler WH (1989) Comparative physiology of the vertebrate kidney. Springer, Berlin Heidelberg New York

Davis JO, Freeman RH (1976) Mechanisms regulating renin release. Physiol Rev 56: 1–56

deCaro G, Epstein AN, Massi M (1986) The physiology of thirst and sodium appetite. NATO ASI Ser, Ser A: Life Sci Vol 105. Plenum Press, New York

Denton D (1982) The hunger for salt. Springer, Berlin Heidelberg New York

Dzau VJ (1988) Circulating versus local renin-angiotensin system in cardiovascular homeostasis. Circulation 77: I4–I13

Fitzsimons JT (1979) The physiology of thirst and sodium appetite. Cambridge Univ Press, Cambridge

Fitzsimons JT (1993) Renin in thirst and sodium appetite. In: Robertson JIS, Nicholls MG (eds) The renin-angiotensin system, vol. 1. Gower Medical Publishing, London, pp 32.1–32.8

Ganten D, Printz M, Phillips MI, Scholkens BA (1982) The renin angiotensin system in the brain. Springer, Berlin Heidelberg New York

Gibbons GH, Dzau VJ, Farhi ER, Barger AC (1984) Interaction of signals influencing renin release. Annu Rev Physiol 46: 292–308

Grossman SP (1990) Thirst and sodium appetite. Physiological basis. Academic Press, San Diego

Harding JW, Wright JW, Speth RC, Barnes CD (1988) Angiotensin and blood pressure regulation. Academic Press, San Diego

Henderson IW, Deacon CF (1993) Phylogeny and comparative physiology of the renin-angiotensin system. In: Robertson JIS, Nicolls MG (eds) The renin-angiotensin system, vol. 1. Gower Medical Publishing, London, pp 2.1–2.28

Johnson JA, Anderson RR (1980) The renin-angiotensin system. Plenum Press, New York

Lind RW, Swanson LW, Ganten D (1985) Organization of angiotensin II immunoreactive cells and fibers in the rat central nervous system. Neuroendocrinology 40: 2–24

Mendelsohn FAO (1985) Localization and properties of angiotensin receptors. J Hypertension 3: 307–316

Nishimura H (1980a) Evolution of the renin-angiotensin system. In: Pang PKT, Epple A (eds) Evolution of vertebrate endocrine systems. Graduate Studies Texas Tech Univ, Texas Tech Press, Lubbock, pp 373–404

Nishimura H (1980b) Comparative endocrinology of renin and angiotensin. In: Johnson JA, Anderson RR (eds) The renin-angiotensin system. Plenum Press, New York, pp 29–77

Peach MJ (1977) Renin-angiotensin system: biochemistry and mechanisms of action. Physiol Rev 57: 313–370

Persson AEG, Boberg U (1988) The juxtaglomerular apparatus. Elsevier, Amsterdam

Peters G, Fitzsimons JT, Peters-Haefeli L (1975) Control mechanisms of drinking. Springer, Berlin Heidelberg New York

Phillips MI, Speakman EA, Kimura B (1993) Tissue renin-angiotensin system. In: Raizada MK, Phillips MI, Summers C (eds) Cellular and molecular biology of the renin-angiotensin system. CRC Press, Boca Raton, pp 97–130

Ramsay DJ, Booth DA (1991) Thirst. Springer, Berlin Heidelberg New York

Robertson JIS, Nicholls MG (1993) The renin-angiotensin system, vols. 1 and 2. Gower Medical Publishing, London

Rolls BJ, Rolls ET (1982) Thirst. Cambridge Univ Press, Cambridge

Saavedra JM (1992) Brain and pituitary angiotensin. Endocrine Rev 13: 329–380

Saavedra JM, Israel A, Plunkett LM, Kurihara M, Shigematsu K, Correa FMA (1986) Quantitative distribution of angiotensin II binding sites in rat brain by autoradiography. Peptide 7: 679–687

Smith JB (1986) Angiotensin-receptor signaling in cultured vascular smooth muscle cells. Am J Physiol 250: F759–F769

Smith RD, Chiu AT, Wong PC, Herblin WF, Timmermans PBMWM (1992) Pharmacology of nonpeptide angiotensin II receptor antagonists. Annu Rev Pharmacol Toxicol 32: 135–165

Sokabe H, Ogawa M (1974) Comparative studies of the juxtaglomerular apparatus. Int Rev Cytol 37: 271–327

Taugner R, Hackenthal E (1989) The juxtaglomerular apparatus. Springer, Berlin Heidelberg New York

Wilson JX (1984) The renin-angiotensin system in nonmammalian vertebrates. Endocrine Rev 5: 45–61

Wright JW, Harding JW (1992) Regulatory role of brain angiotensins in the control of physiological and behavioral responses. Brain Res Rev 17: 227–262

Yang ZF, Epstein AN (1991) Blood-borne and cerebral angiotensin and the genesis of salt intake. Horm Behav 25: 461–476

Chapter 2

Comparative Morphology
of the Juxtaglomerular Apparatus (JGA)

The structure of the JGA is different among the classes of vertebrates (Table 2.1). In order to obtain a general model for the function of the JGA, phylogenetic studies are indispensable. Since studies of this apparatus have been generated by clinical interests, most investigations have been carried out in mammals (see Persson and Boberg 1988; Taugner and Hackenthal 1989 for books); studies of the nonmammalian JGA are rather limited. However, some excellent reviews of the comparative anatomy of the apparatus have been published by Sokabe et al. (1969), Capelli et al. (1970), Ogawa et al. (1972), Sokabe and Ogawa (1974), Ogawa (1977), and Nishimura (1980a,b, 1985). Although these reviews were written some 20 years ago and are now classical, they are still useful. Recently, Henderson and Deacon (1993) described the phylogeny and comparative physiology of the renin–angiotensin system.

2.1 Mammals

2.1.1 Components of JGA

In 1925, Ruyter first reported that the smooth muscle cells of the wall of the renal afferent arterioles of the glomeruli differentiate into granulated cells in mice and rats (Fig. 2.1). He named these "epithelioid cells" and thought they play a role in the local regulation of blood flow in the glomerulus. However, he could not find these cells in man, monkey, dog, cat, rabbit, and guinea pig. Oberling (1927) independently found granulated epithelioid cells in the tunica media of the afferent arteriole of human kidney, and he thought that they are involved in the regulation of glomerular circulation and related to hypertension. Okkels (1929) described these cells in a frog. Goormaghtigh (1932) described two groups of cells in the vascular pole region of the glomeruli in the kidneys of man, cat and rabbit. The cells of one group are located in the media of the afferent arterioles close to the glomeruli. These modified myocytes, devoid of myofibrils and containing granules, were called afibrillar cells; they are now called juxtaglomerular cells (JG cells). The other group consists of small flattened agranular cells with clear cytoplasm and small nuclei. They are located near the distal tubular portion in close contact with the vascular pole. Goormaghtigh compared these cells to the tactile corpuscles of the skin and to

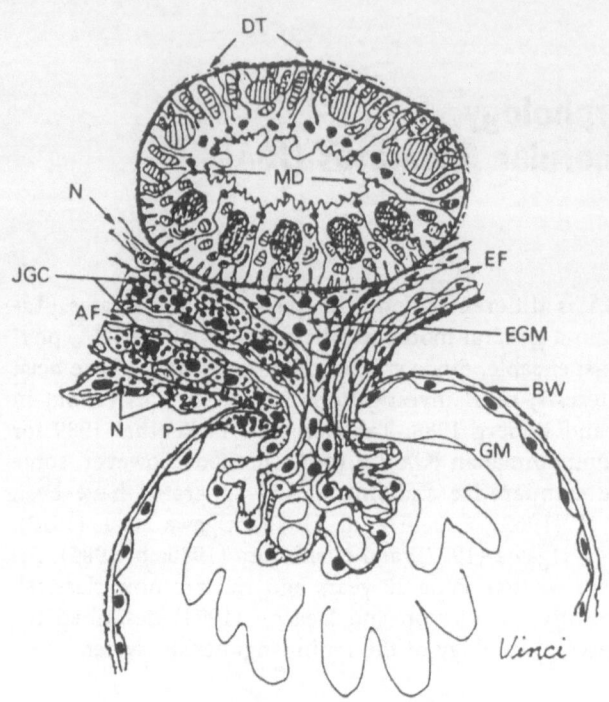

Fig. 2.1. Schematic drawing of the mammalian juxtaglomerular apparatus. *AF* Afferent arteriole; *BW* Bowman's capsule; *DT* distal tubule; *EF* efferent arteriole; *EGM* extraglomerular mesangium; *GM* glomerular mesangium; *JGC* juxtaglomerular cell; *MD* macula densa; *N* nerve terminal, *P* peripolar cell, added by the present authors to the original figure with permission. (Mizuhira 1986)

the Schwannian elements, giving rise to the term "pseudo-Meissnerian cells". Later, these cells were named lacis cells (Oberling and Hatt 1960a) or extra-glomerular mesangium (EGM) (Barajas 1970), terms now commonly used. The term "Polkissen" (Zimmermann 1933) included both the dense aggregate of small cells situated between the arterioles and the modified myocytes of the afferent arteriole. Therefore, "Polkissen" originally included the JG cells and EGM. However, the term is now used synonymously with EGM.

Peter (1970) observed, in rabbit, cat and man, that a portion of the distal tubule that runs toward the parent glomerulus comes in contact with the vascular component. The tubular cells on one side, toward the glomerulus, are taller and their nuclei lie closer together than the cells on the opposite side. Earlier, Zimmermann (1933) had made similar observations and called these taller cells "macula densa" (MD).

Confusion over naming the JG tissues has been resolved. Investigators now agree that the JGA of the kidney in mammals is composed of the afferent arterioles including JG cells and efferent arterioles, and of EGM and MD cells of the convoluted distal tubule (Barajas and Latta 1967; Hatt 1967; Barajas 1970:

Table 2.1. Phylogenetic distribution of JGA components, renin activity, angiotensins and peripolar cells

	JG cells	Renin activity	ANG	MD	EGM	Peripolar cells
Mammals	+	+	+	+	+	+
Birds	+	+	+	+	+	+
Reptiles	+	+	+	+[d]	−	+
Amphibians	+	+	+	+[e]	−	+
Sarcopterygains	+	+		−	−	
Teleosts	+	+	+	−	−	+
Holosteans	+	+		−	−	
Chondrosteans	−	+[a]		−	−	
Holocephalans	+	+		−	−	
Elasmobranchs	+	+	+	+	+	+
Cyclostomes	−	−	+	−	−	−
Cephalochordates			+[b]			
Annelids			+[b, c]			

[a] Renal renin-like activity was detected in *Polypterus*, but uncertain in *Acipenser* (Sect. 2.5.4).
[b] ANG II-like substance was detected immunohistechemically in the amphioxus and the earthworm (Sect. 1.1).
[c] ANG II amide and an ANG I-like molecule were identified chemically in the leech (Sect. 1.1).
[d] A primitive form of MD was found in one species (Sect. 2.3).
[e] An MD-like structure was found in a few species (Sect. 2.4).

Sokabe 1974; Sokabe and Ogawa 1974; Barajas and Müller 1980; Nishimura 1980a,b).

Apart from the known components of the JGA, Ryan et al. (1979) found granulated "peripolar cells" in the sheep JG region (Figs. 2.1, 2.7). These secretory cells are located on each side of the origin of the glomerular tuft at the junction between the parietal and podocytic epithelium. The presence of peripolar cells has been confirmed in the axolotl and toad (Hanner and Ryan 1980), rats (Ryan et al. 1982) and humans (Gardiner and Lindop 1985; Gardiner et al. 1986). It is not known whether the peripolar cells are related to the RAS.

2.1.2 Juxtaglomerular (JG) Cells in Afferent Arterioles

The JG cells originate from smooth muscle cells of the afferent arterioles (Ruyter 1925). The presence of secretory granules in these cells indicates their endocrine nature, and electron microscope studies show that they occasionally retain attachment bodies and myofibrils (see Barajas and Müller 1980). Thus, they have often been called epithelioid or myoepithelioid cells.

The granules in the JG cells can be stained by a variety of methods (see Sokabe and Ogawa 1974). McManus (1948) showed that the granules of human JG cells became bright red in sections treated with the periodic acid–Shiff (PAS) reaction. Wilson (1952) obtained striking results using Bowie's (1935–1936) ethyl violent Biebrich scarlet modification of Benslau's neutral gentian method, which was used for granules in pancreatic acini and islet cells. The granules in JG cells were stained a deep purple and could be detected even under lower powers of the microscope (Fig. 2.2). Since Bowie's method gives reliable results in mammalian JG cells, it has been widely used to determine quantitatively the degree of granularity of the JG cells in experimental animals and embryos (see Barajas and Latta 1967; Barajas and Müller 1980). Biava and West (1966) warned that lipofuscin-like granules were occasionally observed in the JG cells and that these granules also stained with Bowie's method in the rat, mouse and man. However, Barajas and Müller (1980) stated that these granules are not stained with Bowie's method and that, therefore lipofuscin granules most likely do not interfere with the quantitative evaluation of granularity using Bowie's stain. The degree of staining of JG granules varies with different fixatives (Oguri et al. 1969) and between animal species of different classes (Sokabe et al. 1969).

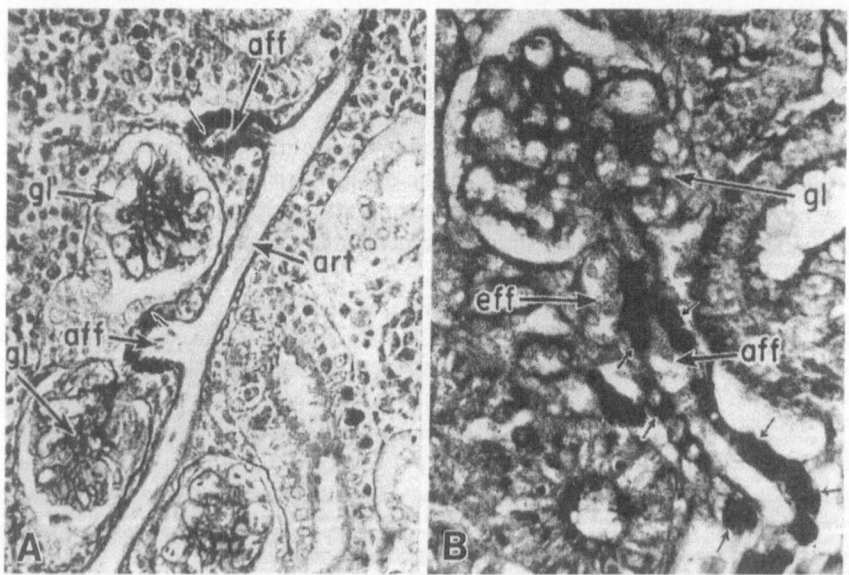

Fig. 2.2A,B. Kidney of the teleost, *Atherinopsis californiensis*, showing the renal artery (*art*), afferent arteriole (*aff*), glomerulus (*gl*). Note the JG cells (*small arrows*) stained darkly with Bowie's stain on the afferent arterioles (type 1 in the text). A × 450; B × 650. (Krishnamurthy and Bern 1969)

As mentioned previously, renin was first observed as a renal pressor factor in the rabbit by Tigerstedt and Bergman (1898). Later renin was found to be an enzyme which reacts with angiotensinogen in plasma, resulting in production of active peptides: angiotensins. The bioassay of renin consists of determining the pressor activity of incubates of tissue extracts with homologous plasma in vitro. Since there is species specificity in the renin–angiotensinogen reaction, homologous plasma should be used as a source of angiotensinogen, especially when tissue extracts of nonmammalian origin are tested. Animals commonly used for blood pressure measurement are the rat, dog, cat or rabbit. (see Sokabe and Ogawa 1974 for details).

To localize renin in the JGA, numerous studies were conducted using different techniques: (1) histological or histochemical examinations of granulation in JG cells stained mostly with Bowie's stain following various experimental procedures; and (2) examination of the correlation between the renin content of the kidney, measured by blood pressure, and granulation in the JG cells following injection of kidney extracts of animals under various experimental conditions (see Bing and Kazimierczak 1963; Cook 1971). Goormaghtigh (1939, 1945) observed that granulation in the afibrillar cells (JG cells) increased in man and experimental animals with arterial hypertension. He proposed that the endocrine activity of these granular cells is related to the production of a hypertensive substance (renin) and that the JG cells are the source of renin. This conclusion was supported by the observation that granulation and the renin content of the rat kidney increased or decreased in parallel (Pitcock et al. 1959; Tobian et al. 1959a).

More direct methods to localize renin in JG cells have also been applied. Separation of the glomeruli revealed that only the pieces containing the vascular pole region of the glomeruli had renin (Cook 1971). Further, several investigators performed a more detailed microdissection to separate the different cell types using frozen or freezedried sections of glomeruli from rabbit, cat, mouse and pig; renin was then bioassayed in each tissue fragment. Bing and Kazimierczak (1962), using frozen sections, found that the most renin was in the preparations that included the tubular portion containing MD. In the kidney of newborn pigs, they (1963) found that renin was present in the subcapsular zone, which was devoid of afferent and efferent vessels. Thus, they concluded that renin was produced in the MD. However, using freezedried sections of cat kidneys, Faarup (1967, 1968) dissected various parts of the JGA with fine techniques and found that renin was mostly in the tissue with many JG cells. Cook (1971) separated granules in arteries with two pairs of fine forceps and found that most renin of the glomerules was in the granules but not in the MD.

Nairn et al. (1959) performed immunofluorescence studies using a partially purified extract of pig kidney as antigen and found that the glomerular capillary and capsular epithelium were immunoreactive. Hill et al (1953) used antibody raised against crude extracts of rat kidneys in rabbits, and observed immunoreactivity in convoluted tubules. However, Edelman and Hartroft (1961) and Hartroft et al. (1964) demonstrated that canine fluorescence-labeled antibodies to partially purified rabbit and pig renin immunoreacted to JG granules of rabbit and pig. Their observations seemed to provide convincing evidence for the storage site of renin in the kidney. However, the localization of renin revealed by those studies was disputable because the renin preparations used as antigens were of questionable purity.

Cohen et al. (1972) isolated stable and pure renin from the submaxillary gland of adult male mice. Menzie et al. (1978) localized renin in JG cells in the mouse kidney using an antiserum raised against the submaxillary gland renin. Taugner et al (1979, 1982a,b; Mizuhira 1986; Taugner and Hackenthal 1989) localized renin in the JG cells of the afferent arterioles and in the wall of some of the efferent arterioles in the mouse kidney, using an antibody raised against pure mouse submaxillary gland renin (Fig. 2.3). Immunoreactivity of renin was also found in some cells of the EGM and most of the interlobular arteries and in the apical part of proximal tubule cells. Taugner et al. (1982a) explained the immunoreactivity of these portions as due to reabsorptive pinocytosis of the filtered hormone. The same pattern of renin distribution was observed by Camilleri et al (1980) in human kidney. However, Tanaka et al. (1980) using an antibody against highly purified submandibular gland renin, demonstrated that renin immunoreactivity was restricted to the granules of epithelioid cells

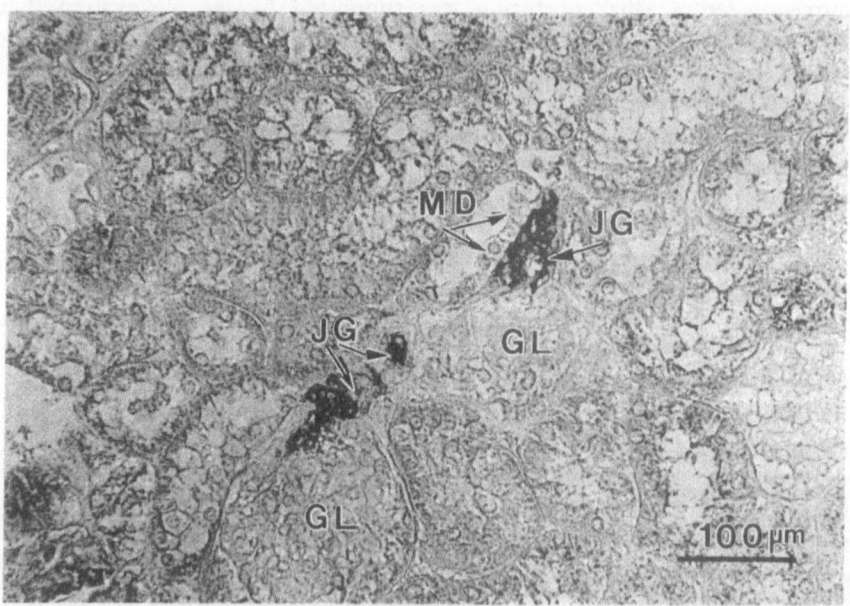

Fig. 2.3. Immunoreactive renin granules in the mouse juxtaglomerular cells (*JG*). The antiserum was raised against renin obtained from the mouse submaxillary gland. *GL* glomerulus; *MD* macula densa; × 175. (Mizuhira 1986)

of the mouse kidney. Human renin was purified from normal kidney (Yoko-sawa et al. 1978) and from a kidney containing a JG cell tumor (Galen et al. 1979). Using a specific antibody to human renin, Celio and Inagami (1981a) observed renin immunoreactivity in the JG cells in the wall of the afferent arterioles and rarely in the cells of the efferent arterioles in human kidney. Celio and Inagami (1981b) localized renin immunohistochemically in the epithelioid granular cells of rat kidney, using goat and rabbit antisera raised against a pure preparation of rat kidney renin. Faraggiana et al. (1982) used an antiserum against purified human renal renin for immunocytochemical detection of renin in the human kidney. Using protein A-gold colloid electron microscope procedures, they showed that renin was localized in the JG cells and was contained in the secretory granules. They also found that immunoreactive renin was often present along the afferent arteriole at some distance from the glomerulus in human kidney.

Celio and Inagami (1981b) found immunoreactive ANG II coexisting in the same cells showing renin immunoreactivity. Coexistence of ANG II and renin was also observed in JG cells and ANG I was detected in rats after treatment with captopril, a converting enzyme inhibitor, (Naruse et al. 1982). Inagami et al (1986) suggested the presence of an intracellular renin angiotensin system in the JG cells (see Chap. 4). Lindop and Lever (1986) observed coexistence of ANG II and renin, and supported the concept of an intrarenal action for renin. The coexistence of renin and ANG II in the same granules of the epithelioid cells has also been demonstrated immunohistochemically by several investigators (Bührle et al 1984; Cantin et al. 1984; Taugner et al. 1982b, 1984b). Kawamura et al. (1985) isolated kidney cortex granules by density gradient fractionation in rats and identified both ANG I and ANG II in a granule fraction rich in renin, indicating the intracellular formation of ANG II. Further, Morris and Johnson (1976) found angiotensinogen-containing particles in rat kidney cortex. Ingelfinger et al. (1990) demonstrated angiotensinogen mRNA in proximal renal tubules of rats. Angiotensin converting enzyme was detected immunocytochemically on the luminal surface and in JG cells in rats (see Inagami 1993 for review). These findings show the presence of RAS in JG cells (see Chap. 4).

2.1.2.3 Electron Microscopy

Early electron microscope studies on mammalian JG cells showed that granules in the cells appeared relatively homogeneous, dense and osmiophilic (Dalton 1951; Bohle 1954). Since 1960, electron microscopy on mammalian JG cells has become more common and the cytological characteristics of JG cells have been further elucidated (see Hatt 1967; Gorgas 1978a,b; Mizuhira 1986 for reviews) (Fig. 2.4). The granules are polygonal or ovoid, membrane-bound, more irregular and larger than those seen in most endocrine cells (Bohle 1959; Oberling and Hatt 1960a,b; Hartroft and Newmark 1961; Latta and Maunsbach 1962; Bucher and Reale 1962). The granules are formed in the Golgi apparatus

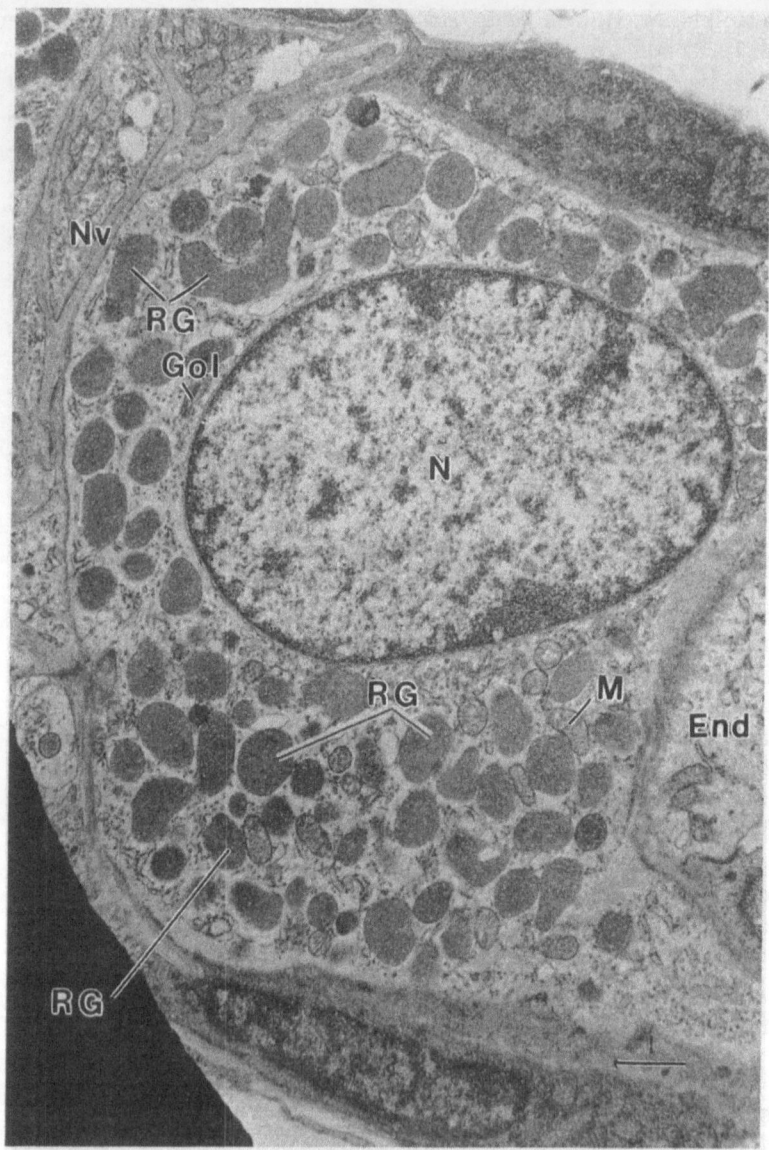

Fig. 2.4. Juxtaglomerular cell, including many renin granules (*RG*), of the mouse kidney. *End* Endothelial cell of the afferent arteriole; *Gol* Golgi apparatus; *M* mitochondria; *N* nucleus; *Nv* nerve terminal. × 9000. (Mizuhira 1986)

(Barajas and Latta 1963; Chandra et al. 1965). Small granules in the Golgi apparatus often contain a crystalline structure. Small granules seem to con-glomerate into large granules, but mature granules rarely contain visible crystalline patterns (Barajas and Müller 1980; Mizuhira 1986). These large granules contain renin, as detected by immunoelectron microscopy (Faraggiana et al. 1982). Some of the small granules may contain other peptides, since the coexistence of two or three peptides in the same cell has often been observed (see Sofroniew et al. 1984). Taugner et al. (1984b) found that renin and ANG II coexist in the same granules. Lipofuscin-like granules are occasionally found in these cells, especially in humans (Biava and West 1966; Barajas and Müller 1980). In the rat, the granules are more numerous in JG cells of kidneys with increased renin production caused by constriction of renal arteries (Barajas and Latta 1967).

Taugner et al. (1985a) determined immunoelectron microscopically that renin and cathepsin B coexist in the secretory granules in JG cells. They also found coexistence of renin with cathepsin D (Taugner et al. 1986). They suggested that cathepsin B may be involved in the conversion of prorenin into the active secretory product, renin, within the secretory granules. In contrast, they hypothesized that cathepsin D may be involved in the intragranular breakdown of renin by nonspecific hydrolysis as a mechanism for the down regulation of the store available for secretion. The secretory granules have characteristics of lysosomes (Taugner et al. 1985b) and show macrophagic phenomena (Taugner et al. 1988).

There have been few studies on the morphology of renin secretion from the epithelioid cells. Peter (1976) observed deep channel-like invaginations of the plasma membrane into the interior of the epithelioid cells in rats. The plasma membrane invaginations were in close contact with secretory granules or, in some cases, contained granule-like material. He suggested that renin release occurs through deep invaginations of the plasma membrane. This is an unusual form of exocytosis. Ryan et al. (1982) and Mizuhira (1986) noted similar invaginations or canaliculi in the epithelioid cells for renin in the rat. However, Taugner et al. (1984a) observed ordinary exocytosis, which is common in other endocrine glands, in *Tupaia belangeri* and mice, using thin sections and freeze-fracture replicas. Granule content is released from omega-shaped cavities into the extracellular space. They observed extrusion of several closely apposed granules, reminiscent of compound exocytosis. Taugner et al. (1984a) interpreted the invagination differently from Peter (1976), Ryan et al. (1982) and Mizuhira (1986). Taugner and colleagues argued that (1) the membrane-bound saccule that persisted after granule depletion collapsed first in the region of its neck, thus resembling an invagination; and (2) processes of epithelioid cells often closely approached each other or the perikaryon, thus forming clefts which mimicked invaginations in transverse sections. They believed, therefore, that invaginations of plasma membrane were not related to renin secretion.

Skøtt and Taugner (1987) reported that superfusion with hypotonic solutions caused both stimulation of renin release from rat epithelioid cells adherent to isolated glomeruli and swelling of the secretory granules. They hypothesized that the swelling might markedly increase the probability of pre-exocytotic fusions between the granule and cell membrane, followed by increased frequency of exocytosis. Whatever the mechanism, renin seems to be secreted mainly into the interstitium near the outer aspects of the vessel, and gains access to the general circulation via the peritubular capillaries (Lindop and Lever 1986).

2.1.3 Extraglomerular Mesangium (EGM) and Macula Densa (MD)

Electron microscope observations of these organs as well as JG cells are well documented in reviews by Hatt (1967); Davis and Freeman (1976); Barajas and Müller (1980); Mizuhira (1986); and Taugner and Haeckenthal (1989; Figs. 2.1, 2.5).

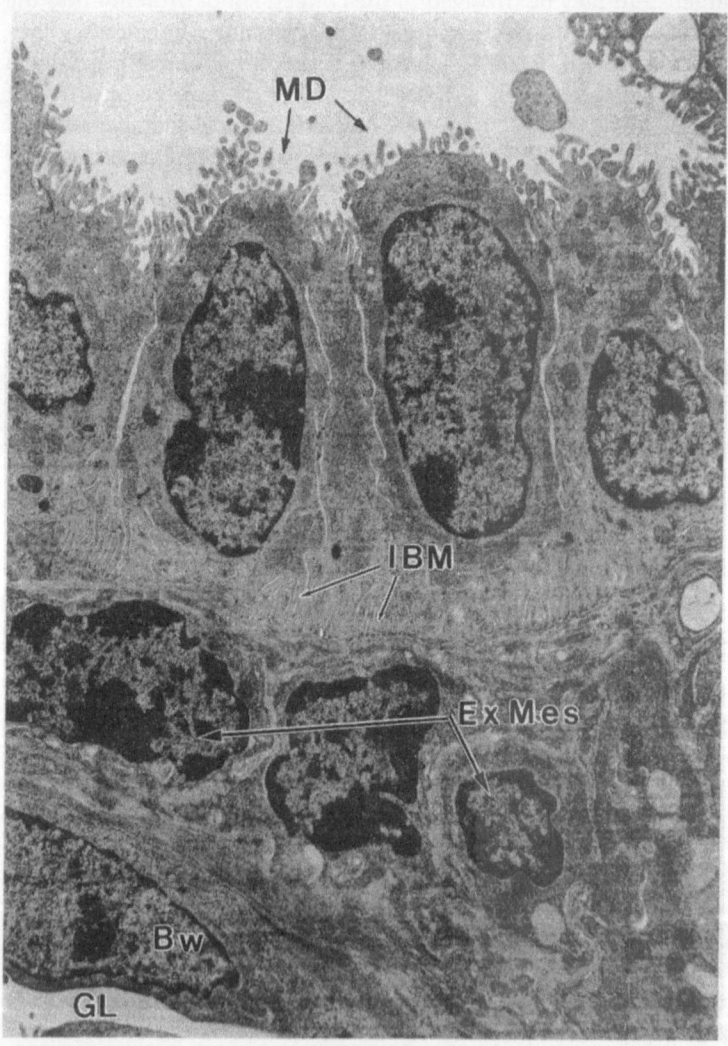

The EGM is a mass of small cells between the MD and the glomerulus; given its continuity with the glomerular mesangium, the mass is called the extra-glomerular mesangium (Barajas 1970; Fig. 2.1). In this cell mass, agranular cells predominate. As mentioned before, Zimmermann (1933) originally included agranular small cells and the modified myocytes of the afferent arterioles in the term "Polkissen". Now this term is synonymous with the EGM. These cells were named the lacis by Oberling and Hatt (1960a,b) and the pseudo-Meissnerian cells by Goormaghtigh (1932). In the rat, the cells have vesicular cytoplasm and no granules. Mitochondria are small, ovoid and have rectilinear cristae. Microvilli are not particularly rich on the cell surface. Occasionally, an intracytoplasmic fibrillar structure is discernible (Hatt 1967).

The following descriptions of the MD are mostly based on observations in the rat (Figs. 2.1, 2.5, 2.6). The renal tubule returns to its own glomerulus and one of its sides attaches to the wall of the afferent arteriole at the vascular pole. The epithelial cells on the attached side are cuboidal and cytoplasm is scant. The microvilli are considerably developed on the cell surface. The nuclei are electron dense and lie closer together than in the cells on the opposite side. Mitochondria are short and irregularly distributed, in contrast to the characteristic palisade arrangement of mitochondria in most of the distal tubule. The Golgi apparatus is developed and is located on the basal side of the nuclei, although it is usually apical to the nuclei in the adjacent tubule cells. The most characteristic features of the MD are that the infolded basal membranes and interdigitating basal processes are much less developed than in the adjacent portions of the distal tubule. Therefore, mitochondria are not included in the basal processes, in contrast to the inclusion of mitochondria in the basal processes of the adjacent distal tubule (Fig. 2.6). The basal membrane may be irregularly thickened or folded under the MD. Intracellular canaliculi are often observed in the MD cells. In addition to the close juxtaposition of the MD to the JG cells, special contact is formed by cytoplasmic projections of MD cells toward the granular cells (Barajas and Latta 1967). In some instances, distinct separations between JG cells and MD cells are difficult to find (Hartroft and Newmark 1961).

2.1.4 Efferent Arterioles

Renin has been detected immunohistochemically in the wall of the afferent arteriole and the efferent arteriole near the glomerulus in several mammalian species (Taugner et al. 1979; Celio and Inagaki 1981a; Taugner et al. 1982b; Lindop and Lever 1986; Taugner and Hackenthal 1989). However, the immunoreaction in the efferent arteriole was moderate compared with that in the afferent vessel.

◄───

Fig. 2.5. Macula densa (*MD*) and extraglomerular mesangium (*ExMes*) of the juxtaglomerular apparatus of the mouse. *Bw* Bowman's capsule; *GL* glomerulus; *IBM* infolded basal membrane. × 12000. (Mizuhira 1986)

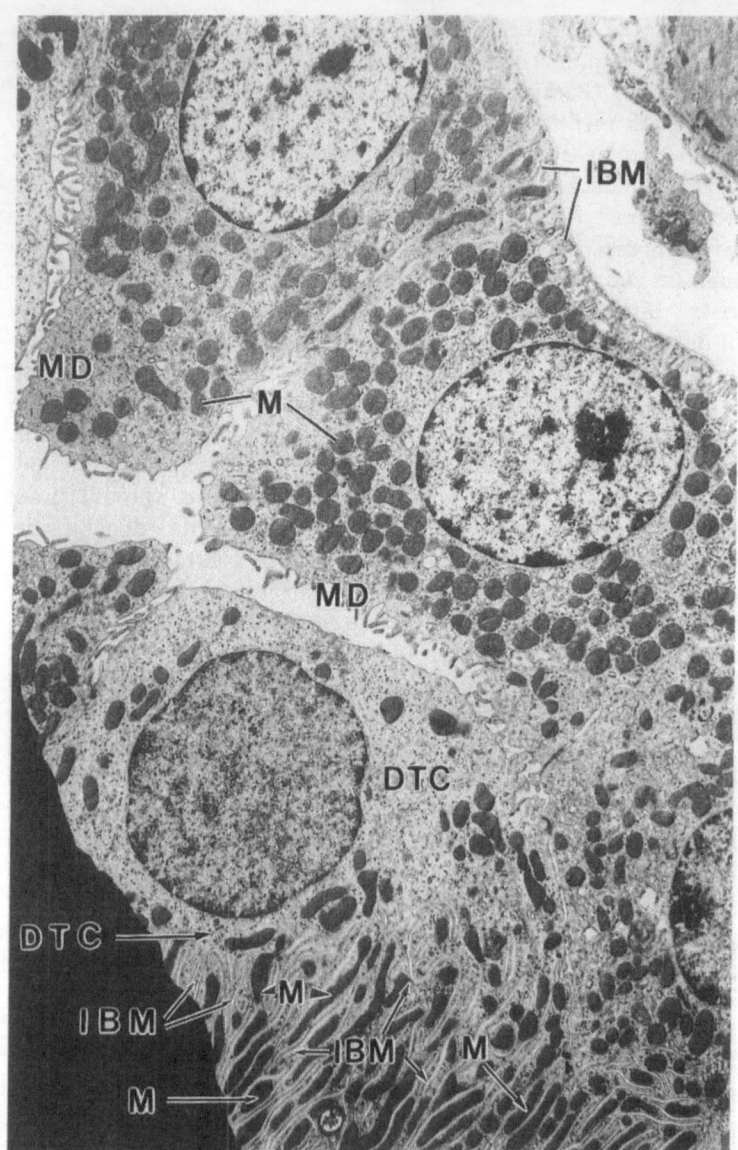

Fig. 2.6. Macula densa (*MD*) cells and the adjoining distal tubule cell (*DTC*) in the rat kidney. The infolded basal membranes (*IBM*) or processes in the MD cells are much less developed than in the DTC. Mitochondria (*M*) are not included in the basal processes in the MD cells, in contrast to the DTC. × 6000. (Courtesy of Dr. Mizuhira)

2.1.5 Peripolar Cells

Ryan et al. (1979) found new, distinctive types of cells which contain multiple cytoplasmic granules closely resembling those of arteriolar myoepithelial cells

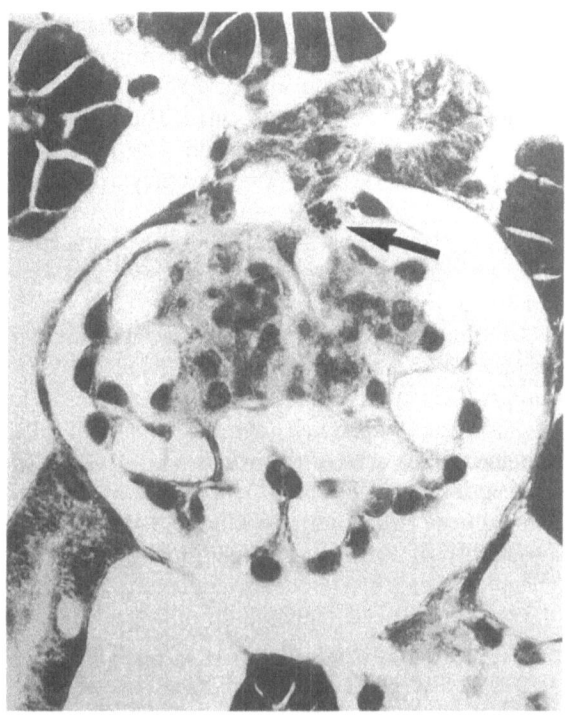

Fig. 2.7. Chicken glomerulus showing a heavily granulated peripolar cell (*arrow*) at the glomerular hilus. MSB trichrome. × 830. (Morild et al. 1988)

in sheep kidney. These cells are in an epithelial position, encircling the origin or polar region of the glomerular tuft. They named them "peripolar cells" (Figs. 2.1, 2.7). It is not known whether these cells are associated with the RAS. One surface of the cells is attached to the basement membrane of Bowman's capsule and the other is directly exposed to the urinary space. Ryan et al. (1979) observed up to four such peripolar cells surrounding the origin of any one glomerular tuft. The granules react positively with methylene blue, PAS, brilliant crystal scarlet and Bowie's stain. Electron microscopically, most granules range from 100 to 500 nm in diameter. Since more glomeruli are found within the superficial renal cortex, more peripolar cells are distributed in the superficial zones of the cortex. The percent of vascular poles which contain one or more peripolar cells is from 3.6 to 100 in seven sheep examined. No cilia or microvilli are identified (Kelly et al. 1990). Granulated peripolar cells have been observed in sheep, man, rat, and mouse (Ryan et al. 1979), monkey, guinea pig, rabbit, dog, giraffe, koala, and platypus (Ryan et al. 1982) and antelopes and goats (Mbassa 1989). Peripolar cells are most prominent in the ovine kidney (Ryan et al. 1979); they are often scanty but variable in number in the human kidney (Gardiner and Lindop 1985). Gall et al (1986) observed peripolar cells in 16 mammalian species and noted their close anatomical relationship with renin-containing myoepithelioid cells.

In the rat, Gibson et al. (1989) performed detailed scanning electron microscope studies. According to these investigators, the peripolar cells are situated in the annular groove at the root of the glomerulus, between the parietal epithelial cells and the podocytes. The cells are dendritic cells with long processes embracing the glomerular arterioles. Up to three peripolar cells are present at each vascular pole, and they are distributed mainly in the glomeruli of the outer third of the renal cortex. Some peripolar cells have a smooth or ruffled surface and are often covered by microvilli (Gibson et al. 1989). Newborn rats and lambs show more cells than their adult counterparts (Gall et al. 1986).

In sheep, the contents in the granules are released by exocytosis into the urinary space (Ryan et al. 1982; Hill et al. 1983, 1984); in human peripolar cells, no examples of typical exocytosis were found, but there were many complex invaginations of the cell membrane, similar to those previously described in the epithelioid cells of the afferent arterioles (Gardiner et al. 1986). There is a close anatomical relationship between the peripolar cells and the renin containing cells (Gardiner and Lindop 1985; Gardiner et al. 1986), suggesting the possibility of a functional relationship between these two kinds of cells. In sheep that had been sodium-depleted by carotid cannula drainage for 2 or 7 days, peripolar cell granules discharging their contents into the urinary space were observed (Ryan et al. 1982). Ryan et al. (1982) speculated that the granules might be the source of an intrarenal sodium-retaining hormone. Further, there is some evidence of increased activity of peripolar cells in sodium-depleted sheep (Hill et al. 1983) and chickens (Morild et al. 1988). Alcorn et al. (1984) observed that peripolar cells and their granules are very much larger in newborn sheep than in fetal lambs or adult sheep. They also observed that similar peripolar cell hypertrophy was triggered in fetal lambs treated in utero with intraperitoneal injections of dexamethasone. They suggested that the enlargement of peripolar cells during the newborn period might contribute to the important functional adaptations that affect water and electrolyte homeostasis in immediate postnatal life.

Several attempts have been undertaken to explore the composition of peripolar cell granules. Granules in peripolar cells in the human kidney did not show immunoreactivity to renin antiserum (Gardiner and Lindop 1985). Trahair et al. (1989) demonstrated in sheep that some glomerular peripolar cells were immunoreactive for neuron-specific enolase (NSE), but nonreactive for renin and kallikrein antisera. Since NSE immunoreactivity was found in non-neuronal and non-neuroendocrine cells in a wide variety of human tissues (Haimoto et al. 1985; 1986), the presence of this enzyme has no specific significance with respect to the peripolar cells. Cytoplasmic granules of peripolar cells of newborn lamb and sheep contain immunoreactive albumin and immunoglobulin. The origin of these proteins is mostly filtered plasma proteins. Some leakage of plasma proteins normally occurs across the glomerular filtration barrier in sheep, particularly in newborn lambs. Uptake of the proteins by the peripolar cells remains unexplained (Trahair and Ryan 1988). Nakajima et al. (1989) showed with an immunogold labeling technique, that droplets in the peripolar cell contain immunoglobulins, C_{3c}, fibrin/fibrinogen and albumin in the same distribution pattern as that in visceral or parietal epithelial cells. They suggested that the peripolar cells might have the same characteristics as mesangial cells and epithelial podocytes. Morild et al. (1988) observed that the droplets in the peripolar cells do not differ morphologically from those in the visceral or parietal epithelial cells of the avian kidney (Fig. 2.7). Further, they found that the droplets do not show immunocytochemically cathepsins B, D, L and H and renin. Thus, it is still uncertain whether the peripolar cell is a specific cell type which possesses a secretory role.

2.2 Birds

Avian JG cells take up Bowie's stain. They have been observed in the wall of the afferent arteriole near the glomerulus of the kidney in the following birds: the budgerigar (*Branta canadensis*) and *Anas platyrhynchos* (Sutherland 1966), pigeon (Smith 1966; Miller 1967), Japanese quail (Sokabe et al. 1969; Ogawa and Sokabe 1971), chicken (Edwards 1940; Smith 1966; Sokabe et al. 1969; Ogawa and Sokabe 1971; Taylor et al. 1970; Kon et al. 1984) and Pekin duck (Sutherland 1966).

In the chicken, fine granules with high electron density were observed in the cytoplasm of JG cells located in the tunica media of the afferent arteriole, vascular pole and mesangial region. These granules are bound by a limiting membrane, vary in size and contour and contained homogeneous material. JG cells contain well-developed Golgi apparatus, elongated endoplasmic reticulum, a few myofibrils and characteristic attachment bodies (Kon et al. 1984). These data are well summarized in reviews by Sokabe and Ogawa (1974) and Wilson (1984a).

Immunohistochemical studies have revealed that renin reaction is observed mostly in cells within the tunica media of the afferent arteriole, and frequently in cells within the glomerular mesangial region. Only a few renin-containing cells are localized in the efferent arterioles in the chicken, though not in the duck (Kon et al. 1986). The presence of renin in the avian kidney has been found in many avian species. Pressor activity was measured by incubating kidney tissue extracts with homologous angiotensinogen preparations in vitro. Bean (1942), Schaffenburg et al. (1960), Weichert (1965), Taylor et al. (1970) and Nishimura et al. (1981) demonstrated renin activity in the chicken kidney; Chan and Holmes (1971) in *Anas platyrhynchos* and *Columba livia*; Nolly and Fasciolo (1972, 1973) in *Rhea americana, Gallus domesticus, Meleagris gallopavo, Anser vulgaris, Cairina moschatta, Columba livia, Zenaida auriculata, Passer domesticus, Nothura mendozensis*; Pagnan et al. (1978) in the broadbreasted white turkey. Bean (1942) and Haas et al. (1954) showed renin activity in the chicken kidney using heterologous substrates from duck and from dog. Homologous plasma for incubation with renin tissue extracts is recommended, because of species specificity (see Sokabe and Ogawa 1974; Wilson 1984a).

There is some disagreement about the presence of MD in the avian kidney (Table 2.1). Ogawa and Sokabe (1971) and Sokabe and Ogawa (1974) reported that avian MD is structurally transitional between the typical MD cells in mammals and the ordinary cells of the distal tubule, since the Golgi apparatus is located apically to the nucleus, unlike in mammals where its location is basal. Observations in the chicken and Japanese quail showed that the cells of avian MD are not tall. Wideman et al. (1981) also found no evidence of typical MD cells in fowl. However, the presence of MD in the avian kidney was reported by Edwards (1940), McKelvey (1963), Berger (1966), Sutherland (1966), Johnson and Mugaas (1970), Taylor et al. (1970), Siller (1971), Johnson (1979) and Kon et al. (1984). Christensen et al. (1982) showed that, in domestic fowl, the MD is distinct and the cells have protrusions on the luminal surface, which are not seen in the tubular cells of the opposite wall. The MD cells often appear taller, and the nuclei often lie closer to each other than in the other tubular cells. Christensen et al. (1982) reported that the MD cells are taller, with smaller internuclear distances than other distal tubule cells in chicken fed 2 g sodium/ kg food. The MD cells have protrusions on their luminal surface and an extensively branched basement membrane. Observations on the MD in birds should be carefully done under appropriate physiological circumstances.

The presence of EGM has been reported in chicken (Siller 1971; Johnson 1979; Christensen et al. 1982; Table 2.1), although Sokabe et al. (1969) and Ogawa and Sokabe (1971) could not find it. Edwards (1940) noted that EGM cells (periarteriolar pad) were observed infrequently in birds, and developed very poorly. Christensen et al. (1982) reported in chicken that the extraglomerular cells are clearly visible in the space among the distal tubule, the afferent and efferent anterioles, and the glomerular hilus. These cells are connected to the mesangial cell mass. The nuclei are elongated and oriented parallel to the MD basement membrane. The cells are surrounded by basement membrane material that is continuous with the basement membranes of the MD, the hilar arterioles and the mesangial cells.

Peripolar cells resembling the visceral or parietal epithelial cells have been demonstrated in the avian kidney, as described in Section 2.1.5 (Morild et al. 1988; Fig.2.7).

2.3 Reptiles

McKelvey (1963) found granulated JG cells, using Bowie's stain, in the media of the afferent arteriole in reptiles (Table 2.1). Later, JG cells with granules were demonstrated, using Bowie's stain, in *Eumeces latiscutatus* (lizard), *Agkistrodon blomboffii, Elaphe quadrivirgata, Natrix tigrina* (snakes) and *Clemmys japonica* (turtle) (Sokabe et al. 1969). Studies demonstrating the presence of JG cells in other species of reptiles are summarized in a review by Sokabe and Ogawa (1974). The presence of renin was shown following incubation of kidney extracts, mostly with homologous plasma, in approximately 20 species of reptiles (Kaley and Donshik 1965; Weichert 1965; Sokabe et al. 1969; Capelli et al. 1970; Nothstine et al. 1971; Nolly and Fasciolo 1972; see Sokabe and Ogawa 1974 for review). In the reptile kidney, the MD is absent in the lizard, snake, alligator and turtle (Edwards 1941; McKelvey 1963; Kaley and Donshik 1965; Sutherland 1966; Sokabe et al. 1969; Sokabe and Ogawa 1974: Table 2.1). Ogawa and Oguri (1978) reported that the distal tubule frequently is located near the vascular pole of the glomerulus; however, the epithelial cells that are in contact with the vascular pole do not differ in appearance from cells within the remainder of the distal tubule. The authors could not identify the EGM (see Sokabe et al. 1969; Sokabe and Ogawa 1974; Ogawa 1977; and Oguri 1978). In the lizard, *Lacerta agilis*, Koval'Chuk (1987) observed JG cells in the wall of the afferent arteriole, primitive forms of the MD in some nephrons and the JG island, and peripolar cells situated on the basal membrane of the external part of the glomerular capsule near the vascular pole. Further detailed studies are needed using more species.

2.4 Amphibians

As early as 1929, Okkels found granulated cells in the vascular hilus of the glomerulus in the frog (Table. 2.1). This observation followed closely the finding of granulated cells in the afferent arteriole of rats and mice (Ruyter 1925) and in humans (Oberling 1927). Later, JG cells with granules were observed with Bowie's stain and other stains (see Sokabe and Ogawa 1974) in the afferent arteriole close to the glomerulus in a number of species of toads, frogs and newts (Okkels 1929; McKelvey 1963; Hartroft 1966; Sutherland 1966; Sokabe et al. 1969; van Dongen and van der Heijden 1969; Capelli et al. 1970; Lamers et al. 1974; Sokabe and Ogawa 1974). Electron microscopically, the presence of granules in the JG cells was established by Hartroft (1966) and Bellocci et al. (1971). According to Lamers et al. (1974) many of the granules in JG cells contain lamellar material. The myofilaments are situated near the vascular lumen. Other cell organelles that include granules are located at the opposite pole of the cell in the neighborhood of the MD cells (see below) and adventitial nerve fibers. In these regions the JG cells have many pinocytotic vesicles. The nerve fibers innervating the JG cells are nonmyelinated and contain small dense cored vesicles. The basement membranes of the JG cells and adjacent MD cells occasionally fused, which might indicate a functional relationship between these cells (Lamers et al. 1974).

The distal tubule of *Bufo vulgaris* returns to the vascular pole of the glomerulus infrequently. However, the epithelial cells on the side toward the glomerulus are not differentiated into the MD (Ogawa 1977). In *Rana nigromaculata, Rana japonica, Rana catesbeiana* and *Triturus pyrrhogaster*, the distal tubule is not near the glomerulus (Sokabe and Ogawa 1974; Ogawa and Oguri 1978). However, Edwards (1940) reported the presence of MD in frog (species not specified), Sutherland (1966) in *Rana catesbeiana* and *Rana pipiens*, and Capelli et al. (1970) in *Rana pipiens*. Lamers et al. (1973) reported that in the part of the tubule adjacent to the afferent arteriole of *Bufo bufo* an accumulation of nuclei is present and that this structure is similar to the MD of the mammalian JGA. The EGM was not observed in amphibians by the investigators mentioned above. Hanner and Ryan (1980) could not find a clearly defined MD structure nor EGM in *Ambystoma mexicanum* and *Bufo marinus* (Table 2.1).

Renin or a renin-like substance was demonstrated immunohistochemically in the JG cells, but not in MD and efferent arterioles, in three *Bufo* species (Lamers et al. 1985b). Using an immunogold staining method for electron microscope immunocytochemistry, Lamers et al. (1985a) also shown that renin immunoreactivity is confined to lamellated granules in the JG cells of afferent arterioles and to the media cells of some larger arteries in *Bufo bufo*. Renin activity was demonstrated by bioassay in a medium after incubation of kidney extracts with homologous plasma in more than ten species of amphibians (see Sokabe and Ogawa 1974). Hanner and Ryan (1980) found peripolar cells in *Ambystoma mexicanum* and *Bufo marinus*, although such cells were difficult to find in the toad.

2.5 Fishes

Before 1960, few studies were performed on the fish renin–angiotensin system. In 1940, Edwards reported that structures resembling mammalian epithelial plaque (macula densa) and the periarteriolar pad (extraglomerular mesangium) are absent in fish. In 1942, Friedman and Kaplan, using Mallory's triple stain, were able to find clusters of cells which they thought might be JG cells, but were unable to identify any granules. Friedman and Kaplan reported the absence of renin activity in aglomerular kidneys of midshipman fish (Batrachoidiae, *Porichthys notatus*). They also reported that renin activity is absent in extracts of the glomerular kidneys of cod and sole but present in carp and catfish kidneys (Friedman et al. 1942). They examined renin activity of extracts in vivo by measuring the blood pressure of nephrectomized and anesthetized dogs. Bean (1942) could not detect renin activity after intravenous injection of kidney extracts of shark into anesthetized dogs. However, these earlier studies are not reliable, since the investigators were unaware of the species specificity of the renin–angiotensin reaction. Further studies of the fish RAS were not performed until the early 1960s. Based on recent data, the phylogenetic distribution of each component of the JGA in fishes is summarized in Table 2.1.

2.5.1 JG Cells in Teleosts with Glomerular Kidney

In the 1960s, numerous histological and pharmacological studies were performed in nearly 100 species of teleosts (see Sokabe and Ogawa 1974). The presence of granules of JG cells was established primarily by Bowie's stain (Fig. 2.2) and periodic acid Schiff stain. Aldehyde–fuchsin–trichrome and Mallory's hematoxylin and Mallory's triple stains also showed granules, but stainability varied with the fixatives (Oguri et al. 1969) and species (Krishnamurthy and Bern 1969). The granules are usually minute in size and stain homogeneously in teleosts. The JG cells were demonstrated in both glomerular and aglomerular kidneys of fish (Bohle and Walvig 1964; Oguri and Sokabe 1968; Capréol and Sutherland 1968; see Sokabe et al. 1969; Sokabe and Ogawa 1974). Capréol and Sutherland (1968) found JG granules in about 20 species, but not in *Salmo gairdneri*. Krishnamurthy and Bern (1969) examined 26 species including marine and freshwater fishes, using several stains and identified JG cells in all, except *Salmo gairdneri, Oncorhynchus kisutch, O. tshawytscha* and *O. nerka.* Oguri and Sokabe (1968) found JG cells in 34 teleostean species, both marine and freshwater, and both glomerular and aglomerular. They discovered JG cells in *O. keta* and *Salmo gaidneri irideus*, although the JG cells were not abundant. JG cells in marine teleosts are more clearly stained with Bowie's stain and are more numerous than in freshwater teleosts (Bohle and Walving 1964; Capréol and Sutherland 1968; Meyer et al. 1967; Oguri and Sokabe 1968; Olivereau and Lemoine 1969; Krishnamurthy and Bern 1969), although renin activity is higher in freshwater teleosts than in marine teleosts (Mizogami et al. 1968).

The JG cells are distributed along renal afferent arterioles and arteries. Six types of distribution of JG cells have been classified by Krishnamurthy and Bern (1967 Fig. 2.8).

Type 1. JG cells occur on afferent arterioles entering a glomerulus directly from arteries; none occur on the efferent arterioles: *Carassius auratus, Atherinopsis californiensis, Roccus saxatilis, Tilapia mossambica, Hyperprosopon ellipiticum, Artedius notospilotus, Citharichthys sordidus* and *Eopsetta jordani.*

Type 2. JG cells are found on the preglomerular arterioles originating from the arterial branches and on the afferent arterioles, but not on the arteries: *Atherinopsis californiensis* and *Sebastodes auriculatus.*

Type 3. JG cells occur both on afferent arterioles and on adjacent arterioles: *Hyperprosopon ellipticum, Hypsurus caryi, Phanerodon furcatus, Rhacochilus vacca, Sebastodes auriculatus* and *Platychthys stellatus.*

Type 4. JG cells occur only on arteries and never on afferent arterioles: *Cymatogaster aggregata, Phanerodon furcatus, Hexagrammos decagrammus, Ophiodon elongatus, Scorpaenichthys marmoratus* and *Citharichthys sordidus.*

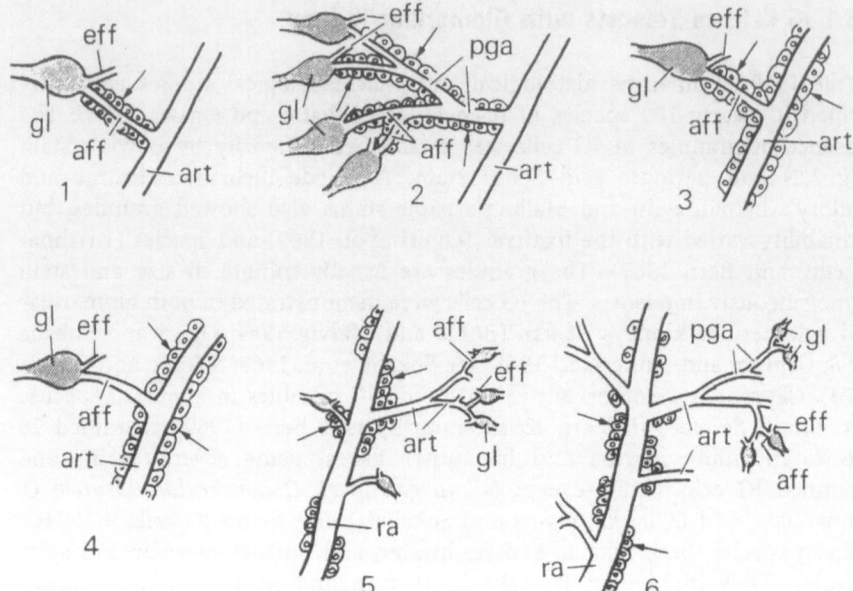

Fig. 2.8. Diagram showing six types (1–6) of distribution of juxtaglomerular cells (arrows) in relation to arteries, arterioles, and glomeruli (gl) in the kidneys of fishes. aff Afferent arteriole; art arterial branch; eff efferent arteriole; pga preglomerular arteriole; ra renal artery. (Krishnamurthy and Bern 1969)

Type 5. JG cells are found at the junction of the main and secondary branches of the renal artery. Although most afferent arterioles have no JG cells, afferent arterioles branching directly off the main artery may occasionally have them: *Gillichthys mirabilis.*

Type 6. JG cells occur primarily in the wall of the main renal artery; some secondary arterial branches also have them. Glomeruli are always distant from the JG cells: *Leptocottus armatus.*

More than one distribution type can occur in a single species: *Atherinopsis californiensis* shows types 1 and 2, and *Sebastodes auriculatus* shows types 2 and 3.

Immunocytochemical studies on the carp kidney, using antiserum raised against mouse submandibular gland renin in rabbit, revealed that numerous immunoreactive JG cells were located in the tunica media of the afferent arterioles or the small arteries. Occasionally, the cells were found in the boundary region between the tunica media and the tunica adventitia of the arterial vessels (Kon et al. 1987).

2.5.2 JG Cells in Teleosts with Aglomerular Kidney

The kidney of aglomerular fishes contains cells including granules that stain with Bowie's stain. These cells are customarily called JG cells (JG), although the fish do not have glomeruli. Bohle and Walvig (1964) observed JG cells in the aglomerular teleosts, *Pleuronectes microcephalus* and *Lophius piscatorius*. Christensen et al. (1987) found granulated epithelioid cells in the walls of the arteriolar networks in the caudal half of the *Pleuronectes* kidney. The cells were immunostained with antisera directed against murine and human renin. Capréol and Sutherland (1968) reported the presence of JG cells in the walls of arterial branches near the kidney surface, but not in the central region of the kidney of the American goosefish, *Lophius americanus*. In the Japanese goosefish, *Lophius litulon*, Oguri and Sokabe (1968) and Oguri and Sokabe (1974) found JG cells located in the walls of the remarkably developed arterial branches in the peripheral subcapsular regions and also in the central regions of the kidney. In *Lophius litulon*, a few glomeruli were present (Ogawa et al. 1972), which seemed to be nonfunctional. Similar structures had been named "pseudoglomeruli" in *L. piscatorius* (Grafflin 1929). In the frogfish, *Phrynelox tridens*, small groups of JG cells were located in the walls of arterial branches in the peripheral subcapsular region of the kidney. In *P. nox*, JG cells formed well developed clusters. In the sargassum fish, *Histrio histrio*, a large number of JG cells were detected in the walls of the arterial branches but only in one limited region near the kidney surface, as in the frogfishes. In the toadfish, *Opsanus tau*, Oguri et al. (1972) found JG cells distributed diffusely in the walls of small arteries located in the interstitial hematopoietic tissue of the kidney.

As mentioned above, the distribution of JG cells in aglomerular kidneys varies markedly among species. Four types of distribution of JG cells were observed in aglomerular fish (Ogawa et al. 1972; Ogawa 1977; Ogawa and Oguri 1978). The simplest type seen in the toadfish kidney had JG cells distributed in the walls of small arteries. In the second type, found in the goosefish, JG cells were distributed in several regions within the kidney. In the third type, observed in the frogfish and the sargassum fish, JG cells were concentrated in a limited area. In the fourth type, found in the seahorses, *Hippocampus aterrimus*, *H. japonicus*, *H. spinosissimus* and *H. whitei*, JG cells were in the walls of the renal arteries located just outside the renal connective capsule.

2.5.3 Renin Activity, MD, EGM and Peripolar Cells in Glomerular and Aglomerular Fishes

The presence of renin has been demonstrated, primarily through an indirect in vitro method (Sokabe and Ogawa 1974) using homologous substrates and kidney extracts, in both glomerular and aglomerular teleosts of more than 30 species (see Mizogami et al. 1968; Nishimura and Ogawa 1973; Sokabe and Ogawa 1974). Arillo et al. (1981) identified renin activity by incubating kidney extracts of rainbow trout with pig angiotensiogen and measuring ANG I radioimmunologically. Further, renin was found in plasma of eel and toadfish (Nishimura et al. 1976, 1979), *Tilapia mossambica,* two species of tuna (Malvin and Vander 1967), *Anguilla anguilla* (Henderson et al. 1976), *Anguilla japonica* (Sokabe et al. 1966) and *Salmo gairdneri* (Bailey and Randall 1981).

There was no clear evidence for the existence of MD. Distal convoluted tubules do not always return to their parent glomeruli. Tubule cells which contacted JG cells were never histologically differentiated from other tubule cells (Krishnamurthy and Bern 1969). The EGM was not found (see Sokabe and Ogawa 1974; Ogawa 1977; Ogawa and Oguri 1978). Cells resembling mammalian peripolar cells were described in *Cyprinus carpio* and *Carassius carassius* (Koval'Chuk and Likhacheva 1990; Table. 2.1).

2.5.4 Sarcopterygians and Primitive Bony Fishes

In the crossopterygian coelacanth, *Latimeria chalumnae,* granules stainable with Bowie's stain were not found in the kidney (Nishimura and Ogawa 1973), although renin activity was found in renal extracts of this species (Nishimura et al. 1973). However, granules were found electron microscopically in large arteries (Lagios 1974). No anatomical proximity was observed between afferent arterioles and distal tubules, indicating the absence of MD and EGM in the coelacanth (Nishimura et al. 1973; Lagios 1974). Two dipnoan species, the lungfish, *Protopterus aethiopicus* (Ogawa et al. 1972; Nishimura et al. 1973) and *Lepidosiren paradoxa* (Nishimura et al. 1973) contain cells which resemble granulated JG cells. They are located in the media of small arteries and arterioles distant from the glomeruli. Distal tubules do not return to their parent glomeruli. These fishes lack MD and EGM (Nishimura et al. 1973). The presence of renin was examined by measuring pressor activity of incubates of kidney extracts and homologous angiotensinogen. Renin activity was found in extracts of kidneys of the two dipnoans (Nishimura et al. 1973). Plasma renin was demonstrated in *Neoceratodus fosteri* (Blair-West et al. 1977).

Of two holostean species, the longnose gar, *Lepisosteus osseus,* possessed kidneys containing granules near the vascular poles of the glomeruli, while the bowfin, *Amia calva,* did not show granules stained with Bowie's stain (Ogawa et al. 1972; Nishimura et al. 1973). However, the kidneys of both species had renin activity (Nishimura et al. 1973). In the bowfin, ANG II increased blood pressure at high doses and captopril, an inhibitor of converting enzyme, inhibited the increase. The cardiovascular system seems to be regulated by the RAS (Butler et al. 1995). Further histological, electron microscope and chemical studies on renal arteries and arterioles of *Amia* are needed.

The chondrosteans, *Acipenser brevirostris, Calamoichthys calabaricus* and *Polypterus senegalus,* did not show granules that stained with Bowie's stain in their kidneys (Ogawa et al. 1972; Nishimura et al. 1973). No granulated JG cells were found in *Acipenser transmontanus* (Krishnamurthy and Bern 1969).

However, *Polypterus senegalus* showed renin activity in kidney extracts. The presence of renin in *Acipenser* kidney is uncertain (Nishimura et al. 1973). In holosteans and chondrosteans, renal tubules do not contact the vascular pole of the glomeruli. The MD and EGM are not present (Nishimura et al. 1973; Sokabe and Ogawa 1974).

2.5.5 Holocephalans

Granulated cells stainable with Bowie's method were detected in the walls of the afferent arteries adjacent to glomeruli in the ratfish, *Hydrolagus colloeo* (Nishimura et al. 1973; Oguri 1978) and rabbitfish, *Chimaera monstrosa* (Oguri 1978, 1980). The granules were more coarse than those of teleosts and, in contrast to teleostean JG granules, were negative to PAS reaction (Oguri 1978, 1980). Oguri (1978, 1980) observed Bowie-positive granules in epithelioid cells of the arteries distant from the glomeruli and in the glomeruli near the vascular pole in both species. A pressor substance biologically similar to angiotensin was produced when *Hydrolagus* kidney extracts were incubated with homologous plasma. Kidney extracts of the teleost, *Opsanus tau*, did not form a pressor substance when incubated with ratfish plasma (Nishimura 1985). The angiotensinogen content in plasma of the ratfish seems to be low (Nishimura et al. 1973; Nishimura 1985). Although the distal convoluted tubules return to their parent glomeruli to contact afferent arterioles at the vascular poles, structures resembling mammalian MD were not observed in ratfish (Nishimura et al. 1973) nor in rabbitfish (Oguri 1980).

2.5.6 Elasmobranchs

No granulated JG cells were found in the walls of renal arteries and arterioles of around 20 elasmobranch species, including sharks and rays (Bohle and Walvig 1964; Capréol and Sutherland 1968; Nishimura et al. 1970; Oguri et al. 1970; Ogawa et al. 1972; Crockett et al. 1973; Nishimura et al. 1973; see Sokabe and Ogawa 1974; Wilson 1984a; Nishimura 1985 for reviews). Recently, however, JG cells have been found in elasmobranchs (Lacy and Reale 1990; see below).

No pressor activity in rats was found in incubated homologous plasma with kidney extracts of sharks (Bean 1942). In *Triakis sacyllia, Heterodontus japonicus, Orectolobus japonicus, Dasyatis akajei* and *Squalus acanthias*, renin was not detected (Nishimura et al. 1970; Nishimura 1985). Recently, renin-like activity has been estimated in *Scyliorhinus canicula* by measuring ANG I generated in incubated renal extracts with synthetic and porcine angiotensinogen (Uva et al. 1992). Further, these authors found an ANG I converting enzyme-like activity, which was inhibited by captopril, an inhibitor of converting enzyme, in renal extracts using spectrophotometric assay.

In elasmobranchs, the glomeruli are well developed and the distal convoluted segments contact the vascular pole of their parent glomeruli (Kempton 1943; Borghese 1966). However, there is no special structure comparable to mammalian MD cells. Thus, it has been believed that the MD is absent in elasmobranchs (Sokabe and Ogawa 1974; Sokabe et al. 1969). Similar observations were made in the freshwater stingray, *Potamotrygon magdalenae* (Ogawa and Hirano 1982). In contrast, Ghouse et al. (1969) observed that the portion of the distal renal tubule in close contact with JG cells contains a group of cells resembling MD cells in the dogfish, *Squalus acanthias*. The EGM was not observed in elasmobranchs (Sokabe et al. 1969; Ogawa 1977).

Opdyke and Holcombe (1976) reported that ANG I and ANG II exhibited strong pressor activity and that SQ 20881, an ANG I-converting enzyme inhibitor, blocked the activity of ANG I in the spiny dogfish shark, *Squalus acanthias*. Further, Henderson et al. (1981) discovered that an intravenous injection of renal extract of the dogfish, *Scyliorhinus canicula*, or an incubated extract with rat renin substrate, produced a pressor response similar to that of angiotensin in the nephrectomized rat bioassay. Hazon et al. (1989) demonstrated that exogenous ANG II induced drinking and pressor activity in the elasmobranch, *Scyliorhinus canicula*, and suggested the presence of a RAS-like system in elasmobranchs. In fact, elasmobranch ANG I was recently isolated from the incubated renal extract of *Triakis scyllia* with homologous plasma and its amino acid sequence was determined (Takei et al. 1993a; Table 4.2). Therefore, it is now clear that the RAS is present in the cartilaginous fish.

Based on these physiological or pharmacological experiments, Lacy and Reale (1990) reinvestigated the anatomy of the vascular pole of the renal corpuscle in the elasmobranchs, *Squalus acanthias, Mustelus canis, Raja erinacea* and *Rhinoptera bonasus*. They showed, for the first time, JG cells containing few granules by staining with toluidine blue, although the granules were not stained with Bowie's stain. The MD structure was also identified with electron microscopy (Table 2.1). The EGM was not extensive, but the extraglomerular mesangial cells had abundant microfilaments and inconspicuous and discontinuous lacis; they possessed numerous vacuoles, probably containing lipids. Lacy and Reale (1989) found peripolar cells in the kidneys of the elasmobranchs, *Raja erinacea, Mustelus canis, Rhizoprionodon terraenovae, Sphryna lewini* and *Rhinoptera bonasus*. These cells lay close to the granulated cells of the glomerular afferent arteriole. In *Squalus acanthias*, peripolar cells could not be identified. Recently, Galli-Phillips (1991) detected immunoreactive ANG II in plasma (30–80 pg/ml) in the nurse shark, *Gynclymostoma cirratum*, and reported that hemorrhage and transfer to 25% sea water induced an increase of plasma ANG II. ANG II was also detected in the brain, kidney, pituitary and rectal gland tissue of this shark, suggesting the presence of tissue RAS.

2.5.7 Cyclostomes

No JG cells containing granules stainable with Bowie's stain were found in the lamprey, *Lampetra japonica*, nor in the hagfish, *Paramyxine atami* (Nishimura et al. 1970; Oguri et al. 1970; Table 2.1). Sutherland (1966) did not find JG granules in adult or larval lampreys from fresh water, nor in the hagfish, *Myxine glutinosa* (cited in Sokabe and Ogawa 1974). No structure like MD or EGM was found in cyclostomes (Sokabe et al. 1969). Incubated renal extracts from *Paramyxine* or *Lampetra,* using homologous, rat or carp plasma, did not show pressor activity when bioassayed in anesthesized rats. Kidney extracts from the rat or carp also produced no pressor substance when incubated with cyclostome plasma (Nishimura 1985). However, Henderson et al. (1981) reported that renal extracts of *Lampetra fluviatilis* incubated with canine renin substrate generated pressor substances similar to angiotensin when assayed in the rat. Further, Carroll and Opdyke (1982) showed that ANG II injection elevated blood pressure in *Myxne glutinosa*. These findings suggest the presence of the RAS in cyclostomes. Very recently ANG has been isolated in *Lampetra fluviatilis* (Takei and Rankin, unpubl.). Further investigations of JGA using different fixatives, stains, biochemical and electron microscope techniques are needed in cyclostomes.

2.5.8 Summary

Mammals and birds possess JG cells, MD, EGM and peripolar cells. In reptiles, amphibians and fishes, further studies are needed to determine the presence or absence of each JG component observed in mammals and birds, using new techniques. As these components have been found in elasmobranchs, reinvestigations on cyclostomes would be most interesting. Phylogenetic distribution of each component of JGA is summarized in Table 2.1, based on data obtained by recent investigations.

Chapter 3

Innervation in the JGA

3.1 Light Microscopy

Oberling (1944), using a silver impregnation method, observed a rich nervous plexus along the afferent arteriole of the glomerulus and nerve endings in contact with cells of the arterial wall of the human kidney. deCastro and dela Peña (1952; cited in Barajas 1964) reported the presence of a rich sympathetic innervation in both afferent and efferent arterioles in the human kidney. De Muylder (1952) described in his monograph that nerves are associated with the renal arterioles in the region of the JG cells and MD in the mouse. Since then, extensive investigations have confirmed a nonmyelinated nerve supply to the JGA (see Davis and Freeman 1976; Barajas and Müller 1980; Taugner and Hackenthal 1989 for reviews; Figs. 2.1, 2.4, 3.1).

The availability of a fluorescence histochemical technique for biogenic monoamines (Falck et al. 1962) has made possible the elucidation of the distribution of sympathetic nerves in the kidney. Adrenergic innervation in the glomerular afferent arterioles was demonstrated by Nilsson (1965) in rabbits and rats; McKenna and Angelakos (1968a) in the dog; Doležel (1966) in the rat and dog; Wågermark et al. (1968) and Ljungqvist and Wågermark (1970) in the rat; Munkacsi (1969) in the rat and the desert rat (*Dipodomys merriami*); Müller and Barajas (1972) in the monkey; Doležel et al. (1976) in the dog, guinea pig, rat and mouse; Gorgas (1978a,b) in the rat; and Barajas (1978) in the monkey and rat. Gill and Stephens (1983) found sympathetic nerves associated with arterioles, ending at the glomerulus in the turtle, *Pseudemys scripta*. Adrenergic innervation in the afferent arterioles was observed by Unsicker et al. (1975) in the toad and frog and by Morris and Gibbins (1983) in the toad. In the carp, monoaminergic innervation was reported in the wall of glomerular arterioles (Kuzmina et al. 1986). Dense adrenergic innervation was found along the terminal arterioles and afferent arterioles and was conspicuous at the preglomerular sphincters in *Salmo gairdneri* (Elger et al. 1984). The adrenergic involvement in renin secretion based on pharmacologic studies is discussed in Sections 5.1.3 and 5.1.4.2.

Adrenergic nerve fibers were not seen in association with the efferent arterioles in the canine kidney (Mckenna and Angelakos 1986a). However, monoaminergic innervations were reported in both afferent and efferent arterioles in rat and desert rat kidneys (Munkacsi 1969). Ljungqvist and Wågermark (1970) and Gorgas (1978a,b) observed adrenergic nerve terminals

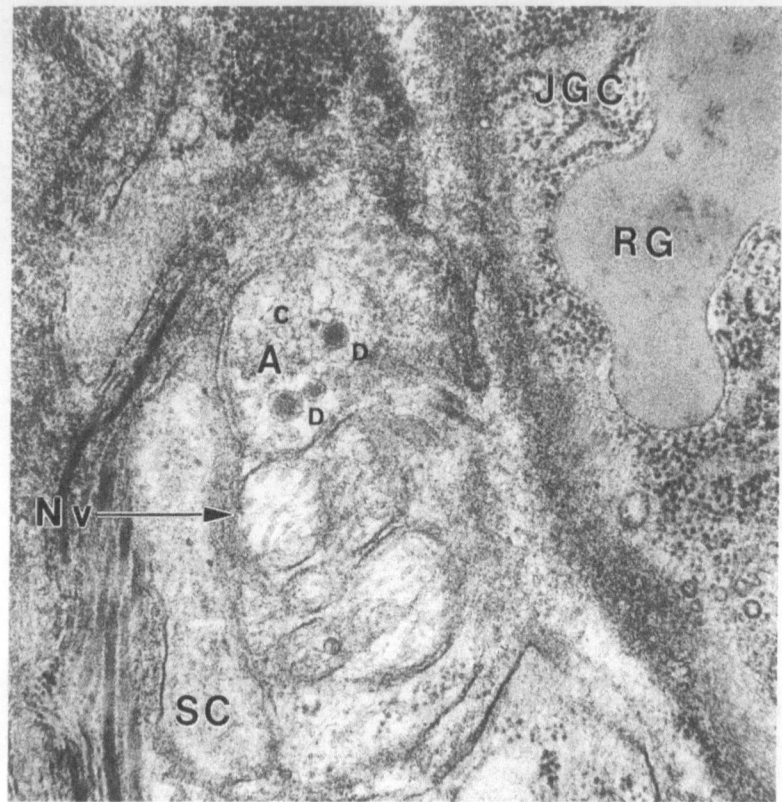

Fig. 3.1. Nerve bundle (Nv) associated with an axon (A) including clear vesicle (C) and dense-cored vesicle (D) in the mouse. RG Granule of JG cell; SC Schwann cell process. × 48000. (Courtesy of Dr. Mizuhira)

along both afferent and efferent arterioles in the rat. Further, adrenergic innervations in the efferent arterioles were described by Doležel et al. (1976) in mice, rats, pigs, dogs and monkeys, but not in guinea pigs and cats. Monoaminergic innervations were found in 3–5% of efferent anterioles in the kidney of the toad, *Bufo marinus* (Morris and Gibbins 1983).

Histochemical demonstration of acetylcholinesterase has been performed by several investigators studying cholinergic innervation in the kidney. McKenna and Angelakos (1968b) found that acetylcholinesterase–containing fibers from ganglion cells in the hilus of the canine kidney innervate the afferent arterioles of the kidney. These fibers were not affected by ablation of the nerves traveling with the renal vessels, which removed all of the cortical noradrenergic nerve fibers. The investigators concluded that cholinergic innervation of the afferent arterioles is independent of the noradrenergic nerve supply. The distribution and extent of acetylcholinesterase-positive innervation of the JGA, however, were found to parallel those observed with catecholamine fluorescence in the rat (Barajas and Müller 1980). After administration of 6-hydroxydopamine, both the fluorescence and the acetylcholinesterase precipitate decreased concomitantly within the glomerular arterioles in rats. It seems that the glomerular arterioles are innervated by adrenergic nerves that display acetylcholinesterase activity (Barajas 1979; Barajas and Müller 1980). Thus, nerves showing acetylcholinesterase reaction may not always be those having only acetylcholine (ACh) as a transmitter. Under certain conditions a single neuron can synthesize and secrete two traditional neurotransmitters, ACh and noradrenaline (Landis 1984). If so, acetylcholine may function as a modulator of the release of noradrenaline. It is still not clear whether cholinergic innervation of the JG cells occurs. Immunocytochemistry of ACh may elucidate this problem, if the procedure can be accomplished.

3.2 Electron Microscopy

The first electron microscope studies on innervation of the JGA were performed by Simpson and Devine (1964) in sheep and by Barajas (1964) in the monkey and rat. Simpson and Devine (1964, 1966) demonstrated the innervation of a bundle of nerve axons or a single axon, covered incompletely by Schwann cells, at the basement membrane located between the axon and the JG cells in sheep kidney. Barajas (1964) observed numerous nonmyelinated nerve fibers associated with the afferent and efferent glomerular arterioles in the monkey and rat. He found dilated vesiculated nerve processes located adjacent to both smooth muscle cells and granular cells in the afferent arteriolar wall. According to Barajas (1978) and Müller and Barajas (1972), single axons established *en passant* synapses with granular and agranular vascular cells and proximal and distal tubular cells in the juxtaglomerular region. Similar *en passant* synapses were less frequently observed in the rat (Barajas and Müller 1973, 1980; Gorgas 1978a,b). Figures 2.1, 2.4 and 3.1 show the innervation of the JG cells.

Four types of granules or vesicles occurred in nerve terminals that contacted afferent arterioles in the monkey : (1) small, densely cored granular vesicles with diameters around 50 nm; (2) small agranular (clear) vesicles of similar size; (3) large granular vesicles with diameters of around 90 nm; and (4) large agranular vesicles (Barajas and Müller 1980). Nerves containing the first type of vesicles were thought to be noradrenergic. In sheep kidney, the vesicles or granules were roughly classified into three types according to their size. The different types of vesicles or granules were usually present together in any single axon endings (Simpson and Devine 1964, 1966). The chemical identity of substances within these granules or vesicles needs to be clarified. They may contain some biologically active peptides in addition to noradrenaline. Coexistence of noradrenaline and biologically active peptides has been demonstrated within the same neuron (see Schultzberg 1984). The peptides may modulate noradrenaline secretion from the endings, resulting in regulation of renin release from the JG cells.

In the lizards, *Lacerta muralis* and *Lacerta sicula*, innervations of afferent arterioles and efferent arterioles were also found. The terminals contained three vesicle types, with a prevalence of the small dense-cored ones, that were thought to be adrenergic; terminals with only small agranular vesicles were rare (Guglielmone and Daneo-Sisto 1978).

No nerves were seen in contact with the MD in the monkey, but nerves did contact distal tubular cells near the MD (Müller and Barajas 1972; Barajas and Müller 1980). Wågermark et al. (1968) also were unable to detect any fluorescent fibers in close relation to the MD or lacis cells in the rat. However, Hartroft (1966) found nerve endings in contact with the MD in the rat; unlike the study of Wågermark et al. (1968). Gorgas (1978a,b) reported that fluorescent nerve endings make contact with peripherally located cells (so-called intermediary cells) of the MD in the rat.

3.3 Summary

A dense adrenergic innervation is observed with fluorescence histochemistry in the juxtaglomerular region. Adrenergic fibers innervate the wall of afferent and efferent arterioles in vertebrates generally. These findings are supported by data from electron microscopy. It is not definite whether adrenergic fibers are in contact with cells of the MD and the EGM. Histochemical demonstration of acetylcholinesterase in the juxtaglomerular apparatus suggests the presence of cholinergic innervation therein, but it is difficult to identify whether or not the fibers are truly cholinergic.

Chapter 4

Tissue Distribution of the RAS

The components of the plasma RAS originate primarily from renal renin, hepatic angiotensinogen, and pulmonary converting enzyme (Fig. 1.2). However, recent progress in biotechnological and immunocytochemical techniques enables us to detect trace amounts of biologically active peptides and their mRNA, which facilitates discovery of each component of the RAS in various tissues which were not thought to be the sites of production. Since some of these tissues such as brain, kidney, heart and blood vessels produce all components of the RAS (see Campbell 1987a; Dzau 1988; Baker et al. 1992; Saavedra 1992 for reviews; Fig. 1.3; Table 4.1), a paracrine or autocrine function of ANG II has to be taken into account when considering its physiological action.

Table 4.1. Presence of the RAS in various tissues demonstrated in mammals. (For references, see Mendelsohn 1985; Campbell 1987a; Dzau 1988)

Tissue	Renin		Angiotensinogen		CE[a]	ANG II
	Protein	mRNA	Protein	mRNA		
Adrenal cortex	+	+	+	+	+	+
Adrenal medulla	+					
Anterior pituitary	+	+	+		+	+
Brain	+	+	+	+	+	+
Endothelium	+	+	+	+	+	+
Gut	+			+		+
Heart	+	+	+	+	+	+
Kidney	+	+	+	+	+	+
Liver	+		+	+		
Lung	+		+	+	+	
Ovary	+	+	+	+		+
Placenta	+	+	+			
Salivary gland	+	+	+	+	+	+
Spleen	+			+		
Testis (Leidig cell)	+	+	+	+	+	+
Thyroid	+					
Uterus	+	+	+			
Vascular smooth muscle	+	+	+	+	+	+

[a]Converting enzyme.

4.1 Renin

Renin, as its name indicates, is a renal enzyme that initiates a cascade of angiotensin-forming reactions. In addition to renal renin, however, renin-like activities have been identified in many extrarenal tissues by radioimmunoassay and immunocytochemical techniques (Hayduk et al. 1970; Ganten et al. 1976; Campbell 1987; Dzau 1988). These extrarenal renins do not originate from renal renin taken up from plasma because renin mRNA is also detected in most of these tissues (Dzau 1988). The presence of extrarenal sources of renin is also supported by the observation that a measurable renin activity persists in plasma after bilateral nephrectomy (Ganten et al. 1976). In order to distinguish these extrarenal renin-like enzymes from renal renin, they are called by various names such as isorenin, tissue renin, pseudorenin, or angiotensinogenase. The term renin is used for extrarenal renins in this chapter although the Nomenclature Committee of the International Society for Hypertension recommends the use of isorenin (Clin Sci Mol Med 55: 113s–115s, 1978).

The presence of brain RAS is of particular interest in view of various central actions of ANG II (Phillips 1978), since the brain tissue is strictly protected from blood-borne ANG II by the blood–brain barrier (Weindl 1973). The local synthesis of ANG II in the brain ensures the paracrine/autocrine action independent of the blood-borne ANG II. The brain RAS has been investigated intensively which resulted in production of several excellent reviews (e.g. Ganong 1984; Ganten et al. 1984, 1988). The brain renin is not derived from plasma because brain renin concentration increases after bilateral nephrectomy (Ganten et al. 1982). The presence of brain RAS is also supported by the fact that intracerebral injection of each component of the RAS induces short-latency drinking in the rat (Fitzsimons et al. 1978a). All components of the RAS are identified in neuroblastoma cells in culture (Fishman et al. 1981; Okamura et al. 1981). Renin was found together with converting enzyme in the synaptosomes from the rat brain, which indicates localization of these enzymes within nerve terminals (Paul et al. 1985). The mRNA that hybridizes cDNA of mouse submaxillary gland renin was identified in the mouse brain (Ganten et al. 1984). These data strongly suggest that the brain synthesizes renin whose structure is similar to that of peripheral renin.

Renin activity was found in the brain of human, sheep, dog, rat, hog and desert rat (Ganten et al. 1976). The highest concentration was measured in the pineal organ followed by the pituitary and choroid plexus in the hog which was bilaterally nephrectomized and exsanguinated by saline perfusion (Hirose et al. 1980). In this study, renin was separated from acid proteases and its concentration determined by direct radioimmunoassay for mouse submaxillary gland renin. The renin concentration of the pineal gland is much lower than that of the kidney cortex, and the neural lobe contains very low renin. Within the brain parenchyma, the hypothalamus showed the highest concentration followed by the cerebellum and amygdaloid nucleus. In the rat and mouse, renin was localized by immunocytochemical techniques in the paraventricular and periventricular nuclei as well as in the supraoptic nucleus, and in nerve fibers of the neurohypophysis (Rix et al. 1982). Immunoreactive renin was also detected in the cerebellum. Renin immunoreactivity was most intense in the cell bodies of oxytocin-containing neurons of normal rats (Calza et al. 1982) or in the magnocellular paraventricular nucleus, supraoptic nucleus and neural lobe of the pituitary of nephrectomized rats (Healy and Printz 1984). It seems that the distribution of renin activity differs among different species and is not always correlated with the target sites for central actions of ANG II. In addition to active renin, prorenin was found in the hog brain and its ratio to the active renin varies among regions (Hirose et al. 1980). Inagaki and Inagami (1984) reported the presence in the rat brain of a new form of inactive renin which is different from prorenin. They termed it latent renin which may be a complex of renin and its binding protein.

High renin activity was measurable in the submaxillary gland of mice (Cohen et al. 1972) and rats (Hackenthal et al. 1972); such activity far exceeds that of the kidney (Wilson et al. 1981). The renin from the submaxillary gland is not distinguishable from renal renin by immunological and physicochemical criteria (Malling and Poulsen 1977), thereby providing a good model for biochemical studies of the renin molecule. With respect to amino acid sequence, renal and submaxillary gland renin are highly homologous but distinctly different. Renin of the mouse submaxillary gland was the first to be isolated in a pure crystalline form (Cohen et al. 1972). It is of interest to note that renin content of the submaxillary gland is greater in the male than in the female and is increased by plasma levels of androgens (Oliver and Gross 1967). However, manipulations that alter renal renin concentration such as Na loading and mineralocorticoid treatment do not affect renin concentration and its mRNA content in the submaxillary gland of rat and mouse (Catanzaro et al. 1985). Renin concentration in the submaxillary gland is higher in spontaneously hypertensive rats than in normotensive controls, but it does not seem to be related to high blood pressure (de Jong et al. 1972). In addition, Pedersen and Poulsen (1983) demonstrated a huge release of renin from the submaxillary gland into saliva in the male mouse after confrontation with another male. This aggression-induced renin release resulted in salivary renin concentration six orders of magnitude higher than that in the plasma. The salivary gland renin may be injected into the opponent on biting.

Renin activity was measurable in the uterus and placenta of several species of mammals (Table 4.1; Ganten et al. 1976). The uterine renin activity of pregnant rabbits appears to exceed that of the kidney (Bing and Faarup 1966; Ferris et al. 1967). However, renin activity was low in the uteri of rat, goat, pig and mouse (Ganten et al. 1976). The biochemical nature of uterine renin is similar but not identical to that of renal renin (Anderson et al. 1968; Ryan 1970). It is suggested that JG cells of the kidney take up renin from circulating blood (Taugner et al. 1985b), but this is not the case in the uterus as uterine cells in culture synthesize renin (Symonds et al. 1968). The renin activity in rabbit plasma that remained after bilateral nephrectomy may be explained for the most part by the release of renin from the uterus (P. Gordon et al. 1967). Release of renin was reported from the perfused uterus of pregnant rabbit after histamine, angiotensin or norepinephrine treatment (Ryan and Ferris 1967). A possible local function of uterine RAS may be the discharge of endometrium and amnion, regulation of myometrial tonicity, interference with prostaglandin action, and regulation of local blood flow (see Ganten et al. 1976 for review).

Renin activity was identified in the arteries and veins of pig (Dengler 1956), rat (Rosenthal et al. 1969), and dog (Hayduk et al. 1970). Renin was detected in the cultured vascular smooth muscle cells and endothelial cells of cat, calf and rat (Re et al. 1982; Dzau 1984). The vascular renin has a molecular mass of 38 kDa and a pH optimum at 6.5–6.8. Changes in plasma renin activity scarcely influence the vascular renin concentration, and nephrectomy seems to stimulate renin release from the mesenteric artery of dog (Ganten et al. 1976). Renin-specific immunostaining can be demonstrated in vascular smooth muscle cells as well as endothelial cells even after 2 days in serum-free culture (Dzau 1988). A part of renin in endothelial cells may be taken up from blood (Loudon et al. 1983), but the local synthesis is apparent because of the accumulation of renin mRNA and incorporation of ^{35}S-methionine into the renin molecule after pulse labeling in both endothelial and smooth muscle cells (Dzau et al. 1987). Intracellular localization of other components of the RAS, i.e., angiotensinogen, converting enzyme and angiotensins, was also demonstrated in cultured vascular cells (Lilly et al. 1983) and in the extracts of vascular tissues (Rosenthal et al. 1984). Furthermore, ANG II is secreted from the endothelial cells (Kifor et al. 1987). Based on these observations, Dzau (1987) proposed that ANG II secreted from the endothelial cells may act on themselves (autocrine) or on the adjacent vascular smooth muscle cells (paracrine) to regulate arterial pressure/arterial tone or angiogenesis. ANG II receptors are identified in the endothelial cells which may mediate the secretion of endothelium-derived vasoactive factor. In the spontaneously hypertensive rats with normal plasma renin activity, the systolic blood pressure correlates positively with the aortic renin (Asaad and Antonaccio 1982). In an experimental model of hypertension in the rat (Riegger et al. 1977) or essential hypertension of the human (Haber et al. 1983) with normal plasma renin activity, blockade of the RAS decreases blood pressure probably through blockade of the arterial system. The renin activity is extremely high in the choroid plexus of dog, rat and human (Erdös 1975). Renin in this vascular organ may regulate ANG II levels in the cerebrospinal fluid. The cardiac action of ANG II has attracted attention of clinical researchers in relation to its possible role in ischemic and congestive heart failure (Dzau 1988). Not only renin but also angiotensinogen and converting enzyme are produced by the heart as demonstrated by the mRNA production (Table 4.1). However, nothing is yet known as to the biochemical nature of cardiovascular renin molecules.

Since the adrenal gland is the target of ANG II and ANG III (Aguilera et al. 1978), their local synthesis in the organ is of particular importance. In fact, renin was identified in the adrenal glands of rat (Naruse and Inagami 1982), man (Naruse et al. 1985), rabbit (Ryan 1967), and dog (Hayduk et al. 1970). Although adrenal renin has not been isolated and its chemical structure is undetermined, it satisfies all criteria for positive identification of renin (Naruse et al. 1984a). The renin content in the adrenal gland increases after ne-phrectomy in spontaneously hypertensive rats, indicating its independence from renal renin (Naruse and Inagami 1982). In the rat, Na deficiency increases adrenal renin concentration (Ganten et al. 1974), and Na loading decreases it in the zona glomerulosa where renin as well as aldosterone are preferentially synthesized (Doi and Mulrow 1984). Pottasium loading increases renin and aldosterone concentrations in the zona glomerulosa of normal rats, and ACTH increases these concentrations in hypophysectomized rats. Changes in renin concentration in the zona glomerulosa are positively correlated with changes in aldosterone concentration but not with plasma renin activity in the rat (Doi and Mulrow 1984). ANG II receptor blockade by saralasin produces dose-dependent inhibition of aldosterone production by the rat adrenal cortical cells (Williams et al. 1974). Thus, adrenal RAS might function as a local regulator of aldosterone production in the rat.

ANG II is known to regulate the release of hypophysial hormones as sum-marized in section 8.6. Renin activity was detected in the luteinizing hormone cells of the rat pituitary (Naruse et al. 1981) and in prolactin cells of the human pituitary (Mukai et al. 1984), but not in the neural lobe. Since anterior pituitary cells do not originate from neural ectoderm, pituitary renin may be different from the brain enzyme. The pituitary renin shares some biochemical char-acteristics with renal renin such as molecular weight, optimum pH, and pre-sence of a trypsin-activatable inactive form (Mizuno et al. 1985). Renin extracted from the bovine anterior pituitary seems to be slightly different from those of human and hog kidney and mouse submaxillary gland (Hirose et al. 1982).

Immunoreactive renin was identified in Leidig cells of the rat testis (Par-mentier et al. 1983). The testicular renin activity is not correlated with plasma renin activity, and is increased by gonadotropin administration and decreased by hypophysectomy as measured by a direct radioimmunoassay for rat renal renin (Naruse et al. 1984b). The rat Leidig cells contain inactive (latent) renin as well as active renin, and the former is activated by the sulfhydryl reagents (Pandey et al. 1984b). The rat Leidig cells seem to synthesize renin, since these cells contain mRNA which hybridizes with the cDNA of rat submaxillary gland renin (Pandey et al. 1984a). The molecular weights of active and latent renin of rat testes are estimated to be 39 and 48 kDa, respectively. All components of the RAS are found in rat Leidig cells as examined by radioimmunoassay in combination with reverse-phase HPLC, which indicates intracellular formation of ANG II in testicular cells (Pandey et al. 1984c).

Other organs that exhibited renin activity are the skeletal muscle, liver, lung and spleen of dog (Hayduk et al. 1970), thyroidal follicular cells and prostate gland of human (Naruse et al. 1985), small intestine and colon of mouse (Dzau 1984), and pancreas, thymus and duodenum of rat (Skeggs et al. 1969). In the case of pancreas, thymus and duodenum, however, the renin activity may be attributable to cathepsin D since the enzymes from these tissues had very low pH optima. Recently, renin mRNA was detected in rat ocular tissues (Brandt et al. 1994).

4.2 Angiotensinogen

The presence of angiotensinogen in brain tissue was first suggested by the observation that ANG I was liberated after incubation of brain extract with renal renin in the dog, rabbit and sheep (Table 4.1; Ganten et al. 1976). The highest concentration was found in the circumventricular organs such as area postrema, organum vasculosum lamina terminalis and median eminence in the rat (Lewicki et al. 1978). Adrenalectomy decreases angiotensinogen content in plasma and in several brain regions such as preoptic area, anterior hypothalamus and area postrema in the rat, and corticosterone treatment restores the decrease (Wallis and Printz 1980). The localization of angiotensinogen appears to be intracellular (Printz et al. 1982). Angiotensinogen activity was also noted in the cerebrospinal fluid of rat, dog, sheep and man. The concentration of angiotensinogen in the cerebrospinal fluid is variable among species but is generally greater than plasma. Heterogeneity was noted in the brain angiotensinogen of rat (Murakami et al. 1984). Several investigators suggest that angiotensinogen from the liver or plasma differs biochemically and immunologically from that of brain or cerebrospinal fluid (Printz et al. 1978; Ito et al. 1980). However, a study using cell-free translation of mRNA showed that angiotensinogen from the brain tissue is identical with that of the liver (Campbell et al. 1984). They also showed that the translatable mRNA of angiotensinogen in the liver increases several fold after nephrectomy or dexamethasone (glucocorticoid) treatment, whereas the increment is less than twofold in the brain. Thus, the expression of the angiotensinogen gene seems to be regulated differently between the liver and the brain. The neuroglial astrocytes are known as a source of angiotensinogen production, but pure rat neuronal cells in culture also synthesize and secrete angiotensinogen (Thomas et al. 1992). Angiotensinogen synthesis and secretion are stimulated by estradiol and synthetic mineralocorticoid in the rat liver (Klett et al. 1992), but its synthesis is inhibited by cortisol (Olson et al. 1991). Angiotensinogen synthesis is also stimulated by estrogen in the pituitary of ovariectomized rat (Healy et al. 1992).

Tissue distribution of angiotensinogen mRNA has been investigated in the rat using a cloned cDNA of rat angiotensinogen as a probe (Ohkubo et al. 1986). Angiotensinogen mRNA was found in the brain, kidney, adrenal gland, ovary and lung (Table 4.1). The highest concentration was noted in the brain, but it is one-third to one-fourth of that of the liver. The mRNA content in the kidney, adrenal gland and ovary is 1/20–1/30, and that of the lung only 1/1000 of that of the liver. Angiotensinogen was also found in rat proximal tubules by immunohistochemistry, but it appears to be a material reabsorbed from the glomerular filtrate rather than locally synthesized (Richoux et al. 1983). However, angiotensinogen was detected in cultured rabbit proximal tubule cells by a specific radioimmunoassay (Yanagawa et al. 1991).

4.3 Converting Enzyme

Converting enzyme, which requires Cl⁻ ions for its full activity, was first identified in plasma (Skeggs et al. 1954). However, the activity in plasma was too low to account for the immediate conversion of ANG I to ANG II. It was later shown that the lung contains high levels of converting enzyme which may be responsible for generation of ANG II in the blood (Ng and Vane 1967). The enzyme is concentrated in endothelial cells as shown by electron microscopic immunocytochemistry (Ryan et al. 1975). It is suggested that the vascular endothelial cells of the lung and peripheral blood vessels, and the epithelial cells of renal tubules, are major sources of this enzyme. The converting enzyme activity was also identified in a variety of peripheral tissues (see Erdös 1976), among which brain converting enzyme is the subject of the most intensive research.

Converting enzyme activity was identified in the caudate nucleus of man (Poth et al. 1975), in the choroid plexus of man, dog and rabbit (Defendini et al. 1983), and in the pituitary gland, locus caeruleus, substantia nigra and hypothalamus (Yang and Neff 1972), striato-nigral pathway (Strittmater et al. 1984), and circumventricular organs such as the subfornical organ, area postrema and choroid plexus (Weindl et al. 1977) of rat. The converting enzyme activity is enriched in brain microvessels compared to homogenates of intact cerebral cortical gray matter in cattle and rat (Gimbrone et al. 1979). Since Paul et al. (1985) suggested that converting enzyme and renin coexist in the same nerve endings of the rat, intraneuronal synthesis of ANG I and II is likely. It seems that converting enzymes of striatal neurons and brain microvessels are different as a result of differential glycosylation (Williams et al. 1991). Converting enzyme activity is detectable in the cerebrospinal fluid when assayed by the hydrolytic activity of Hip-His-Leu (Rix et al. 1981). Brattleboro rats with hypothalamic diabetes insipidus have a greater converting enzyme activity in the supraoptic nucleus and periventricular hypothalamic nuclei than normal rats, and the increased activity is reversed by vasopressin treatment (Saavedra and Chevillard 1982). It seems that vasopressin regulates the formation of brain ANG II by modulating the converting enzyme activity.

Endogenous converting enzyme may have a regulatory function for the activity of the RAS in some tissues. Evered et al. (1980) showed that intracerebroventricular injection of SQ14225 (Captopril) inhibits drinking induced by hog renin, synthetic renin substrate and ANG I but does not inhibit drinking induced by ANG II. Rosivall et al. (1984) showed that intrarenal arterial infusion of ANG I manifests its effects after conversion to ANG II. This may be achieved by the endothelial cell-bound enzyme in preglomerular and glomerular vessels (Caldwell et al. 1976). Addition of ANG I to the aortic ring preparation resulted in contraction, which is partially inhibited by SQ20881 (Saye et al. 1984). The removal of endothelium blocked initial conversion to ANG II. Thus the converting enzyme in the endothelium plays a role in ANG I-induced contraction. More recently, however, it has been shown that aortic smooth muscle is also capable of converting ANG I to ANG II as the conversion occurs in the endothelium-denuded aortic ring and in primary culture of aortic smooth muscle cells (Andre et al. 1990).

4.4 Angiotensins

ANG I was extracted from the brain of nephrectomized rat, rabbit and monkey (Ganten et al. 1984). ANG I was also found in the cerebrospinal fluid of man, dog and rat in quantities ranging from 4–28 pg/ml (Ganong 1984). Brain ANG I concentration varies from 45 to 157 pg/g in the rat (Ganten et al. 1983). The concentration is highest in the hypothalamus, and it increases significantly after nephrectomy. An immunohistochemical study revealed that ANG I is localized in the circumventricular organs such as the area postrema, subfornical organ and median eminence (Changaris et al. 1977).

ANG II was identified immunohistochemically in the brain, but the result is variable depending on the antisera used (Ganong 1981). A high density of immunoreactive ANG II was found in the median eminence, neural lobe of the pituitary, central nucleus of amygdala, spinal nucleus of the trigeminal nerve and substantia gelatinosa (Kilcoyne et al. 1980). Radioimmunoassay measures variable concentrations of ANG II in the brain of dog, rat and rabbit (see Reid 1977 for review). The variability seems to be due to the degradation of the peptide during extraction and incubation for radioimmunoassay. Sirret et al. (1981) measured concentrations ranging from 45 pg/g in the cerebral cortex to 360 pg/g in the hippocampus after inhibition of degradation. However, Simonnet et al. (1984) found that immunoreactive ANG II measured by radioimmunoassay has a molecular weight of ca. 10 kDa, and the radioimmunoassay always gives greater values than the radioreceptor assay. In fact, radioimmunoassay may not be a good tool for measurement of ANG II, because ANG II is measurable in human plasma even though no ANG II is detected by HPLC (Nussberger et al. 1985). Despite these observations, Ganten et al. (1983) finally isolated ANG II from the brain of nephrectomized rats by HPLC. ANG II immunoreactivity was also found in cultured brain cells of neonatal rat (Phillips et al. 1979) and in human brain cells (Quinlan and Phillips 1981). The immunoreactive ANG II from cultured brain cells of neonatal rats comigrated with authentic [Asp^1, Ile^5] ANG II on HPLC (Raizada et al. 1984b). Phillips and Stenstrom (1985) examined the regional distribution of ANG II in the nephrectomized rat brain after purification of ANG II by HPLC followed by radioimmunoassay. The levels are higher in the hypothalamus (125 pg/g), pituitary (190 pg/g) and spinal cord (199 pg/g) than in the cortex (60 pg/g). The hypothalamic levels of spontaneously hypertensive rat are not higher than those of normotensive controls.

ANG II-(2–7), -(3–7) and -(3–8) were identified in the canine hypothalamus (Welches et al. 1991), whereas ANG II-(1–7) was predominant in the rat brain (Chappell et al. 1989). ANG II-(1–7) is as effective as ANG II for vasopressin release (Schiavone et al. 1988) and neurogenic hypertension (Campagnole-Santos et al. 1989) but not for dipsogenic action (Schiavone et al. 1990). Since ANG II-(1–7) is localized in nerve fibers and terminals (Block et al. 1989), this peptide may have a neuromodulator function in the rat. However, ANG II-(1–7) reported previously in the rat hypothalamus could be artifactual (Lawrence et al. 1992), because brain tissues have high activity of prolyl endopeptidase that cleaves phenylalanine from the C-terminus of ANG II (Welches et al. 1991). ANG II and III are released from the neurosecretory paraventricular nucleus of rat after water deprivation (Harding et al. 1992). The release of ANG II from fetal rat brain is regulated by intracellular Ca ions influxed mostly through N-type calcium channels (Gadbut et al. 1991).

Although sometimes undetectable, ANG II concentration in the cerebrospinal fluid was usually as high as that in plasma (Ganong 1984). Immunoreactive ANG I, II and III were extracted from the rat and human cerebrospinal fluid which migrated identically with their standards on two different HPLC systems (Herman et al. 1982). ANG II is formed in vivo and in vitro when renin is added to the cerebrospinal fluid (Husain et al. 1983). It seems that ANG II is formed intracellularly in the brain (Re 1984). ANG II in the cerebrospinal fluid is not derived from plasma, since its levels remain low after intravenous infusion of ANG II in the dog (Mikami et al. 1985). It was also shown that dehydration increased plasma ANG II levels but decreased cerebrospinal fluid levels in cattle (Bell et al. 1985). ANG II was identified by immunohistochemistry in the gonadotrophs but not in the lactotrophs in the rat anterior pituitary (Deschepper et al. 1985). Renin was found in both gonadotrophs and lactotrophs in the rat as mentioned above. Tonin (Kondo et al. 1980; Schiffrin and Genest 1983; Fig. 1.2) and cathepsin G (Wintroub et al. 1981) in the brain seem to be capable of cleaving ANG II directly from angiotensinogen.

Since most ANG II antisera used for radioimmunoassay and immunohistochemistry recognize its C-terminus and thus cross-react with ANG III and other N-terminally truncated ANG II, the concentration and distribution of ANG II reported should have included these peptides if not separated by HPLC. In fact, ANG III and other fragments are present in the cerebrospinal fluid (Semple et al. 1980). Approximately 10% of immunoreactive ANG II extracted from the brain is true ANG II (Ganten et al. 1983). ANG II-immunoreactive neurons innervate both magnocellular and parvicellular neurons in the paraventricular nucleus (Lind et al. 1984a), but ANG III-immunoreactive neurons innervate only parvicellular neurons (Ganten et al. 1984).

Immunohistochemical studies showed that ANG II and renin coexist in the JG cells of intact and adrenalectomized rat (Taugner et al. 1984b). Virtually all mature JG cells contain both renin and ANG II, and the content increases after adrenalectomy. However, since ANG I, angiotensinogen and converting enzyme are not found in the JG cells, ANG II is thought to be taken up from the extracellular fluid by pinocytosis into the secretory granules. More recently, however, colocalization of renin, ANG I and II were immunoelectron microscopically demonstrated in the cultured JG cells of newborn rat (Inagami et al.

49

1991). They also showed that in the perfused rat kidney ANG I and II are released into the perfusate in quantities which account for a large part of the intrarenal formation of ANG II observed in vivo (Inagami et al. 1992b). In addition, Kohara et al. (1991) reported that ANG II-(1–7) is generated in the blood by cleavage of ANG I through a pathway independent of converting enzyme in dogs. They claimed that ANG II-(1–7) is a member of circulating ANG peptides. In fact, ANG II-(1–7) is a potent antidiuretic peptide in rats (Santos and Baracho 1992), and displays different effects on fluid absorption in the isolated proximal straight tubule depending upon doses (Garcia and Garvin 1994).

4.5 Nonmammalian Vertebrates

A complete set of the RAS components is present in all vertebrate classes because incubation of kidney extract with homologous plasma produces ANG I in selected species from cyclostomes to mammals and because ANG I, but not ANG II no longer functions as a vasopressor substance after inhibition of the converting enzyme (Chaps. 6 and 8). However, little is known about the tissue distribution of components of RAS in nonmammalian vertebrates.

The presence of JG granules as shown by Bowie's staining or modified Movat's silver staining has been demonstrated in the kidney of avian, reptilian, amphibian and piscine species so far examined (Table 1.2). Lamers et al. (1985b) applied immunohistochemistry to the toad kidney using antisera raised against purified mouse submandibular gland renin, and found immunoreactive renin in the JG cells of *Bufo* species. Electron microscopy revealed that so-called JG granules in the toad appear to be lysosomes (Lamers et al. 1985a). Bowie-positive granules were not identified in the elasmobranch kidney, and the incubation of elasmobranch kidney extract with rat or carp plasma or incubation of rat or carp kidney extract with elasmobranch plasma did not produce any vasopressor substance when assayed in the rat (Nishimura et al. 1973). However, JG granules were identified electron-microscopically in the kidney of four elasmobranch species, *Squalus acanthias, Mustelus canis, Raja erinacea* and *Rhinoptera bonasus* (Lacy and Reale 1990), and incubation of the kidney extracts of *Scyliorhinus canicula* with rat plasma produced a vasopressor substance when assayed in the rat (Henderson et al. 1981). The renin-like activity was also identified in the corpuscles of Stannius of teleost fishes (Sokabe et al. 1970). Incubation of the extracts with homologous plasma yielded ANG I identical to that formed by renal renin in *Lophius litulon* (Hasegawa et al. 1984b) and *Oncorhynchus keta* (Takemoto et al. 1983), but the incubation produced different ANG I in other fish species (Sokabe and Nakajima 1972). Immunoreactive ANG II was demonstrated in the corpuscles of Stannius (Yamada and Kobayashi 1987) and ovarian follicle cells (Mandich and Massari 1994) of the rainbow trout. The presence of renin in other extrarenal organs, including the brain, has not been reported.

Converting enzyme activity, as assessed by hippurate production from Hip-His-Leu, was highest in the kidney followed by the lung in the chicken (Polanco et al. 1990). However, the substrate should be Hip-Ser-Leu for chicken converting enzyme although this enzyme rather nonspecifically cleaves off the C-terminal dipeptide. Converting enzyme activity is concentrated in the gill and corpuscles of Stannius of the rainbow trout as determined by the rate of Hip-His cleavage (Gallardy et al. 1984). Converting enzyme activities in these organs are 30-fold higher than in plasma on a weight basis. Since the gills receive the entire cardiac output, they are, like mammalian lungs, ideally situated to regulate plasma titers of circulating hormones. Olson et al. (1986) showed that ANG II genesis occurs in the respiratory pathway of the gill, whereas degradation of ANG II occurs in the venous pathway. During stress, elevated plasma catecholamines may reduce venous perfusion (Olson 1984), thereby helping maintain elevated circulating ANG II levels. In African lungfish, *Protopterus aethopicus,* and six species of air-breathing teleosts, the highest converting enzyme activity was found in the gill and the accessory respiratory organs (Olson et al. 1987). The converting enzyme activity was also detected in the heart and kidney, and occasionally in the liver of these fishes.

Chapter 5

Regulation of Renin Release

Despite an abundance of studies on the mechanisms regulating renin release in mammals, studies in nonmammalian species are rather scant. The initial part of this chapter will therefore be devoted to reviewing the present state of knowledge about the regulation of renin release in mammals. The details of the mechanism are extensively reviewed by Davis and Freeman (1976), Keeton and Campbell (1981), Gibbons et al. (1984), Fray et al. (1987) and Hackenthal et al. (1990). Increasing evidence shows that the kidney secretes both active and inactive renin, and the activation of inactive renin significantly contributes to changes in plasma renin activity in humans (Sealey and Laragh 1975), dogs (James and Hall 1974), and pigs (Bailie et al. 1979; Okamura and Inagami 1984). This activation process, however, has not been studied sufficiently in nonmammalian species.

In order to quantify renin release, plasma renin activity is usually used as an index. Plasma renin activity is measured by the rate of ANG I formation after incubation of plasma under inhibition of converting enzyme and angiotensinase activities (Boucher et al. 1967). However, the activity determined by this method is influenced by changes in plasma angiotensinogen concentration, although angiotensinogen is usually in excess in plasma, and its concentration fluctuates less than the renin concentration (Keeton and Campbell 1981). In a strict sense, renin release should be expressed by the difference in renin activity between renal artery and vein. It is not yet clear to what extent extrarenal renin contributes to plasma renin activity.

5.1 Factors Affecting Renin Release

5.1.1 Baroreceptor

The involvement of blood pressure in the control of renin release was first indicated by Goldblatt and colleagues in 1934. They made an experimental model of hypertension in the dog by constricting the renal artery, and found that a vasopressor substance was released from the kidney in response to hypovolemia and/or hypotension. The substance was later identified as renin released from the afferent arteriole of the glomerulus. Since then, attention has been focused on the role of renal perfusion pressure in the control of renin release. Tobian et al. (1959b) demonstrated an inverse relationship between renal perfusion pressure and renin release (Fig. 5.1). This observation was further confirmed by Skinner et al. (1963) who found increases in renin release in proportion to decreases in blood supply to the renal artery within the range of autoregulation. In these pioneering studies, however, changes in other factors that affect renin release were not considered. It is apparent that renal hypotension and hypovolemia decrease glomerular filtration rate (GFR), which in turn decreases NaCl delivery to the MD. Additionally, these changes stimulate sympathetic nerve activity, which results in increases in renal sympathetic nerve activity and catecholamine release from the adrenal medulla (Kopp and DiBona 1984). All of these factors stimulate renin release. Blaine et al. (1970) sequentially removed each factor in the dog by (1) making a nonfiltering kidney model to remove the influence of MD; (2) renal denervation to remove the influence of renal nerve activity; and (3) adrenalectomy to remove the influence of catecholamines and corticosteroids. Even after removal of these factors, a decrease in renal perfusion pressure clearly stimulates renin release (Fig. 5.1).

Baroreceptor regulation is also supported by in vitro studies using isolated kidney preparation of the dog and rat (Kaloyanides et al. 1973; Churchill et al. 1974; Hofbauer et al. 1974). The baroreceptor is localized in the afferent arteriole by experiments using papaverine (Blaine et al. 1971; Davis et al. 1972; Witty et al. 1972). This smooth muscle relaxant is known to dilate the renal

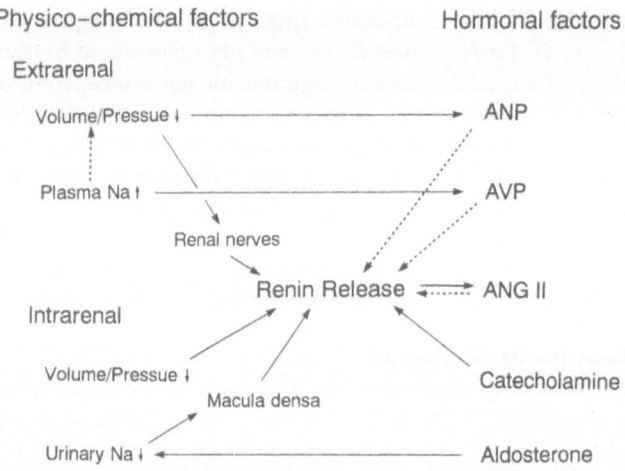

Fig. 5.1. Schematic drawing showing interactions of factors regulating renin release in mammals. Major physicochemical factors affecting renin secretion at the peripheral level are blood pressure, blood volume and plasma osmolality, all of which alter activity of the pressure/volume receptor at the afferent arteriole of the renal glomerulus, the Na sensor at the macula densa, and renal sympathetic nerves. Established hormones affecting renin secretion are atrial natriuretic peptide (*ANP*), arginine vasopressin (*AVP*), ANG II and catecholamines. Endothelin seems to be another possible inhibitor as suggested recently (Sect. 5.1.4.3). The effect of aldosterone may be indirect. *Solid line* indicates stimulation and *broken line* inhibition

afferent arteriole to block renal autoregulation (Thurau and Kramer 1959). The baroreceptor may perceive (1) intraluminal pressure of the afferent arteriole; (2) transluminal pressure at the arteriolar wall (difference of intraluminal and interstitial pressure); or (3) tension of the arteriolar wall (product of trans-luminal pressure and the arteriolar radius) (Vander 1967). Renal blood flow may also be involved when the change in renal perfusion pressure exceeds the range of autoregulation (Abe et al. 1973).

In addition to the renal baroreceptor, baroreceptors in the cardiopulmonary circulation and carotid sinus are implicated in the control of renin release (Fig. 5.1; Thames et al. 1978). However, involvement of the carotid baror-eceptor remains controversial (Jarecki et al. 1978; Rocchini and Barger 1979). Lee et al. (1984) failed to increase renin release after reduction in the renal perfusion pressure if right atrial pressure was elevated. The authors ascribed this result to the powerful inhibitory signal arising from atria. They thought this signal to be of nervous origin, but humoral signals such as atrial na-triuretic peptide are also candidates (Fig. 5.1).

5.1.2 Macula Densa

As early as 1939, Goormaghtigh suggested a role of the macula densa (MD) in the control of renin release from its morphological characteristics and topo-graphical proximity to JG cells (Fig. 5.1). Low plasma Na levels were later shown to correlate positively to high JG cell indices of granulation and active cellular profiles of the MD (Fisher 1961; Capelli et al. 1968). Vander and Miller (1964) observed that an increase in renin release caused by aortic constriction is prevented by administration of diuretics. The authors attributed this pre-vention to increased Na load to the MD. More convincingly, Shade et al. (1972) showed that hypernatremia decreases renin release in dogs with intact filtering kidneys but not in those with nonfiltering kidneys. Blair-West et al. (1977) showed that increased plasma renin concentration caused by water deprivation is depressed when sheep are allowed to drink more than 50 mM NaCl solution, but the depression is minute when 20 or 35 mM NaCl solution is provided. Since blood volume is scarcely affected by the Na concentration in drinking water, it seems that plasma renin concentration is more closely related to Na excretion than to blood volume. Vander and Carlson (1969) suggested that increased renin release is a function of decreased Na transport by the MD which occurs in association with decreased Na load. Consistently, intrarenal arterial infusion of ouabain, which blocks Na–K ATPase activity and inhibits Na transport, blunts an increase in plasma renin activity after ureteral occlu-sion or during partial arterial clamping (Churchill and McDonald 1974). Churchill et al. (1979a) also suggested that not renal tubular Na concentration but Na load (Na concentration × volume) to the MD influences renin release. A decrease in Na load also stimulates renin synthesis, since low NaCl diet in-creases both renal renin activity and the renin mRNA content in the mouse kidney (Catanzaro et al. 1985).

There are also a number of reports which indicate stimulation of renin release by increased Na load to the MD. Thurau (1964) examined the effect of changes in Na load to the MD on renin release using a retrograde micro-perfusion technique and observed a positive relationship between Na concentration of the perfusate and degree of collapse of the proximal tubule. Thurau ascribed the collapse to the decreased GFR which is caused by increased intrarenal formation of ANG II, i.e., increased renin release. Furthermore, the renal venous renin activity increases after administration of diuretics in the rabbit (Meyer et al. 1968) and the dog (Cooke et al. 1970), even though Na and volume depletion caused by diuretics are prevented by reinfusion of ureteral urine into the circulation. This result is interpreted as suggesting that an increased Na load to the MD increases renin release. However, Freeman et al. (1974) repeated the in vivo experiment in the dog and observed that renin secretion decreases while renal venous renin activity increases after injection of diuretics.

Kotchen et al. (1976, 1978) provided evidence that Cl ions, rather than Na ions, are important for regulation of renin release. An increase in plasma renin activity caused by dietary Na restriction is attenuated when isotonic NaCl solution is provided for drinking, but it remains elevated when isotonic Na_2CO_3 solution is given. Na balance is identical between both groups, while Cl balance is positive in the former and negative in the latter. Similarly, KCl or choline chloride is effective but K_2CO_3, Na acetate, Na nitrate or Na thiocyanate are without effect. An acute volume expansion with isotonic NaCl solution decreases plasma renin activity to a greater extent than a similar volume expansion established by isotonic Na_2CO_3 solution in anesthetized rats. Among halide ions, Cl and Br ions are effective but I ions are ineffective for inhibition of renin release (Galla et al. 1977). However, infusion of Na lactate or K lactate solutions decreases renin release in anesthetized, thoracic vena cava-constricted dogs although Cl excretion is elevated only in K lactate-treated animals (Stephens et al. 1978). Thus, the role of Cl ions in the control of renin release remains to be determined.

It is suggested that the MD is involved in the intrarenal tubulo-glomerular feedback mechanism (Schnermann et al. 1973; Navar et al. 1974; Thurau 1975). The MD senses changes in composition or flow of renal tubular fluid and influences the single GFR of the same nephron. The RAS appears to mediate this feedback. Since all components of the RAS are present intrarenally (Granger et al. 1972), changes in renin secretion caused by the MD may influence local formation of ANG II, which in turn modifies GFR by regulating constriction of renal arterioles. In addition, there is evidence that ANG II is secreted together with renin from the same granules in JG cells (Inagami et al. 1991). However, there are also several reports which oppose the presence of an intrarenal feedback loop (Morgan 1971; Bartoli and Earley 1973; Maddox et al. 1974; Knox et al. 1975) and involvement of the RAS in the feedback mechanism (Potkay and Gilmore 1973; Ganong et al. 1974).

The route of information transfer from the MD to JG cells has not been fully clarified yet. One possible route is via changes in the JG cell volume (Skøtt

1988). It is known that the variation of Na concentration at the MD affects Na–K-2Cl co-transporters of the cell, which may alter the ambient osmotic pressure of JG cells (Schlatter et al. 1989). Another possible route is via adenosine which may lead to changes in Ca ions in JG cells (see Sect. 5.1.4.6).

5.1.3 Renal Nerve

Innervation of the JG apparatus has been investigated extensively at both light and electron microscope levels (see Chap. 3). A fluoro-histochemical study of biogenic amines in combination with specific staining of JG granules revealed that nerve terminals containing norepinephrine make synaptic contacts with JG cells (Wågermark et al. 1968). Silverman and Barajas (1974) observed that these vesicles are depleted after reserpine treatment, which is accompanied by a decrease in plasma renin activity. The presence of β-adrenergic receptors on the JG cells is demonstrated by histochemical fluorescence study using a β-adrenergic antagonist (Atlas et al. 1977; see Sect. 3.1).

Vander (1965) was the first to demonstrate renin release after electrical stimulation of renal nerves (Fig. 5.1). Assaykeen and Ganong (1971) subsequently reported that the increase in renin release is completely blocked by pretreatment with propranolol, a β-adrenergic antagonist. Johnson et al. (1971) isolated the effects of renal nerve stimulation from other factors such as the renal baroreceptor and MD, and still observed an increase in renin release. Similarly, electrical stimulation of the distal cut end of renal nerves resulted in an increase in renin release in the absence of changes in GFR or renal perfusion pressure (La Grange et al. 1973), and without altering renal blood flow, GFR, or Na excretion (Taher et al. 1976). In the latter study, the effect was blocked by l-propranolol but not by d-propranolol which has no β-adrenergic blocking action.

Experiments using adrenergic agonists and antagonists have assessed the nature of adrenergic mechanisms involved in the control of renin release. The β-adrenergic agonist, isoproterenol, has a potent stimulatory effect on renin release both in vivo (Assaykeen et al. 1974) and in vitro (Vandongen et al. 1973). The effects of the vasodilating drugs, minoxidil and hydrallol, seem to be mediated by the β-adrenergic mechanism (Pettinger and Keeton 1975). Johnson et al. (1974) showed that intrarenal administration of low doses of isoproterenol increases, while propranolol infusion decreases, renin release caused by Na depletion in the dog. The administration of such low doses may not affect β-adrenergic mechanisms outside the kidney. Experiments using different types of β-blockers indicate the involvement of β_2-receptors in the regulation of renin release (Assaykeen et al. 1974). Both active and inactive renins are secreted after β-adrenergic stimulation (Okamura and Inagami 1984).

Although stimulatory effects of β-adrenergic mechanisms are evident, the role of α-adrenergic mechanisms is still a subject of controversy. Assaykeen

(1973) showed that propranolol blocks renin release induced by renal nerve stimulation, while an α-adrenergic blocker, phenoxybenzamine, fails to block the increase in the dog. Pettinger et al. (1972) observed that in conscious rats epinephrine stimulates renin release at low doses but not at high doses, whereas norepinephrine inhibits it at low doses and stimulates it at high doses. These results are explained by a dynamic balance of α-adrenergic depression and β-adrenergic stimulation of renin release. An α-adrenergic agonist, methoxamine, antagonizes isoproterenol-induced renin release in the isolated, perfused rat kidney (Vandongen and Peart 1974) and rat kidney slices (Weinberger et al. 1975). α-adrenergic antagonists potentiate norepinephrine-induced renin release in vitro (Nolly et al. 1974; Vandongen and Greenwood 1975). Meyer and Herrmann (1978) showed that isoproterenol-induced renin release is inhibited by tyramine in conscious rats. This result is interpreted as showing that inhibitory α-adrenergic receptors are a stimulated by norepinephrine released by tyramine. However, it is not yet determined whether the α-adrenergic system is inhibitory for renin secretion because stimulatory effect of α-adrenergic system is also evident (Blair 1983), probably mediated by α_1-adrenergic receptors (Takahashi et al. 1984).

The central nervous system also seems to be involved in the regulation of renin release via renal sympathetic nerves (Ueda 1976). Brain regions implicated in the stimulation of renin release include the pressor areas of midbrain and hindbrain of dogs (Ueda et al. 1967; Passo et al. 1971) and dorsolateral pons of cats (Richardson et al. 1974). Electrical stimulation of the hypothalamus depresses renin release in the conscious dog, and renal denervation or pharmacological blockade of the β-adrenergic system inhibits the response (Zehr and Feigel 1973). The anteroventral third ventricle region (AV3V), which seems to be a center for fluid and pressure homeostasis, is also inferred to be a regulatory center for renin release in the rat (Weekley 1984).

The role of renal nerves in renin release in physiological or pathophysiological conditions is suggested. Renal denervation blocks renin release in response to mild nonhypotensive hemorrhage (Bunag et al. 1966; Weber et al. 1974), but it only slightly diminishes the renin response to more extensive hemorrhage (Zanchetti and Stella 1975). Renal renin content and renin secretion increases after 1–3 weeks of renal denervation (Tobian et al. 1964; Bencsath et al. 1972). Renal nerves may be involved in renin release caused by upright posture, tilting, exercise, and exposure to cold in the human (R.D. Gordon et al. 1967; Hosie et al. 1970; Johnson and Park 1973; Zanchetti and Stella 1975). These stimuli are known to increase sympathetic activity. Furthermore, it is suggested that regulatory signals for renin release from cardiopulmonary volume receptors and carotid sinus baroreceptors are transmitted via renal sympathetic nerves (Fig. 5.1; see Keeton and Campbell 1981).

5.1.4 Humoral Factors

5.1.4.1 Ions

The effect of Na ions is closely related to the MD system. An inverse relationship is noted in plasma Na concentration and renin activity in hypertensive humans and dogs (Fig. 5.1; Brown et al. 1965). The inverse relationship is also observed in vitro in perfused kidneys (Yamamoto et al. 1969) and renal cortex slices (Michelakis 1971) of dogs. Na depletion increases renin release but the increase is nullified by intrarenal infusion of hypertonic NaCl in anesthetized dogs (Nash et al. 1968). Since intrarenal infusion of NaCl solution no longer suppresses renin release in nonfiltering kidneys as mentioned above, Na ions may act via the MD mechanism (Shade et al. 1972). The effects of Na ions on renin release from renal cortical slices (Michelakis 1971; Capponi and Valloton 1976), renal cortical cell suspensions (Lyons and Churchill 1975), and superfused glomeruli (Blendstrup et al. 1975; Frederiksen et al. 1975) are quite variable. A culture system of pure JG cells has to be established to examine the direct action of Na ions on renin release.

It is shown that concentrations of K ions in diet are inversely related to renin release in the human (Maebashi et al. 1968; Brunner et al. 1970), rat (Sealey et al. 1970), and dog (Abbrecht and Vander 1970). Vander (1970) found that intrarenal infusion of KCl suppressed renin release in anesthetized dogs. Since K sulfate and K lactate were as effective as KCl (Shade et al. 1972; Stephens et al. 1978), K ions, rather than accompanying anions, may be responsible for the effect. Shade et al. (1972) found that renin secretion caused by caval constriction is not inhibited by intrarenal infusion of KCl into dogs with nonfiltering kidneys. Thus, the effect of K loading is not a direct action on JG cells. This is further confirmed in vitro where an increase in K concentration in the medium does not inhibit renin release from rat renal cortical slices (Aoi et al. 1974) and isolated glomeruli (Frederiksen et al. 1975) except in rabbit renal cortical slices (Ginesi et al. 1983).

Since Ca ions play an essential role in hormone secretion and muscle contraction, and JG cells are endocrine cells originating from vascular smooth muscle cells, it is reasonable to assume that the Ca ion is an important factor for renin secretion. However, changes in extracellular Ca ions alter urinary Na excretion, catecholamine release, blood pressure and other factors that affect renin release. For example, intrarenal infusion of $CaCl_2$ into anesthetized dogs decreases renal venous renin activity with concomitant increases in Ca and Na excretion (Kotchen et al. 1974), resulting in an influence on the MD mechanism. However, Watkins et al. (1976) infused $CaCl_2$ and Ca gluconate intrarenally in Na-depleted, anesthetized dogs with a single nonfiltering kidney and observed a suppression of renin release. A decrease in renin release was also reported in humans after intrarenal infusion of $CaCl_2$ and Ca gluconate (Weidmann et al. 1972), and an increase was observed after Ca depletion caused by infusion of EDTA (Llach et al. 1974). After renal denervation, however, intrarenal infusion of $CaCl_2$ increased renin secretion in anesthetized dogs (Iwao et al. 1974). Increased endogenous Ca ions caused by para-thyroidism and other aetiologies do not always accompany abnormal plasma renin activity in humans (Brinton et al. 1975). Increases in Ca ions suppress renin release from isolated perfused rat kidney (Fray 1978; Logan et al. 1975) or renal cortical slices (sheep, Park and Malvin 1978; rats, Churchill et al. 1979b; Ginesi et al. 1983), whereas they increase renin release in other animals (dogs, Michelakis 1971; Yamamoto et al. 1984; rats Aoi et al. 1974; Chen and Poisner 1976). Thus, the effect of extracellular Ca ions on JG cells remains undefined (see Sect. 5.2).

Churchill and Lyons (1976) observed an increase in renin release with a concomitant increase in Na excretion when $MgCl_2$ was infused into the renal artery of anesthetized dogs. Wilcox (1978) also observed increases in renin release and arterial pressure after $MgCl_2$ infusion in anesthetized dogs with denervated kidneys. Thus, Mg ions seem to act on JG cells to stimulate renin secretion. A parallel relationship was observed between Mg concentration and renin release in renal cortical slices (rat, Fray 1977; dog, Morimoto et al. 1970). Since Ca ions antagonize the stimulatory effect of Mg ions in vivo (Wilcox 1978) and in vitro (Fray 1977), Mg ions may stimulate renin release by in-hibiting Ca influx across the JG cell membrane.

After high doses of LiCl injection, plasma renin activity increases in the rat (Kierkegaad-Hansen 1974; Gutman et al. 1973). The increase by Li ions is also observed in isolated, perfused rat kidneys. Plasma renin activity increases after oral administration of Li ions in the dog (Beck et al. 1975) and man (Shopsin et al. 1973). The effect of Li ions may be mediated by a decrease in Na load to the MD because Li ions increase proximal reabsorption of Na ions (Gutman et al. 1973).

5.1.4.2 Amines

The effect of catecholamines on renin secretion should be interpreted in connection with renal nerve activities (Fig. 5.1; Chap. 3; Sect. 5.1.3). Briefly, intravascular infusion of isoproterenol, epinephrine and norepinephrine in-creases renin secretion, and the increase is blocked by propranolol and po-tentiated by α-adrenergic antagonists in several species of mammals (Pettinger et al. 1972; Tanigawa et al. 1972; see Keeton and Campbell 1981). Direct action of catecholamines is suggested by in vitro experiments using renal cortical slices (dog, Michelakis et al. 1969; human, Rosset and Veyrat 1971; rat, Weinberger et al. 1975), isolated, superfused rat glomeruli (Morris et al. 1976), and isolated, perfused kidneys (rat, Vandongen et al. 1973; rabbit, Viskoper et al. 1977; cat, Harada and Rubin 1978). The presence of an extrarenal site of action of circulating catecholamines is also suggested in anesthetized dogs which may be β_1-receptor mediated (Johnson 1984).

An intravenous infusion of dopamine increases plasma renin activity in anesthetized dogs (Otsuka et al. 1970) and humans (Wilcox et al. 1974), but direct infusion of dopamine into the renal artery causes a slight decrease (Chokshi et al. 1972). Imbs et al. (1975) observed an increase in renin release when dopamine was infused into the denervated kidney of anesthetized dogs without altering blood pressure, urine volume and Na excretion. The dopaminergic antagonist, haloperidol, but not propranolol, blocked the dopamine effect. Thus, the effect may be mediated by specific dopamine receptors within the kidney. Dopamine increased renin release from rat kidney slices, but the effect was not blocked by haloperidol but by propranolol (Henry et al. 1977).

Intrarenal arterial infusion of serotonin increases plasma renin activity in rats (Meyer et al. 1974) but not in dogs (Bunag et al. 1966). Ganong et al. (1979) found that intravenous injection of serotonin precursors, tryptophan and 5-hydroxytryptophan, causes dose-dependent increases in renin release in dogs. The effect of serotonin seems to be mediated by the renal sympathetic nerves and by prostaglandin production in JG cells (K. Takahashi et al. 1991).

5.1.4.3 Peptides

Among peptide hormones thus far examined, those stimulating renin secretion include glucagon (rat, Vandongen et al. 1973; humans, Fernandez-Cruz et al. 1975; cat, Harada and Rubin 1978; dog, Ueda et al. 1978), ACTH and parathyroid hormone (see below), α-MSH (rat, Hauger-Klevene 1970), bradykinin (humans, Margolius et al. 1974), and kallikrein (rat, Beierwaltes et al. 1985). Indirect evidence suggests that erythropoietin is also stimulatory (rat, Gould et al. 1974). On the other hand, peptides inhibiting renin secretion are ANG II and arginine vasopressin (see below), arginine vasotocin (dog, Vander 1968), somatostatin (rat, Scholkens 1978), and atrial natriuretic peptide (see below). Variable effects are obtained with oxytocin (dog, Bunag et al. 1967; Brooks et al. 1984; rat, Hauger-Klevene 1970), insulin (dog, Assaykeen et al. 1970; human, Hedeland et al. 1972; rat, Campbell and Zimmer 1980; Cohen et al. 1983), and growth hormone (rat, Palkovits et al. 1970; Henderson and Balment 1975; Hauger-Klevene 1976). Calcitonin (dog, Smith et al. 1979) and thyroid stimulating hormone (rat, Hauger-Klevene 1970) are without effect.

Hauger-Klevene et al. (1969) observed a transitory increase in plasma renin activity after administration of ACTH in conscious rats. The short duration of the increase may be due to the delayed suppression by ACTH-liberated corticosteroids (see below), because adrenalectomy prolonged the increase. However, the increase in plasma renin activity may also be due to an increase in plasma angiotensinogen level, as ACTH was shown to increase it in the dog (Haynes et al. 1953). Effects of hypophysectomy on plasma renin activity are not consistent in the rat (Marks et al. 1960; Rojo-Ortega et al. 1972).

Parathyroid hormone increases plasma renin activity in the dog (Smith et al. 1979; Zawada and Johnson 1984), human (Horky et al. 1986) and rat (Helwig et al. 1991). Since parathyroid hormone has vasodepressor and diuretic actions, its effect may be mediated by the renal baroreceptor and MD. However, direct actions are also likely because high affinity binding sites are localized in the preglomerular arteriole (Nikcols et al. 1990), and because parathyroid hormone is known to increase cyclic AMP (cAMP) production (Helwig et al. 1984) and inhibit Ca entry through L-type Ca channels (Pang et al. 1990). Ca ions and cAMP are important intracellular messengers for renin secretion (Helwig et al. 1991; see Sect. 5.2).

Vander and Geelhoed (1965) reported that intravenous infusion of ANG II decreases basal renin release in dogs even if arterial pressure was held constant (Fig. 5.1). Furthermore, subpressor doses of ANG II infusion inhibit renin release caused by renal arterial hypotension or reduced Na excretion in the sheep (Blair-West et al. 1971a). ANG II seems to act directly on JG cells, since it is effective at doses which do not alter arterial pressure, GFR and Na excretion in dogs with a single nonfiltering kidney (Shade et al. 1972). ANG III is equally potent as ANG II in inhibiting renin release in vivo (Freeman et al. 1975). The inhibitory effect of ANG II is reversed by [Sar^1, Gly^8] ANG II (McDonald et al. 1975). Intracerebro-ventricular administration of ANG II decreases plasma renin activity in the conscious goat (Eriksson and Fyhrquist 1976) and anesthetized cats (Lokhandwala et al. 1978). The inhibitory signal appears to be conveyed via renal nerves or vasopressin release. ANG II inhibits renin release in renal cortical slices (dog, Michelakis 1971; rat, Capponi et al. 1977), and isolated, perfused rat kidneys (Vandongen et al. 1974). Naftilan and Oparil (1978) showed that ANG II and N-terminally truncated ANG IIs equally inhibit renin release from rat renal cortical slices, but in isolated, perfused rat kidneys, ANG III is as effective as ANG II, but ANG II-(3–8), -(4–8), and -(1–7) are less effective (Vandongen et al. 1974). SQ14225 is shown to induce renin release in conscious rats by activation of the β-adrenergic system and by interruption of the short feedback loop of ANG II (Schiffrin et al. 1981); it also increases the number of JG cells in rat afferent arteriole (Geary et al. 1992). The specific AT_1 receptor antagonist, Dup753, also stimulates renin release but the specific AT_2 receptor antagonist, PD123177, is without effect in conscious rats (Pals and Couch 1993). The effect was also mediated by the $β_1$-adrenergic system.

Intravenous adminstration of arginine vasopressin decreases renin release in anesthetized dogs (Bunag et al. 1967; Vander 1968), rats and humans (Fig. 5.1; see Keeton and Campbell 1981). The effective doses are within the physiological range, and arterial pressure and plasma Na concentration are not altered (Tagawa et al. 1971). Shade et al. (1973) also induced a suppression of renin release in anesthetized, Na-depleted dogs with a single nonfiltering kidney without affecting renal blood flow and arterial pressure. Interestingly, Brattleboro rats with hypothalamic *diabetes insipidus* have higher plasma renin activity than normal rats of the same strain (Gutman and Benzakein 1971; Balment et al. 1975). In humans, the effect of vasopressin is variable on basal renin release but apparently inhibitory on the elevated release caused by other stimuli (Hesse Nielsen 1977). Vasopressin may play an important role in the inhibition of centrally administered ANG II, because hypophysectomy abolishes the ANG II effect (Malayan et al. 1979). Vasopressin has no effect on the basal renin release from rat renal cortical slices (de Vito et al. 1970) and isolated, perfused rat kidneys (Vandongen 1975; Konrads et al. 1978), but inhibits renin release caused by isoproterenol in the latter preparation.

Atrial natriuretic peptide (ANP) inhibits renin secretion when infused into the circulation of dogs (Fig. 5.1; Burnett et al. 1984; Maack et al. 1984; Salazar et al. 1986). The effect seems to be direct because ANP lowered the renin release from renal cortical slices of rats (Henrich et al. 1986) and monkeys (Henrich et al. 1987). As evidenced by a study using cultured JG cells, cGMP functions as a second messenger of the ANP effect (Henrich et al. 1988) as is the case for other biological actions of ANP (Brenner et al. 1990). The MD mechanism may also contribute to the inhibitory action of ANP because ANP increases Na load to the MD through its natriuretic action, and because its inhibitory action is not evident in the nonfiltering kidney of dogs (Opgenorth et al. 1986). Although ANP inhibits sympathetic nerve activity (Kuchel et al. 1987) and blunts the renin release caused by renal nerve stimulation (Hisa et al. 1989), the effect of ANP on renin release is not totally dependent on its effect on renal nerves (Scheuer et al. 1989). Interaction between renin and ANP has been reviewed recently by Richards and Nicholls (1993) and Schiffrin et al. (1993).

Endothelin-1, a potent vasoconstrictor identified in endothelial cells (Yanagisawa et al. 1988), infused intrarenally has little effect on basal renin release but inhibits renin release stimulated by renal arterial constriction or ureteral occlusion in dogs with nonfiltering kidneys (Lin et al. 1993) and in those with denervated kidneys (Naess et al. 1993). Endothelin-1 also inhibits basal and β-adrenergically stimulated renin release from rat renal cortical slices and isolated JG cells (Moe et al. 1991). Three endothelin isoforms have similar inhibitory potency in cultured JG cells, and the inhibition is marked by cAMP-stimulated renin release (Kramer et al. 1994; Ritthaler et al. 1995). The inhibitory effect seems to be mediated by ET_B receptor and protein kinase C. In contrast to kidney JG cells, endothelin-1 stimulates renin secretion from human decidua via Ca dependent mechanisms (Chao et al. 1994).

5.1.4.4 Steroids

Goodwin et al. (1970) observed that long-term treatment with deoxycorticosterone acetate (DOCA) does not alter plasma renin activity in Na-deplete rats but causes a decrease in Na-replete rats. Pettinger et al. (1971) reported that the inhibitory effect of DOCA was greater in rats offered NaCl solution as drinking water than in those offered tap water. Similar results were obtained in the dog (Robb et al. 1969) and humans (Warren and Ferris 1970; Shade and Grim 1975). These results indicate that the effect of DOCA is closely related to the Na balance of the animal. Nasjletti and Masson (1969) found that adrenalectomy increases plasma renin activity in the rat, and the effect is reversed by corticosteroids. Plasma renin activity in adrenalectomized dogs increases when aldosterone is infused at a rate to restore its plasma levels below normal, but it decreases to undetectable levels when the level is higher

(Young and Guyton 1977). It should be noted that an increase in plasma aldosterone levels is accompanied by increased extracellular fluid volume and decreased Na excretion (Fig. 5.1). Since the latter stimulates renin release via the MD mechanism, the inhibitory effect of this mineralocorticoid may be mediated by volume receptors or baroreceptors. The absence of a direct action on JG cells is suggested by in vitro experiments using renal cortical slices (rat, de Vito et al. 1970; humans, Rosset and Vayrat 1971).

The effect of glucocorticoids on renin release is not as evident as that of mineralocorticoids. According to Nasjletti and Masson (1969), the inhibition of renin release by cortisol or corticosterone is less than 1/100 of that by DOCA in adrenalectomized rats. Methylpredonisolone increases plasma renin activity in normal rats but decreases it in Na-loaded rats (Krakoff et al. 1975). The increase in normal rats may be due to the effect of angiotensinogen on plasma levels (Helmer and Griffith 1951). Effects of dexamethasone, hydrocortisone, predonisolone and predonisone on renin release are variable in humans (Newton and Laragh 1968; Kelsch et al. 1971; Katz et al. 1972).

Testosterone does not alter renin release in ovariectomized rats (Nasjletti et al. 1971) but increases it in orchidectomized rats (Katz and Roper 1977). The orchidectomy itself does not alter plasma renin activity. The plasma renin activity is high during the luteal phase of the menstrual cycle and during pregnancy in the rat (Nasjletti et al. 1971) and human (Brown et al. 1964). The high renin levels during pregnancy or estrogen treatment may be due to the increased release of angiotensinogen (see Sect. 4.2). Progesterone seems to decrease plasma renin activity without altering plasma angiotensinogen levels in the human (Oparil et al. 1975).

5.1.4.5 Prostaglandins

The role of prostaglandins in renin release has been investigated extensively and a few reviews summarize the results (Oates et al. 1979; Henrich 1981; Freeman et al. 1984). Prostaglandins seem to be good candidates for the regulator of renin release, since arachidonic acid, a precursor of prostaglandins, stimulates renin release in vivo and in vitro in several mammalian species (see Sect. 5.2). Furthermore, the increase induced by arachidonic acid is blocked by pretreatment of a cyclooxygenase inhibitor indomethacin. Administration of prostaglandins (PGD_2, PGI_2, PGE_2, and 13, 14-dihydroPGE$_2$) into the renal artery or to the incubation medium of the renal tissue increases renin release in several species of mammals. PGI_2 is apparently the most effective prostaglandin for stimulating renin release (McGiff and Wong 1979). Prostaglandins may also influence renin release via other mechanisms, since they alter both urinary Na excretion and renal arteriolar tone (Tannenbaum et al. 1975; Bolger et al. 1976).

The decrease in renal arterial pressure within the autoregulatory range stimulates renin release via prostaglandin-dependent mechanisms (McGiff et al. 1970), but stimulation by further decreases is mediated by other mechanisms

(Blackshear et al. 1979; Seymour and Zehr 1979). Involvement of prostaglandins in baroreceptor-mediated renin release has been suggested in vivo using dogs with denervated, nonfiltering kidneys (Freeman et al. 1982) and in vitro using isolated, perfused rat kidneys (Linas 1984). Geber et al. (1981) and Francisco et al. (1982) reported that prostaglandins may be involved in the renin release caused by manipulation of Na balance in the dog whose renal vascular baroreceptors and renal nerves are made nonfunctional. Linas (1984) found that activation of the MD mechanism increased PGE_2 production and renin release from isolated, perfused rat kidneys, and indomethacin inhibited these responses. More recent data indicate that PGI_2 acts directly on JG cells and PGE_2 acts via the MD to stimulate renin release (Ito et al. 1989). However, prostaglandins may not be synthesized locally at the MD because of the absence of cyclooxygenase activity at this site (Smith and Bell 1978). There is also negative evidence for the interaction between prostaglandins and the MD in renin release (Campbell et al. 1979; Villarreal et al. 1982).

Renal nerve stimulation or infusion of norepinephrine and epinephrine increases release and production of renin and release of renal prostaglandins (Dunham and Zimmerman 1970; Needleman et al. 1974). These studies indicate that β-adrenergic mechanisms stimulate renin release, and α-adrenergic mechanisms stimulate prostaglandin release. As α-adrenergic mechanisms are somewhat inhibitory for renin release (Sect. 5.1.4.2), prostaglandins may not be involved in the renal nerve-mediated renin release. Consistently, indomethacin failed to block the β-agonist-induced renin release both in vivo and in vitro (Berl et al. 1979; Seymour and Zehr 1979; Beierwaltes et al. 1980; Vandongen et al. 1981; Henrich and Campbell 1984; Linas 1984; Osborn et al. 1984). However, there are also reports indicating that cyclooxygenase inhibitors influence β-adrenergically stimulated renin release in the rat (Campbell et al. 1979; Suzuki et al. 1981).

5.1.4.6 Adenosines

Adenosine decreases plasma renin activity with concomitant decreases in GFR and Na excretion (Fig. 5.2) when infused into the renal artery of dogs (Tagawa and Vander 1970) or into the thoracic aorta of Na-deplete rats (Osswald et al. 1978). The renal effect of adenosine is independent of its hemodynamic effects (Arend et al. 1984). Adenosine also reduces renin release from rat kidney slices or isolated rat glomeruli (Skøtt and Baumbach 1985). The inhibitory effect of adenosine seems to be mediated by A_1 receptors (Churchill and Churchill 1985), because A_1 receptor antagonists blocked adenosine's effect in cultured rat JG cells (Kurtz et al. 1988). Based on these observations, Jackson (1991) proposed that adenosine functions as a physiological brake for renin release, probably in part via the MD mechanism.

Cyclic AMP increases renin release when infused into the renal artery of dogs without changing other parameters which are known to affect renin release (Fig. 5.2; Winer et al. 1971). Dibutylyl cAMP is more potent than cAMP in stimulating renin release in dogs (Allison et al. 1972; Okahara et al. 1977) and rat (Hauger-Klevene 1970; Campbell et al. 1979). This may be explained by the higher lipid solubility and greater resistance to degradation enzymes of dibutylyl cAMP. Dibutylyl cAMP is effective also in dogs with nonfiltering kidneys and after pretreatment with indomethacin or Ca ionophore (Bondar et al. 1984). Since renin release is specifically increased by cAMP in vivo, and since cAMP is also effective in vitro in isolated, perfused rat kidneys (Hofbauer et al. 1978) and renal cortical slices (dog, Michelakis et al. 1969; Yamamoto et al. 1973; rats, Saruta and Matsuki 1975; Suzuki et al. 1981), cAMP appears to be an endogenous stimulator for renin secretion which acts directly on JG cells. cAMP increases renin mRNA levels as well as renin release in primary cultures of mouse JG cells (Chen et al. 1993). The former is due to the cAMP action on stabilization of renin mRNA. As will be discussed in Section. 5.2, cAMP is thought to be the only stimulatory intracellular second messenger for renin release.

5.2 Subcellular Mechanisms

Despite a spectrum of extracellular first messengers that regulate renin release, there seem to be only a few intracellular second messengers. Churchill (1985) and Kurtz (1989) suggested in their reviews that Ca ions and cAMP are, respectively, major inhibitory and stimulatory second messengers for renin secretion, and cGMP and protein kinase C are other possible intracellular modulators (Fig. 5.2). It is apparent that the inhibitory effect of Ca ions predominates over the stimulatory cAMP effect. More recently, Kurtz and Bruna (1991) have indicated that renin secretion and synthesis are regulated by different transmembrane transduction systems. They showed that an increase in cAMP and inhibition of the Ca-calmodulin system or the Na^+/H^+ exchange system enhances renin secretion, whereas an increase in cGMP and activation of protein kinase C lead to an inhibition. For renin synthesis, however, cAMP appears to be the sole intracellular modulator for stimulation.

The concentration of intracellular Ca ions is controlled by efflux and influx of Ca ions across the cell membrane, and by the release and uptake of Ca ions by intracellular Ca reservoirs (Fig. 5.2). The Ca efflux is controlled by active Ca transport and by Na–Ca exchange which passively follows the active Na–K exchange. The involvement of the Na–Ca exchange in the regulation of renin release is suggested by the fact that both basal and stimulated renin secretion are inhibited by inhibition of Na–K ATPase activity (Schwaltz et al. 1975; Churchill and Churchill 1980a; Fray 1980; Park et al. 1981; Lopetz-Novoa et al. 1982; Churchill et al. 1983). Conversely, renin secretion is stimulated by activation of Na–K ATPase, and this effect is dependent on extracellular Ca ions (Churchill et al. 1979b; Migdal et al. 1980; Phillis and Wu 1981). The involvement of active Ca transport in regulation of renin release is suggested by indirect evidence that vanadate, which inhibits Ca-ATPase activity, inhibits

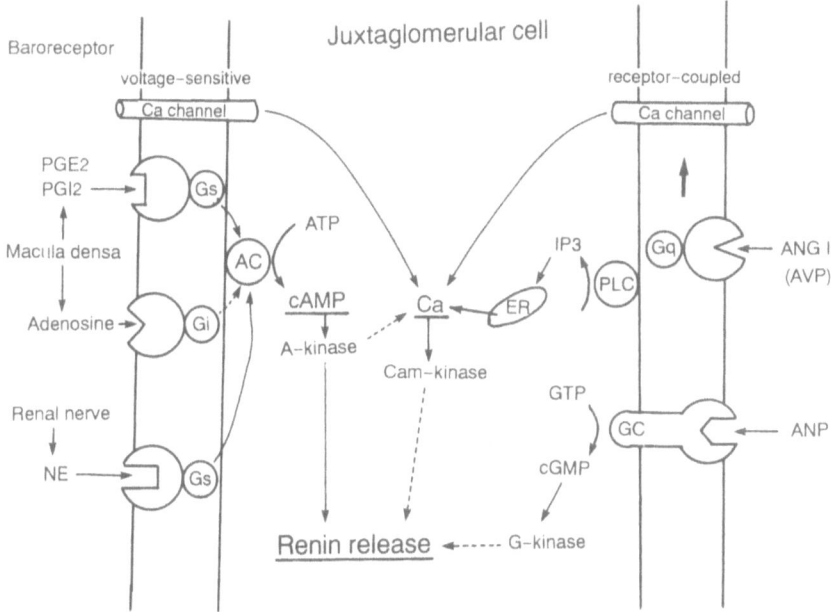

Fig. 5.2. Schematic drawing of intracellular mechanisms regulating renin release from JG cells. Major intracellular messengers are Ca ions for inhibition and cAMP for stimulation. An increase in perfusion pressure at the afferent arteriole activates voltage-sensitive Ca channels, and ANG II and AVP activate the receptor-coupled Ca channel, resulting in an increased influx of Ca ions. ANG II and AVP also activate phospholipase C (*PLC*) via GTP binding protein (*Gq*), and a subsequent increase in inositol triphosphate (*IP3*) stimulates Ca release from the endoplasmic reticulum (*ER*). The resultant increase in Ca ions stimulates calmodulin (*Cam*)-kinase and inhibits renin release. Changes in Na load to the macula densa stimulate prostaglandin production (*PGI2* or *PGE2*) or adenosine production, which modulates adenylate cyclase (*AC*) activity via GTP binding proteins (*Gs* or *Gi*). Norepinephrine (*NE*) released from the adrenal medulla and liberated from renal sympathetic nerves stimulates *AC* via *Gs* protein. The *AC* increases cAMP production and stimulates renin release via protein kinase A (A-kinase). *ANP* increases cGMP production by activation of guanylate cyclase (*GC*) and inhibits renin release via protein kinase G (G-kinase). *Solid line* represents stimulation and *broken line* inhibition (Modified from Jackson 1991)

both basal and stimulated renin secretion in vivo and in vitro (Churchill and Churchill 1980b), and that β-adrenergic agonists, which stimulate Ca-ATPase activity via cAMP production (Van Breemen et al. 1979), increase renin secretion. All the observations mentioned above support the notion that Ca ions are inhibitory second messengers for renin release.

The influx of Ca ions seems to be controlled by voltage-sensitive Ca channels and by receptor-operated Ca channels. Since JG cells are derived from vascular smooth muscle cells (Barajas 1979) and many endocrine cells and vascular smooth muscle cells have voltage-sensitive Ca channels (Rosenberger and Triggle 1978), it is reasonable to postulate that depolarization of JG cells activates voltage-sensitive Ca channels and inhibits renin release. In fact, depolarization of JG cells by excess K ions inhibits renin release (Fray 1978;

Churchill and Churchill 1982), and the inhibition is antagonized by Ca channel blockers such as verapamil and diltiazem (Churchill et al. 1981). Fray (1980) suggested that distortion of the afferent arteriolar baroreceptor leads to depolarization of JG cells, resulting in inhibition of renin release. Inhibitory effects of ANG II, vasopressin, α-adrenergic agonists and adenosine appear to be mediated by receptor-operated Ca influx, because chelation of extracellular Ca ions nullified the inhibition. The depolarization-induced Ca influx may play a minor role in the ANG II and vasopressin effects, since methylverapamil and diltiazem did not block these effects (Churchill et al. 1981). ANG II also activates receptor–coupled phospholipase C, and the resultant product, inositol triphophate (IP_3), increases intracellular Ca. Since intracellular Ca ions are generally stimulatory for hormone secretion, Baumbach and Leyssac (1977) suggested that the mode of renin secretion may be different from exocytosis as observed in general endocrine cells.

The notion that cAMP is a stimulatory second messenger for renin secretion originates from the fact that renin secretion is stimulated by several substances that are known to activate adenylate cyclase such as β-adrenergic agonists, A_2 adenosine receptor agonists, dopamine, glucagon, parathyroid hormone and forskolin (Fig. 5.2), whereas renin secretion is inhibited by the substances that inhibit adenylate cyclase activity such as α-adrenergic agonists and A_1 adenosine receptor agonists (Churchill and Churchill 1985). Renin secretion is also stimulated by inhibitors for phosphodiesterase that metabolizes cAMP, and by exogenous cAMP and dibutylyl cAMP as mentioned above (see Keeton and Campbell 1981; Jackson 1991).

5.3 Nonmammalian Vertebrates

Since renin has not been isolated in nonmammalian species and homologous radioimmunoassays for nonmammalian ANG I are not available, their plasma renin is quantified by the activity determined by the rate of ANG I formation after incubation of its plasma with mammalian angiotensinogen. However, it is not known how efficiently nonmammalian renin can cleave ANG I from heterologous angiotensinogen. Thus, plasma renin activity determined in this way only represents relative values as pointed out by Nishimura et al. (1977).

5.3.1 Birds

Plasma renin activity increases after hemorrhage in the pigeon, and the increase is proportional to the volume of blood lost (Chan and Holmes 1971). Plasma renin activity also increases after hypophysectomy with a concomitant decrease in blood pressure, although plasma angiotensinogen concentration, determined by incubation of plasma with excess porcine renin, decreases after hypophysectomy. When hypophysectomized pigeons are treated for 1 week with ACTH, both renin and angiotensinogen concentrations in plasma return to normal, but blood pressure remains subnormal. Mild hypotension caused by hemorrhage of 10% total blood, or greater hypotension caused by injection of papaverine, fails to increase plasma renin activity in the chicken, but sustained hypotension caused by hemorrhage of 40% total blood or by infusion of nitroprusside causes an increase (Nishimura and Bailey 1982). Papaverine may have blocked the renal baroreceptor which may account for the lack of its stimulatory effect. Plasma renin activity of anesthetized chickens is higher than that of conscious chickens (Nishimura et al. 1981a), probably due to the lower blood pressure in the former. Plasma renin activity, determined by the differences of activity between renal vein and systemic artery, increased as renal arterial perfusion pressure was reduced without changing mean systemic arterial pressure (Wideman et al. 1993). Thus, renal baroreceptors may be present in the chicken and may directly regulate renin secretion.

Plasma ANG II concentration is higher and blood pressure is lower in saltwater-adapted Pekin ducks than in freshwater-adapted ducks (Simon-Oppermann et al. 1984). Gray and Simon (1985a) showed that infusion of a slightly hypotonic NaCl solution into the salt-loaded ducks decreased plasma ANG II levels. The degree of the decrease is proportional to an increase in blood volume as indicated by a decrease of hematocrit. Plasma ANG II levels increase in proportion of the period of water deprivation in the Japanese quail, which is associated with a decrease in blood volume (Kobayashi and Takei 1982; Takei et al. 1988a). The increased plasma ANG II levels in water-deprived birds rapidly return to normal after birds drink. The water intake is shown to accompany restoration of blood volume. An acute hemorrhage of 12.5% total blood increases plasma ANG II levels in the quail (Fig. 5.3; Takei et al. 1989). Since plasma Na concentration and osmolality do not change but blood pressure and blood volume decrease after hemorrhage, the increase in plasma ANG II level may be attributable to hypotension and/or hypovolemia. The blood volume quickly returned to normal after hemorrhage in the quail, which may explain the short duration of the hemorrhage-induced increase in plasma ANG II level (Takei and Hatakeyama 1987). It seems that the baroreceptor mechanism also functions for regulation of renin release in birds (Table 5.1).

Table 5.1. Possible role of afferent arteriolar baroreceptor, macula densa and renal nerves in regulation of renin release

Animals	Stimulus for renin release		
	Hypotension	Urinary Na $\downarrow$	β-Stimulation
Mammals	++	++	++
Birds	+	+	±
Reptiles	?	?	?
Amphibians	−	?	+
Fishes	+	±	±

+, stimulatory; −, inhibitory; ±, no effect; ?, undetermined or conflicting

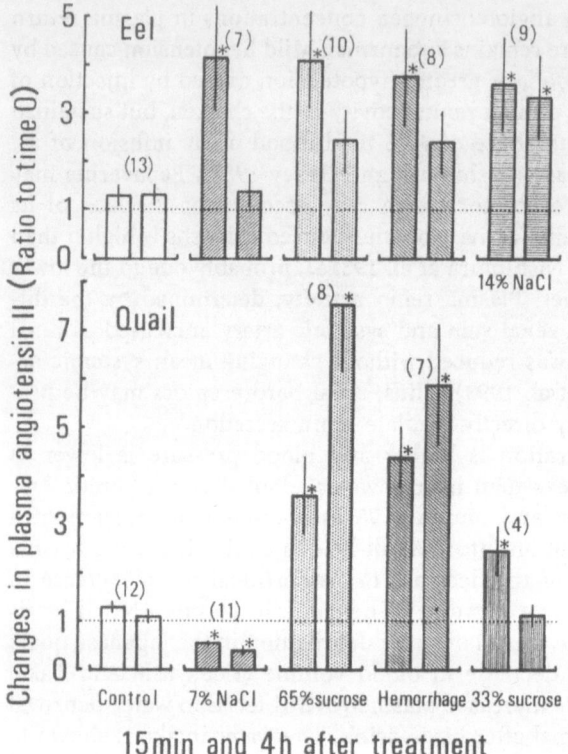

15 min and 4 h after treatment

Fig. 5.3. Changes in plasma ANG II levels after injection of 0.9% NaCl (control), 7% NaCl or 65% sucrose (isosmotic to 7% NaCl) solution, or after 1 ml of hemorrhage in the eel (200 g) and quail (100 g). Blood volumes of both species are similar (ca. 7 ml). Injections of 14% NaCl to eel and 33% sucrose to quail are also made. Values 15 min and 4 h after injection are expressed in terms of ratio to the value before injection. *Vertical lines* represent standard errors of the mean; * $p<0.05$

Na-depletion caused by dietary Na restriction or mercurial diuresis increases granulation in JG cells and renal renin concentration as determined by the rat vasopressor bioassay in chickens (Taylor et al. 1970). Plasma ANG II levels of seawater-adapted ducks are greater than those of freshwater-adapted birds even though plasma Na concentration is higher in seawater-adapted birds (Simon-Oppermann et al. 1984). A positive relationship is oberved between plasma ANG II levels and plasma osmolality in salt-loaded ducks (Gray and Simon 1985a). It is not known whether plasma Na concentration is parallel to urinary Na excretion, i.e. Na load to the MD, because of the presence of a Na-excreting nasal salt gland. In the quail, plasma ANG II level increases after water deprivation with concomitant increases in plasma Na concentration and os-molality (Takei et al. 1988a). Urinary Na concentration increases but Na load (concentration × volume) to the MD can be decreased in water-deprived birds because of oligouria.

Gray and Simon (1985a) showed that slow infusion of hypertonic NaCl solution into salt-loaded ducks increased plasma ANG II concentration. It is possible that increased plasma NaCl is excreted via the nasal salt gland without increasing urinary Na excretion. In the anesthetized chicken, however, slow infusion of hypertonic NaCl solution directly into the renal portal vein decreases plasma renin activity without significant changes in blood volume and blood pressure (Nishimura and Bailey 1982). Plasma ANG II levels decrease profoundly after intravenous injection of hypertonic NaCl solution in the quail (Fig. 5.3; Takei et al. 1988b). The decrease is accompanied by increases in urinary Na excretion, blood volume, and blood pressure. In contrast to NaCl, quail injected with hypertonic sucrose solution have an increased plasma ANG II level (Fig. 5.3). Blood volume and blood pressure increased as in NaCl-loaded birds, but urinary Na excretion may be smaller in sucrose-loaded birds. It has been postulated that the volume receptor responsible for renin secretion is located in the extracellular compartment outside the vascular space in the duck (Gray and Simon 1985b). A similar extravascular volume receptor may be present in the quail because hypertonic sucrose may have caused transitory dehydration in the extravascular, interstitial space because of slow penetration into the extravascular space. Thus the volume-sensing mechanism in the extravascular space may play important roles in the control of renin release in birds (Table 5.1). Renal nerves seem to play a minor role in the physiological regulation of renin secretion in birds because β-adrenergic antagonists fail to alter basal renin release in the chicken (Table 5.1; Nishimura et al. 1981b). Atrial natriuretic peptide reduces plasma aldosterone level and does not affect plasma ANG II levels in normally hydrated Pekin duck (Gray et al. 1991). However, it depresses high plasma ANG II levels caused by 24-h water deprivation.

5.3.2 Reptiles

Acute hemorrhage of 23% blood volume induces a transitory increase in plasma renin activity and a more prolonged decrease in blood pressure in conscious freshwater turtle, *Pseudemys scripta* (Stephens and Creekmore 1984). Plasma renin activity was determined by utilizing radioimmunoassay for human ANG I, because this turtle has ANG I similar to mammals (see Sect. 6.3). Infusion of SQ20881, a converting enzyme inhibitor, prolongs hypotension and further increases plasma renin activity. The same degree of hypotension caused by diazoxide increases plasma renin activity but papaverine is without effect probably because of the blockade of the baroreceptor system. However, cumulative hemorrhage of 30% blood volume fails to increase plasma renin activity and plasma angiotensinogen level in the anesthetized *Pseudemys*, although arterial pressure is reduced to 50% (Cipolle and Zehr 1985). Plasma renin activity was measured by rat vasopressor bioassay in this experiment. A more severe hemorrhage of 60% blood volume or hypotension induced by nitroprusside infusion in the conscious turtle also failed to increase plasma renin activity. In addition to the difference in assay method, plasma was incubated under angiotensinase inhibition in the former study while dialyzed plasma was used in the latter according to the method of Boucher et al. (1967). Because inactive renins seem to be present in the turtle plasma (Cipolle and Zehr 1984), the difference in the two results could be due to the activation of inactive renins. Thus it remains undetermined whether or not the baroreceptor mechanism is present in reptiles (Table 5.1).

Hypovolemia induced by frusemide increases renal renin content in the water snake, *Natrix taxispolota* (LeBrie and Bolecskevy 1979). If the increase is due to a decreased release, the Na sensing mechanism similar to that of the MD may predominate over the pressure/volume sensing mechanism in this snake. Depression of plasma Na concentration caused by frusemide increases renal renin activity in the tortoise, *Testudo hermanni*, whereas dietary Na loading or intraperitoneal injection of hypertonic NaCl solution decreased the activity (Uva and Vallarino 1982), although both frusemide and Na loading should have increased Na excretion. In *Pseudemys*, frusemide administration in conscious animals increases plasma renin activity which is accompanied by reduction in plasma Na and K levels (Cipolle and Zehr 1985). In the same species, however, Stephens and Robertson (1985) found that frusemide and ethacrynic acid did not alter plasma renin activity, plasma electrolyte concentration and blood pressure in the conscious animal in spite of a large increase in urinary Na and K excretion. It needs to be determined whether or not the tubular Na sensing mechanism is involved in regulation of renin release in reptiles (Table 5.1).

Infusion of a β-adrenergic agonist, isoproterenol, into the jugular vein of anethetized *Pseudemys* fails to increase plasma renin activity (Cipolle et al. 1986). Infusion of acetylcholine increases plasma renin activity, but the effect may be mediated by catecholamines because the increase is not blocked by atropine but by propranolol. Epinephrine increases plasma renin activity which is blocked by propranolol but not by phenoxybenzamine. However, it is possible that the effect of propranolol is not specific to the β-blocking action (Cipolle et al. 1986). Thus it is not yet known whether renal nerves are involved in regulation of renin release in reptiles (Table 5.1).

5.3.3 Amphibians

The presence of baroreceptor mechanisms for regulation of renin release is not clear in amphibians. Plasma renin activity is slightly greater in the toad, *Bufo marinus*, adapted to hyperosmotic media than in those is distilled water (Garland and Henderson 1975). Renal renin activity increases and body weight decreases in the toad, *Bufo arenarum*, after exposure to more than 250 mM NaCl solution for several days (Nolly and Fasciolo 1971). It is not known whether the increase in renal renin activity is due to increased synthesis or decreased release. Plasma renin activity decreases after dehydration (exposure to air) and increases after isotonic expansion of blood volume by NaCl or glucose solution in the bullfrog, *Rana catesbeiana* (Sokabe et al. 1972). Therefore, regulation of renin release by the volume sensing mechanism seems to be reversed in this frog compared to mammals and birds (Table 5.1).

Renal renin activity increases in the leopard frog, *Rana pipiens*, maintained in Na-free media (Capelli et al. 1970), but the activity is not altered in the toad, *Bufo arenarum*, after exposure to distilled water, peritoneal dialysis, and frusemide adminstration (Nolly and Fasciolo 1971). When the toad was kept in increasing concentrations of NaCl solution, plasma Na concentration increased in parallel, but renal renin concentration increased only at the highest concentration. Renal renin concentration and plasma and urinary Na levels increase as the exposure to hypertonic saline is prolonged. However, the increase in renal renin concentration is due to the decrease in kidney size and the total amount of renal renin is not altered (Nolly and Fasciolo 1971). The MD is present in some species of frogs (Capelli et al. 1970; Lamers et al. 1974), but its role in the regulation of renin release has not yet been defined (Table 5.1).

The β-adrenergic agonist, isoproterenol, increases plasma renin activity in the tiger salamander, *Ambystoma tigrinum*, kept at 20 °C but not in those kept at 5 °C (Table 5.1; Corwin et al. 1984). Intensive innervation of adrenergic nerve fibers is reported in the granulated JG cells of the toad (Lamer et al. 1985b). In the organ-cultured kidney of *Amphiuma means*, Ca ionophore Br-X537A stimulates synthesis and release of renin, when renin activity is measured by radioimmunoassay after incubation with mammalian angiotensinogen (Worley et al. 1978). The stimulatory action of intracellular Ca ions on renin release is converse to the inhibitory action in mammals (Churchill 1985).

5.3.4 Fishes

Transferring the euryhaline *Tilapia mossambica* from seawater to freshwater does not change plasma renin activity when determined by the rat vasopressor bioassay, although plasma Na concentration decreased (Malvin and Vander 1967). Plasma renin activity decreases in the eel, *Anguilla rostrata*, after transfer from seawater to freshwater (Nishimura et al. 1976). Plasma volume increases and plasma Na concentration decreases slightly after transfer. Plasma and kidney renin activity decrease and increase, respectively, in the toadfish, *Opsanus tau*, after transfer from 50% seawater to 5% seawater (Nishimura et al. 1976). Plasma Na concentration decreases and plasma angiotensinogen concentration does not change after transfer. Plasma renin activity and plasma Na concentration are greater in seawater-acclimated *Anguilla anguilla* than in freshwater fish (Henderson et al. 1976). Plasma renin activity decreases within a day after transfer back to freshwater. The JG cells of *Cymatogaster aggregata* showed a degenerate profile after transfer from seawater to hyposmotic media (Lagois 1968). Transferring eels (Sokabe et al. 1973; Henderson et al. 1976) or rainbow trout (Jackson et al. 1977) from freshwater to seawater increases plasma renin activity within a few hours. However, plasma renin activity returned to the freshwater lever after acclimation to seawater in Japanese eels (Sokabe et al. 1973) but not in European eels (Henderson et al. 1976). Renal renin activity, determined after incubation with hog angiotensinogen, increased in proportion to the increase in external unionized ammonia concentration in the rainbow trout (Arillo et al. 1981). Plasma ANG II levels increase within 4 h after transfer of freshwater eels to seawater (Takei et al. 1985), but the changes were not observed for a day or two after transfer in another study (Okawara et al. 1987). The JG cells of *Tilapia* increase in number and size after transfer from fresh water to seawater (Krishnamurthy and Bern 1973). Renal renin activity is higher in several teleost species acclimated to fresh water than in those acclimated to seawater (Sokabe et al. 1966; Capelli et al. 1970). However, Nolly and Fasciolo (1972) could not find any difference in renal renin activity between seawater and freshwater fishes.

It has been shown that cumulative hemorrhage induces stepwise increases in plasma renin activity in the toadfish acclimated to 50% seawater (Table 5.1; Nishimura et al. 1979). Single massive hemorrhage of 50–60% blood also induces time-dependent increases in plasma renin activity for 1 h and a decrease in arterial pressure for more than 3 h. Plasma angiotensinogen levels decreases after hemorrhage of more than 40% blood. Papaverine induces a transient decrease in blood pressure for 30 min, and unlike mammals, birds and reptiles, this drug increases plasma renin activity in the toadfish. Plasma renin activity of the toadfish is determined by radioimmunoassay because toadfish ANG I cross-reacts with antisera raised against human ANG I (Nishimura et al. 1977). Another aglomerular fish, *Lophius litulon*, has histidine at position 9 as in mammals (see Sect. 4.2). Bailey and Randall (1981) showed that in the rainbow trout, plasma renin activity is linearly correlated with the degree of hemorrhage. In the isolated, perfused, nonfiltering kidney preparation, they also found that a decrease in renal perfusion pressure resulted in an increase in renin release. Therefore, a renal baroreceptor mechanism similar to that of mammals may exist in this fish (Table 5.1). The increase in renin release in response to decreased renal perfusion pressure is not affected by either α- or β-adrenergic blocking agents, whereas ANG II apparently inhibits renin release in this in vitro system. Plasma ANG II concentration increased after acute hemorrhage of 14.3% blood volume in the Japanese eel (Fig. 5.3; Takei et al. 1985). Arterial blood pressure decreases after hemorrhage, and decreases in blood pressure and blood volume continue for more than 4 h (Takei 1988). When isosmotic blood volume expansion is induced by intravenous saline infusion in the Australian lungfish, *Neoceratodus forsteri*, plasma renin activity increases as measured by rat vasopressor bioassay (Blair-West et al. 1977). However, blood volume expansion with hyposmotic saline does not change the activity. This result indicates the involvement of Na-sensing mechanisms in the regulation of renin release in the lungfish.

There is no clear relationship between plasma renin activity and plasma Na concentration in teleost fish after transfer from fresh water to seawater or vice versa. When hypertonic NaCl solution is injected into the conscious eel, which theoretically increases plasma Na concentration by 15%, plasma ANG II concentration paradoxically increases even though blood volume increases by 5–15% for 4 h postinjection (Fig. 5.3; Takei et al. 1988c). This result is opposite to those of mammals and birds in which hypernatremia suppresses renin secretion (Fig. 5.3). A double concentration of hypertonic NaCl solution causes similar and prolonged increases (Fig. 5.3). The lack of the MD in piscine kidneys may be one of the causes of this unexpected result (Sokabe and Ogawa 1974). Injection of hypertonic sucrose also increases plasma ANG II level (Takei et al. 1988c). Since Na excretion is increased after sucrose injection, this result also supports the lack of the MD mechanism in eels (Table 5.1).

Isoproterenol induces a decrease in arterial pressure and an increase in plasma renin activity in the toadfish, and propranolol abolishes both responses (Nishimura 1980b; Nishimura and Madey 1989). Subdepressor doses of isoproterenol also cause slight and transient increases in plasma renin activity. Norepinephrine increases blood pressure but does not alter plasma renin activity in the same species. The α-adrenergic blocker, phentolamine, injected alone into the intact fish increases plasma renin activity. ANG II may be responsible for the physiological inhibition of renin release in the toadfish because administration of SQ14225 (captopril), another converting enzyme inhibitor, increases plasma renin activity as in mammals (Nishimura and Bailey 1982). The stimulatory effect of isoproterenol may be indirect since its effect is absent in renal slices in vitro and adrenergic innervation to JG cells is also absent in the toadfish (Table 5.1; Nakamura et al. 1992). An increase in intracellular Ca ions may be inhibitory for renin release in the toadfish as in mammals because stimulation of Ca influx via the voltage-sensitive Ca channel suppresses renin release, and the effect is reversed by the channel blocker, nifedipine (Nishimura and Madey 1989). It seems that prostaglandins do not play a role in baroreceptor-mediated renin release in the toadfish because indomethacin does not inhibit hemorrhage-induced renin release (Nakamura et al. 1992).

Chapter 6

Biochemistry of the RAS

Renal renin activity, which is demonstrated by ANG I production after incubation of kidney extract with homologous plasma under inhibition of converting enzyme activity, has been detected in selected species of all classes of vertebrates thus far examined (Henderson et al. 1981). Thus, it has become increasingly convincing that the RAS is an old hormonal system that first appeared in the most primitive jawless fish. More recently, [Asp^1, Ile^5] ANG II-amide was isolated from the brain of a leech, *Erpobdella octoculata,* which exhibited a potent diuretic effect in this species (Salzet et al. 1995). Furthermore, non-amidated ANG II and an ANG I-like molecule were identified in another leech, *Theromyzon tessulantus* (Laurent et al. 1995).

6.1 Renin

Recent progress in biotechnological techniques enables us to determine the primary structure of high-molecular-weight proteins such as renin. The history of purification and sequence determination of renin using complementary DNA (cDNA) has been reviewed (Corvol and Ménard 1991; Inagami 1991). The first step for cloning, extraction of renin messenger RNA (mRNA), was achieved successfully for the first time in the mouse submaxillary gland, because this tissue produces 1000-fold more renin mRNA than does the kidney (Rougeon et al. 1981). The cDNA library was screened with an oligonucleotide probe whose sequence was predicted from the partially purified renin (Panthier et al. 1982). The complete amino acid sequence of preprorenin from the mouse submaxillary gland was deduced from the cDNA sequence (Fig. 6.1). Active renin was then produced after removal of the signal sequence and pro sequence at the N-terminus and finally processed into two-chain renin with 35 and 3-kDa molecules. The amino acid sequence of prorenin has a striking similarity to aspartyl proteases such as pepsinogen, prochymosin and pencillopepsin. Unlike other aspartyl proteases, however, renin has uniquely high substrate specificity to angiotensinogen and an optimal pH at neutral. The mRNA that hybridized the cDNA of submaxillary gland renin was detected in all major renin-containing tissues such as kidney, adrenal, brain, heart, pituitary, blood vessels and reproductive organs (Table 4.1; Campbell and Habener 1986; Dzau et al. 1987). Pratt et al. (1983) proposed a mode of renin processing in the mouse submandibular gland. The preprorenin synthesized in the cytoplasm is

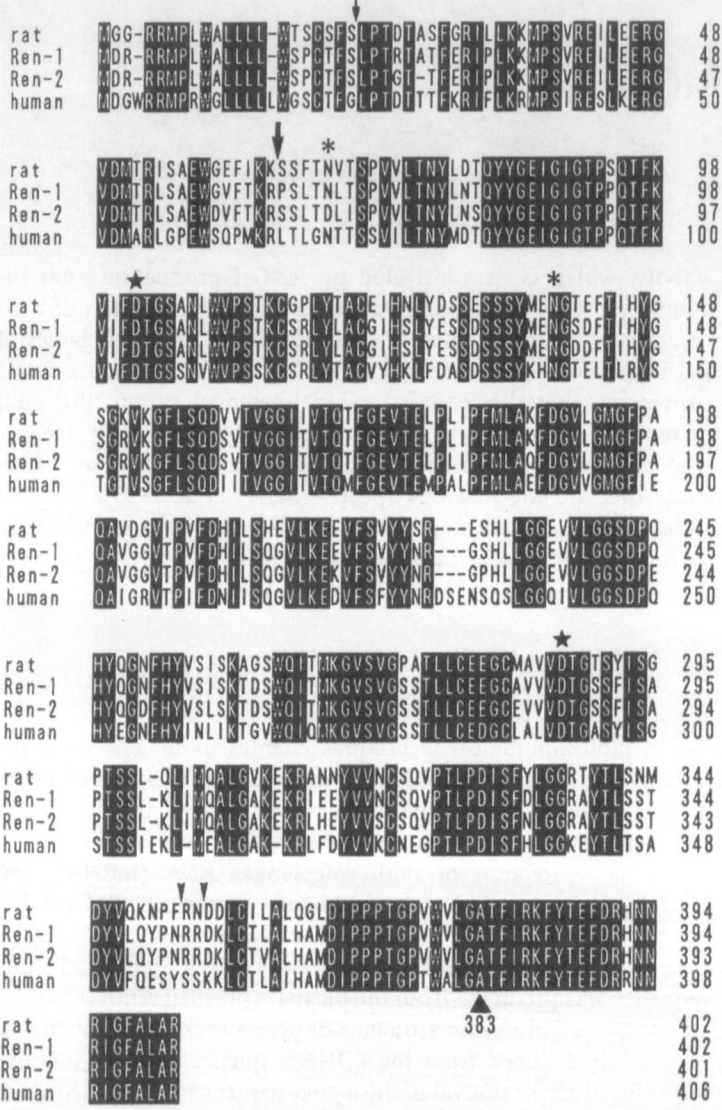

```
rat     MGG-RRMPLWALLLL-WTSCSFSLPTDTASFGRILLKKMPSVREILEERG    48
Ren-1   MDR-RRMPLWALLLL-WSPCTFSLPTRIATFERIPLKKMPSVREILEERG    48
Ren-2   MDR-RRMPLWALLLL-WSPCTFSLPIGI-TFERIPLKKMPSVREILEERG    47
human   MDGWRRMPRWGLLLLLWGSCTFGLPTDITTFKRIFLKRMPSIRESLKERG    50

rat     VDMTRISAEWGEFIKKSSFTNVTSPVVLTNYLDTQYYGEIGIGTPSQTFK    98
Ren-1   VDMTRLSAEWGVFTKRPSLTNLTSPVVLTNYLNTQYYGEIGIGTPPQTFK    98
Ren-2   VDMTRLSAEWDVFTKRSSLTDLISPVVLTNYLNSQYYGEIGIGTPPQTFK    97
human   VDMARLGPEWSQPMKRLTLGNTTSSVILTNYMDTQYYGEIGIGTPPQTFK   100

rat     VIFDTGSANLWVPSTKCGPLYTACEIHNLYDSSESSSYMENGTEFTIHYG   148
Ren-1   VIFDTGSANLWVPSTKCSRLYLACGIHSLYESSDSSSYMENGSDFIIHYG   148
Ren-2   VIFDTGSANLWVPSTKCSRLYLACGIHSLYESSDSSSYMENGDDFIIHYG   147
human   VVFDTGSSNVWVPSSKCSRLYTACVYHKLFDASDSSSYKHNGTELILRYS   150

rat     SGKVKGFLSQDVVTVGGIIVTQTFGEVTELPLIPFMLAKFDGVLGMGFPA   198
Ren-1   SGRVKGFLSQDSVTVGGITVTQTFGEVTELPLIPFMLAKFDGVLGMGFPA   198
Ren-2   SGRVKGFLSQDSVTVGGITVTQTFGEVTELPLIPFMLAQFDGVLGMGFPA   197
human   TGTVSGFLSQDIITVGGITVTQMFGEVTEMPALPFMLAEFDGVVGMGFIE   200

rat     QAVDGVIPVFDHILSHEVLKEEVFSVYYSR---ESHLLGGEVVLGGSDPQ   245
Ren-1   QAVGGVTPVFDHILSQGVLKEEVFSVYYNR---GSHLLGGEVVLGGSDPQ   245
Ren-2   QAVGGVTPVFDHILSQGVLKEKVFSVYYNR---GPHLLGGEVVLGGSDPE   244
human   QAIGRVTPIFDNIISQGVLKEDVFSFYYNRDSENSQSLGGQIVLGGSDPQ   250

rat     HYQGNFHYVSISKAGSWQITWKGVSVGPATLLCEEGCMAVVDTGTSYISG   295
Ren-1   HYQGNFHYVSISKTDSWQITWKGVSVGSSTLLCEEGGAVVVDTGSSFISA   295
Ren-2   HYQGDFHYVSLSKTDSWQITWKGVSVGSSTLLCEEGCEVVVDTGSSFISA   294
human   HYEGNFHYINLIKTGVWQIQWKGVSVGSSTLLCEDGCLALVDTGASYISG   300

rat     PTSSL-QLIWQALGVKEKRANNYVVNCSQVPTLPDISFYLGGRTYTLSNM   344
Ren-1   PTSSL-KLIWQALGAKEKRIEEYVVNCSQVPTLPDISFDLGGRAYTLSST   344
Ren-2   PTSSL-KLIWQALGAKEKRLHEYVVSCSQVPTLPDISFNLGGRAYTLSST   343
human   STSSIEKLWMEALGAK-KRLFDYVVKCNEGPTLPDISFHLGGKEYTLTSA   348

rat     DYVQKNPFRNDDLQILALQGLDIPPPTGPVHVLGATFIRKFYTEFDRHNN   394
Ren-1   DYVLQYPNRRDKLCTLALHAMDIPPPTGPVHVLGATFIRKFYTEFDRHNN   394
Ren-2   DYVLQYPNRRDKLCTVALHAMDIPPPTGPVHVLGATFIRKFYTEFDRHNN   393
human   DYVFQESYSSKKLCTLAIHAMDIPPPTGPTWALGATFIRKFYTEFDRRNN   398

rat     RIGFALAR                                            402
Ren-1   RIGFALAR                                            402
Ren-2   RIGFALAR                                            401
human   RIGFALAR                                            406
```

Fig. 6.1. Primary structures of prepro forms of *rat* renal renin, mouse renal renin (*Ren-1*), mouse submaxillary gland renin (*Ren-2*), and *human* renal renin. ★ Aspartate residue in the active center; * possible glycosylation site. ↓ and ⬇ show, respectively, putative cleavage sites for prorenin and mature renin. *Arrowheads* show cleavage sites to produce two-chain renin. (Hirose and Murakami 1992)

rapidly internalized into the rough endoplasmic reticulum and hydrolyzed by the signal protease to produce prorenin. The prorenin is then converted to an active form in the Golgi apparatus. The one-chain renin is further processed to two-chain renin which is one-sixth as active as the one-chain renin. Both one-

and two-chain renins are secreted, and prorenin does not seem to be released from the salivary gland.

The expression of the renin gene is highly tissue-specific. Some strains of mice have two different renin genes (two-gene mouse), *Ren-1* (kidney type) and *Ren-2* (submaxillary gland type), whereas other strains have only the *Ren-1* gene (one-gene mouse). Only the *Ren-2* gene is expressed in the submaxillary gland of the two-gene mouse, whereas both the *Ren-1* and the *Ren-2* genes are expressed in the kidney. The expression in the salivary gland is tenfold greater than that in the kidney. Tronik et al. (1987) introduced the *Ren-2* gene into the one-gene mouse and found that the introduced *Ren-2* gene is expressed in the submaxillary gland. A renin gene is not expressed in the adrenal of the one-gene mouse, but both genes are expressed in the two-gene mouse. Mullins et al. (1989) showed that the *Ren-2* gene introduced into the one-gene mouse was expressed also in the adrenal. Thus, the *Ren-2* gene seems to possess a regulatory sequence in the promoter region which is responsible for the expression in the adrenal. Furthermore, Sigmund and Gross (1991) showed that the regulatory sequence of the *Ren-2* gene responsible for the expression in JG cells resides 4.6 kb upstream in the promoter region. Similarly, Fukamizu et al. (1991) showed that a site 3 kb upstream of the human renal renin gene regulates its specific expression in the JG cells.

Imai et al. (1983) succeeded in cloning the cDNA of human renal renin and analyzed its sequence (Fig. 6.1). Their success was dependent on the use of a surgically removed ischemic kidney in which renin synthesis was increased tenfold, and on the use of cDNA of the mouse submaxillary gland renin as a hybridization probe. The amino acid sequence predicted from the cDNA sequence showed that human preprorenin consists of 406 amino acid residues with a signal sequence of 23 amino acid residues and a pro sequence of 44 amino acid residues at the N-terminus (Fig. 6.1). Two potential glycosylation sites and two aspartate residues in the active center were identified within the molecule. The cleavage site to produce two-chain renin was also identified (Fig. 6.1). The alanine residue at position 383 may be important for optimum activity at neutral pH. The genomic DNAs were also cloned for human, mouse and rat renal renin (Miyazaki et al. 1984; Fukamizu et al. 1988). The genomic DNA of human renin had ten exons and those of the rat and mouse had nine. A high degree of sequence identity was found among the three renal renins and the mouse submaxillary gland renin (Fig. 6.1). However, renal renins are, unlike submaxillary gland renin, thought to be glycoproteins. Glycosylation is important for the hepatic clearance of renin (Kim et al. 1988). Thus submaxillary gland renin was not cleared by the liver but by the kidney. Heterogeneity was observed in renal and plasma renin (Galen et al. 1979; Chang et al. 1981), which may be accounted for by the difference in the degree of glycosylation (Katz et al. 1991). The genomic DNA as well as cDNA of human renin are similar to those of human pepsin (Miyazaki et al. 1984). The three-dimensional structure of human renal renin was deduced in comparison with the structure of penicillopepsin, another aspartyl protease, with the aid of computer graphics (Akahane et al. 1985). The result suggests that the difference in the molecular geometry of their active sites may influence the high substrate specificity of renin, despite a great similarity in overall structure between renin and penicillopepsin molecules. The pre and pro segments of the renin precursor appear to be exposed to the exterior so as to be easily cut during the maturation process. Carlson et al. (1985) also showed that tertiary structures of aspartyl proteases exhibited high degrees of structural similarity when the structures were compared by X-ray diffraction.

In addition to active renin, inactive renins which were activatable by trypsin or acidification were found in plasma and amniotic fluid (Lumbers 1971; Nielsen and Poulsen 1988). The human kidney also contains inactive renin (Day and Luetscher 1975). These inactive renins are termed 'big renin', 'prorenin' or 'preprorenin'. Misono et al. (1976) showed that both active and inactive renins were released from renal cortical slices of rabbits and rats. Inagami and Murakami (1977) extracted three forms of renin from the hog kidney. The enzymatic activities of 140- and 61-kDa molecules were respectively 0.03 and 9% of that of active 42-kDa renin. The 61-kDa form may correspond to preprorenin as predicted from the cDNA sequence (Imai et al. 1983). The ratio of inactive renins released into the circulation ranges between 50 and 62% in the normal subject (Boyd 1977). In the tumoral JG cells, active renins were secreted from the granules after intracellular processing, but inactive big renins were secreted constitutively without being incorporated into the granules (Galen et al. 1984). Doi et al. (1984) showed that active renin was rapidly decreased, but inactive renin was paradoxically increased, after bilateral nephrectomy in rats, indicating a presence of extrarenal sources. However, the increase in inactive renin after nephrectomy was not prevented by removal of other isorenin-producing organs such as the salivary gland, uterus, spleen, pancreas, stomach, intestine, adrenal gland and pituitary. Dzau (1984) suggested that the endothelium has an enzyme capable of activating human plasma prorenin which, together with the surface-bound converting enzyme, constitutes another biochemical cascade for generation of ANG II. The author also suggested that platelets and neutrophils have a renin-activating enzyme, cathepsin G, and serve as a mobile pathway to amplify the renin–angiotension biochemical cascade. Renin-binding protein has been found in the renal tubule of rats, which appears to be responsible for interconversion of active and inactive renin (Ikemoto et al. 1982). This protein-bound renin is another form of regulation of plasma renin activity and is different from big renin or prorenin.

The first report of the presence of brain renin was made in the dog (Ganten et al. 1971) and rat (Fischer-Ferraro et al. 1971). However, Day and Reid (1976) later suggested that the apparent brain renin activity was caused by cathepsin D, and Hackenthal et al. (1978) in fact isolated cathepsin D from the brain which exhibited an angiotensin-forming activity at low pH. Thus, the existence of brain renin remained a subject of considerable debate. However, Hirose et al. (1978) devised an affinity chromatography using a casein–Sepharose gel, and succeeded in separating renin from cathepsin D in the rat brain. This brain renin had an optimal pH at neutral and cross-reacted with the antiserum raised against pig renal renin. This result was further confirmed by the separation of brain renin with another affinity chromatography using a pepstatin–Sepharose gel (Dzau et al. 1980). However, brain renin as well as other tissue renins have not been fully characterized yet.

6.2 Angiotensinogen

Angiotensinogen, or renin substrate, is a glycoprotein synthesized in the liver of several mammalian species. At least three different forms of angiotensinogen were identified according to the electrophoretic profile; sheep, cow, pig and rabbit angiotensinogens comigrated with α_2-globulin, human and dog angiotensinogens with α_1-globulin, and rat and mouse angiotensinogens with albumin. Plasma angiotensinogen concentration increased after treatment with ACTH, glucocorticoid, estrogen or ANG II, after nephrectomy, hemodilution or severe hypoxia, and during pregnancy (see Reid et al. 1978 for review). Plasma angiotensinogen concentration decreased after hepatectomy, adrenalectomy and hypophysectomy. The decrease in angiotensinogen concentration after adrenalectomy may be due to an increased plasma level of renin (Clauser et al. 1985). The hepatocytes which contained immunoreactive angiotensinogen were very few in the liver of normal rat, but the number increased after nephrectomy or colchicine treatment (Richoux et al. 1983). The plasma concentration of angiotensinogen was 0.8 nmol/ml in normal humans, while the concentration was about one-half of this amount in dogs, one-third in rats and one-fifth in horses, cows, cats, hogs and rabbits (see Peach 1977 for review).

A single species of angiotensinogen was identified in the rat plasma, but two different forms were released from isolated hepatocytes as examined by isoelectric focusing (Murakami et al. 1984). Campbell et al. (1985) suggested that the heterogeneity of angiotensinogen secreted from human hepatocytes is due to the difference in post-translational processing such as glycosylation or association with other proteins. According to Bouhnik et al. (1981), rat angiotensinogen purified from plasma had two different molecular forms with molecular mass of ca. 55 kDa. ANG I was attached to its N-terminus by the Leu–Leu bond. In human angiotensinogen, however, ANG I was attached to the mother molecule by the Leu–Val bond (Tewksbury et al. 1981). More recently, Fernley et al. (1986) isolated and sequenced angiotensinogen from sheep plasma and observed microheterogeneity in the chromatographic pattern. The heterogeneity was due to the difference in the carbohydrate moiety.

Ohkubo et al. (1983) and Kageyama et al. (1984) succeeded in cloning cDNAs of rat and human angiotensinogen, respectively. The deduced amino acid sequence indicated that the rat angiostensinogen consisted of 453 amino acid residues with a putative signal sequence of 24 amino acid residues (Fig. 6.2). The deduced sequence and its molecular weight coincided well with those determined from the purified protein. An ANG I moiety was localized at its N-terminus, preceded by the signal peptide and followed by two internally

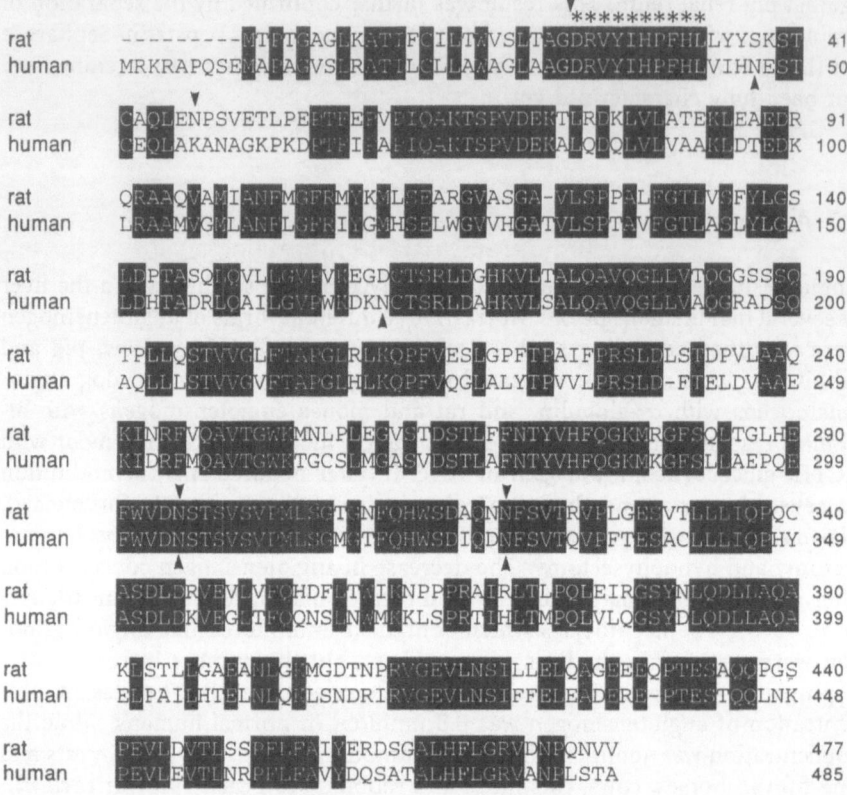

Fig. 6.2. Primary structures of *rat* and *human* preangiotensinogen. *Asterisks* indicate the location of ANG I. *Arrow* shows cleavage site to produce angiotensinogen. *Arrowheads* show possible glycosylation sites

homologous sequences and three potential glycosylation sites. The sequence identity of angiotensinogen molecules of rat and human was 63.6% but the N-terminal portion just distal to ANG I and the location of possible glycosylation sites differed to a greater extent. This may contribute to the known species specificity of the renin–angiotensinogen reaction.

The tissue-specific expression of the angiotensinogen gene has been elucidated. Clouston et al. (1989) introduced into the Swiss-strain mouse an angiotensinogen gene isolated from the Balb/C mouse. The introduced gene was a mini-gene which was constructed by deleting some of the introns and exons of the whole angiotensinogen gene to distinguish it from the host's gene. They showed that both native and introduced genes were expressed in the liver and other angiotensinogen-producing tissues, and that the glucocorticoid and estrogen treatments were also effective for stimulating transcription of the transgene. Thus, the 0.75-kb sequence located upstream of the angiotensinogen gene, which was included in the transgene, may contain a regulatory sequence responsible for the tissue-specific expression and for the response to inducing drugs. Gaillard-Sanchez et al. (1990) demonstrated that the renin gene and the angiotensinogen gene are located in proximity on the same chromosome (No. 1) in the human. Furthermore, Fukamizu et al. (1990) introduced a human angiotensinogen gene into the mouse and found that angiotensinogen mRNA was produced in the kidney as much as in the liver of the transgenic mouse. The expression in the kidney of normal mouse was very low. A possible explanation is that the negative regulatory element which normally inhibits the expression of the angiotensinogen gene in the kidney is located outside the introduced sequence.

6.3 Angiotension I

ANG I is a decapeptide which is cleaved from the N-terminus of angiotensinogen by renin. Since this peptide has little activity in isolated vascular smooth muscles relative to ANG II, it has long been thought to be an inactive precursor of ANG II. However, ANG I seems to act directly on the adrenal chromaffin cells (Peach et al. 1971), central nervous system (Buckley 1972; Swanson et al. 1973), renal vasculature (Itskowitz and McGiff 1974), and probably on the adrenal cortex (Saruta et al. 1972). This notion originates from the fact that pretreatment with converting enzyme inhibitors did not attenuate the biological action of ANG I. The presence of ANG I-specific receptors in some of these tissues supports its direct action (Goodfriend et al. 1972). Two different ANG I molecules were identified in mammals: $[Asp^1, Ile^5, His^9]$ ANG I for the human, rat, mouse, pig, horse, rabbit, guinea pig, and sheep, and $[Asp^1, Val^5, His^9]$ANG I only for the ox (Table 6.1).

6.4 Converting Enzyme

Converting enzyme, like renin, plays an essential role in the expression of various biological actions of the RAS as assessed by the use of its inhibitor (Waeber et al. 1990). Converting enzyme has been purified from the lung, kidney and plasma of several species of mammals, and had molecular weights ranging from 150 to 206 kDa (Peach 1977). It is now clear that converting enzyme is a rather nonspecific dipeptidyl carboxypeptidase that belongs to the class of zinc metalloprotease (Igic et al. 1972). The enzyme is also called kininase II because of its catalytic action on bradykinin (Erdös 1976). Recently, structures of two forms of human converting enzyme have been determined from the cDNA sequence (Soubrier et al. 1988; Lattion et al. 1989; Ehlers et al. 1989; Wie et al. 1991; see Ehlers and Riordan 1990; Hooper 1991 for reviews). A larger 170-kDa form is present in endothelial cells of the lung and other vasculature and epithelial cells of renal tubules, and a smaller 90–110 kDa form is

Table 6.1. Amino acid sequences of ANG I from various vertebrate species

Species	Amino acid sequence										References
	1	2	3	4	5	6	7	8	9	10	
Mammal											
Man, pig, horse, sheep, rat, mouse, rabbit, guinea pig	Asp	-Arg	-Val	-Tyr	-Ile	-His	-Pro	-Phe	-His	-Leu	Akagi et al. (1982) Ferniey et al. (1986)
Ox	Asp-				-Val-				-His-		Akagi et al. (1982)
Bird											
Fowl (*Gallus domesticus*)	Asp-				-Val-				-Ser-		Nakayamam et al. (1973)
Quail (*Coturnix coturnix japonica*)	Asp-				-Val-				-Ser-		Takei and Hasegawa (1990)
Reptile											
Snake (*Elaphe climocophora*)	Asx-				-Val-				-Ser-		Nakayamam et al. (1977)
Turtle (*Pseudemys scripta*)	Asp-				-Val-				-His-		Hasegawa et al. (1984)
Alligator (*Alligator mississipiensis*)	Asp-				-Val-				-Ala-		Takei et al. (1993a)
Amphibia											
Bullfrog (*Rana catesbeiana*)	Asp-				-Val-				-Asn-		Hasegawa et al. (1983a)
Teleost											
Salmon (*Oncorhynchus keta*)	Asn-				-Val-				-Asn-		Takemoto et al. (1983)
Eel (*Anguilla japonica*) (*A. rostrata*)	Asn-				-Val-				-Gly-		Hasegawa et al (1983b) Khosla et al. (1985)
Goosefish (*Lophius litulon*)	Asn-				-Val-				-His-		Hayashi et al. (1978)
Elasmobranch											
Dogfish (*Triakis syllia*)	Asn-		-Pro-		-Ile-				-Gln-		Takei et al. (1993b)

present in the germinal cells of the testis. The mRNAs of the two enzymes are transcribed from the common gene which consists of 26 exons (Hubert et al. 1991). The endothelial mRNA was transcribed from exon 1 to exon 26 except exon 13, and the testicular mRNA from exon 13 to exon 26. Thus the endothelial mRNA contains two homologous sequences of the enzyme in series. Both endothelial and testicular enzymes are integral membrane proteins anchored to the plasma membrane by their hydrophobic C-terminal segment. However, they are also released into the circulation by proteolytic cleavage (Beldent et al. 1993). The cDNAs for endothelial and/or testicular converting enzyme were also cloned in the rabbit (Kumar et al. 1989; Thekkumkara et al. 1992), mouse (Bernstein et al. 1989), and cattle (Shai et al. 1992). The expression of each mRNA is regulated by different promoters; the endothelial enzyme is induced by glucocorticoids whereas the testicular enzyme is induced by androgens (Krulewitz et al. 1984; Velletri et al. 1985). The functional role of testicular converting enzyme has not been determined yet, but in mice carrying an insertional mutation that inactivates both the somatic and the testicular converting enzyme gene, all homozygous female mutants are fertile, but the fertility of the homozygous male mutants is greatly reduced (Krege et al. 1995). Male–female difference is also observed in blood pressure; only males with the heterozygous mutation have lower blood pressure than normal males.

The presence of different types of enzymes with ANG II producing activity was suggested in the human lung, hog and guinea pig plasma, rat submaxillary gland (see Erdös 1976 for review), and human heart (Urata et al. 1990). The enzyme from the submaxillary gland is called tonin (Schiffrin and Genest 1983) and that from the heart belongs to the group of chymases, both of which can produce ANG II directly from angiotensinogen and have a much lower molecular weight than converting enzyme.

The route for ANG III formation is mostly from ANG II by aminopeptidases, but a direct formation from [des-Asp^1] ANG I by converting enzyme is suggested in the pulmonary (Gaynes et al. 1978), mesenteric (Sexton et al. 1979), and hepatic (Britton et al. 1983) circulations.

6.5 Angiotensinases

The peptidases that catabolize angiotensins are termed angiotensinases. These include aminopeptidases, carboxypeptidases and endopeptidases, and play important roles in the termination of angiotensin action. Angiotensinase activity is high in the circumventricular organs such as the subfornical organ and area postrema where its activity far exceeds that in plasma (Schelling et al. 1978). Two aminopeptidases that degrade ANG II are identified in plasma, angiotensinase A_1 selective for an N-terminal asparaginyl residue and angiotensinase A_2 selective for an N-terminal aspartyl residue (Nagatsu et al. 1970). However, these plasma enzymes do not seem to have major roles in ANG II inactivation as was the case for plasma converting enzyme (Peach 1977).

Aminopeptidases are found in the red blood cells, liver, kidney, brain, intestine, spleen, adrenal, pancreas and lung, but in some organs the enzyme activity is localized in lysosomes. The endopeptidase that splits ANG II into two tetrapeptides was found in plasma and kidney homogenates (Regoli et al. 1963), but this enzyme seems to be nonspecific. Similar enzymes are found in the liver (cathepsin C; McDonald et al. 1974) and in red blood cells (Kokubu et al. 1969). Carboxylpeptidases which hydrolyze ANG II are found in the urine, kidney, spleen, liver and brain (Goldstein et al. 1972). The tissue enzymes are localized in lysosomes, thus their physiological function is doubtful. The inactivation of ANG II involves primarily aminopeptidases, since only N-terminal degradation occurred after perfusion of ANG II through the liver and other organs (Leary and Ledingham 1969).

6.6 Nonmammalian Vertebrates

6.6.1 Birds

The presence of the RAS was first implicated in the pigeon, *Columba livia* (Chan and Holmes 1971), in which an angiotensin-like pressor substance was extracted from plasma, and incubation of kidney extracts with nephrectomized rat plasma produced a pressor substance. Plasma ANG II concentration was measured by radioimmunoassay in the Pekin duck, *Anas platyrhynchos*, kept in fresh water (35.3 ± 3.9 pg/ml, $n = 22$; Gray and Simon 1985a) and in the Japanese quail, *Coturnix coturnix japonica*, in normal water balance (61.3 ± 9.9 pg/ml, $n = 8$; Takei et al. 1988a).

The amino acid sequence of ANG I was determined in the chicken (Nakayama et al. 1973) and quail (Takei and Hasegawa 1990; Table 6.1). These two galliform species have identical ANG I. The first and fifth amino acid residues are respectively aspartate and valine as in the ox and other nonmammalian tetrapods, but serine at position 9 is found only in birds. Amino acid sequences of other groups of birds such as anseriforms and columbiforms need to be determined to generalize the structure of avian ANG I.

The presence of converting enzyme is suggested in birds by pharmacological experiments. Intravenous injection of ANG II induces an immediate vasodepressor response which is followed by a delayed vasopressor response in the chicken (Moore et al. 1981a; Nakamura et al. 1982) and quail (Takei and Hasegawa 1990). These vascular effects are blocked by pretreatment with SQ20881 or captopril in both species (Moore et al. 1981b; Nishimura et al. 1982). The effect of converting enzyme inhibitors on the dipsogenic effect of ANG II is described in Sect. 8.3.2. The converting enzyme isolated from the chicken lung is a glycoprotein with zinc in its molecule, and require SC1⁻ ions for the expression of its full activity (Polanco et al. 1992). These characteristics are quite similar to those of mammalian enzymes. However, the chicken enzyme has a molecular mass of 690 kDa which is much greater than those of mammalian enzymes. The difference between mammalian and avian converting enzyme is also suggested by pharmacological experiments showing that [des-Pro^2] bradykinin and bradykinin potentiator B, both of which are potent inhibitors of mammalian converting enzyme, did not display any inhibitory effect in the quail (Takei and Hasegawa 1990).

Recently, a cDNA encoding chicken converting enzyme was cloned from the cDNA library of the lung (Esther et al. 1994). The converting enzyme appears to be a somatic isozyme since the protein deduced from the cDNA sequence consists of two putative zinc binding sites at the center of two homologous domains. A testicular isozyme with a single domain was not identified in the chicken testis by either Northern blot or enzyme assay.

6.6.2 Reptiles

A renin-like activity has been demonstrated in plasma and kidney of the snake, *Elaphe climocophora*, and the activity is lower in the hibernating animals (Seyama et al. 1979). Plasma and kidney renin activities of nonhibernating tortoise, *Testudo hermanni*, are also greater than those of hibernating animals when the activity was determined by radioimmunoassay after incubation of the plasma or kidney extract with porcine angiotensinogen (Vallarino 1984).

Plasma renin activity of normal freshwater turtle, *Pseudemys scripta*, is 0.8 ± 0.1 ng/ml/h when measured by rat pressor bioassay (Cipolle and Zehr 1984) and 2.5 ng/ml/h by radioimmunoassay for human ANG I (Stephens and Creekmore 1984). Both *Pseudemys* and human ANG I have histidine at position 9 (Table. 6.1). Therefore, the difference between the two ANG Is is only at position 5 which is shown to have little influence on the biological and immunological activity of ANGs (Sakakibara et al. 1985). The amino acid sequences of ANG I are determined in the snake *Elaphe climocophora* (Nakayama et al. 1977), turtle *Pseudemys scripta* (Hasegawa et al. 1984a), and alligator *Alligator mississippiensis* (Takei et al. 1993a; Table 6.2). The N-terminal amino acid of turtle and alligator ANG I is aspartate as in other tetrapods, but it is unidentified in the snake due to the N-terminal blockade. The ninth amino acid is variable also in reptiles. The presence of converting enzyme is suggested in *Pseudemys*, because the vasopressor effect of human ANG I, but not of ANG II, is decreased by prior infusion of captopril (Cipolle and Zehr 1984). The trypsinization of the turtle plasma resulted in a slight increase in plasma renin activity, whereas acidification of plasma to pH 3.3 yielded a fourfold increase in the activity, suggesting the presence of inactive renin in the turtle plasma. Captopril decreased the vasopressor effect of human ANG I in *Pseudemys* (Stephens 1981) and the rat snake, *Ptyas korros* (Ho et al. 1984).

6.6.3 Amphibians

The amino acid sequence of amphibian ANG I has been determined only in the bullfrog, *Rana catesbeiana* (Hasegawa et al. 1983b). Since its first amino acid is aspartate, bullfrog ANG I is of the tetrapod type (Table 6.2). Angiotensin-like vasopressor peptides have been extracted from the skin of some frogs, and the amino acid sequence of the one from the Australian frog, *Crinia georgiana*, was determined to be Ala-Pro-Gly-[Ile^3, Val^5] ANG II (Erspamer et al. 1979). This peptide possessed 90% vasopressor potency of human ANG II in the rat (Khosla et al. 1981). Since the vasopressor effect of human ANG I was blocked by captopril in the bullfrog and mudpuppy, *Necturus maculosus* (Fruchter et al. 1980), converting enzyme must also be present in amphibians. Harper and Stephens (1985) produced similar results in the bullfrog. Angiotensinase activity in amphibian plasma may be much lower than that of rat because rat plasma inactivates human and bullfrog ANG II much faster than does bullfrog or mudpuppy plasma (Fruchter et al. 1980).

6.6.4 Fishes

A vasopressor substance is produced after incubation of lamprey kidney extract with purified dog angiotensinogen (Henderson et al. 1981). ANG II increases blood pressure, induces drinking, and stimulates mineralocorticoid release in *Scyliorhinus* (Hazon and Henderson 1985; Hazon et al. 1989; O' Toole et al. 1990), spiny dogfish, *Squalus acanthias* (Opdyke and Holcombe 1976) and hagfish, *Myxine glutinosa* (Carroll and Opdyke 1982). More recently, a fraction of *Scyliorhinus* kidney extracts with a molecular mass of a ca. 48 kDa has been shown to cleave ANG I from the porcine angiotensinogen, and the gill extract very potently cleaved hippuric acid from Hip-His-Leu (Uva et al. 1992). These data suggest the presence of the RAS even in the primitive piscine species. The plasma renin activity and plasma angiotensinogen level of the toadfish, *Opsanus tau*, are 1.7 ± 0.3 ng/ml/h and 4.1 ± 0.4 µg/ml ($n = 8$), respectively, when determined by radioimmunoassay for mammalian ANG II (Nishimura et al. 1979). The ninth amino acid of toadfish ANG I may be histidine as in the mammalian peptide because ANG I of goosefish, *Lophius litulon*, which is systematically close to toadfish, has histidine at this position (Hayashi et al. 1978). ANG I has been isolated from chum salmon, *Oncorhynchus keta* (Takemoto et al. 1983), and eel, *Anguilla japonica* and *A. rostrata* (Hasegawa et al. 1983a; Khosla et al. 1985; Table 6.2). It is of interest to note that all teleost ANG Is have asparagine residues at position 1 instead of aspartate as in all tetrapods, although a proportion of eel ANG Is have aspartate at that position (Hasegawa et al. 1983a; Khosla et al. 1985). Recently, we isolated ANG I from the dogfish, *Triakis scyllia* (Takei et al. 1993a), thereby directly demonstrating the presence of the RAS in elasmobranchs. More recently, we also sequenced ANG II from the lamprey, *Lampetra fluviatilis*

(Y. Takei, and J.C. Rankin, unpubl. data). The first amino acid residue of lamprey and dogfish ANG is asparagine as in teleosts (Table 6.2). Interestingly, the fifth residue is isoleucine in the dogfish as in mammals. Furthermore, the third amino acid residue is unique in that the dog fish has proline instead of valine in all other species.

Captopril appears to be an effective inhibitor in the rainbow trout (Gallardy et al. 1984), rainbow trout and eel (Kenyon et al. 1985), flounder, *Platichthys flesus* (Carrick and Balment 1983), European dogfish (Hazon et al. 1989) and other teleost fishes (see Chap. 8). SQ20881 also inhibited the vasopressor effect of ANG I in the spiny dogfish (Opdyke and Holcombe 1976) and eel (Nishimura et al. 1978; Takei 1987), and blocked seawater-induced drinking in the killifish, *Fundulus heteroclitus* (Malvin et al. 1980). Thus, the converting enzyme in these fishes may be similar to that of mammals.

Recently, a cDNA coding for converting enzyme has been cloned from the cDNA library of the embryo of *Drosophila melanogaster* (Cornell et al. 1995). This insect enzyme has a single catalytic domain with a molecular mass of 67 kD which is reminiscent of the testis isozyme of mammalian converting enzyme. Southern analysis showed that the enzyme was controlled by a single gene which differs from the duplicate genes observed in mammals. The protein expressed in the COS-7 cells is capable of converting mammalian ANG I to ANG II and is inhibited by captopril and trandolaprilat.

Chapter 7

Angiotensin Receptors

ANG II binding sites have been identified by radioligand binding studies or by autoradiography in almost all tissues on which ANG II actions have been described (Table 7.1). Most of these sites satisfy criteria required for positive identification of receptors, i.e. binding is rapid, saturable, reversible, and specific to ANG II peptides, and the binding sites have high affinity with an equilibrium dissociation constant (K_d) comparable to ED_{50} of its in vitro effect.

ANG II receptors have been identified in smooth muscle tissues such as blood vessels, uterus, bladder and digestive tracts, and in cardiac muscles of several mammalian species (Mendelsohn 1985). High affinity ANG II receptors have also been identified in peripheral nonmuscular tissues such as adrenal cortex, adrenal medulla, renal glomeruli, renal tubules, liver, intestinal epithelia, and anterior pituitary. The presence of ANG II receptors is suggested in migratory blood cells and platelets. The central nervous system also contains specific ANG II receptors which are localized in neuronal cells of various brain regions and in spinal cord neurons. All these tissues are indicated as targets of ANG II, and most of them synthesize ANG II locally by the tissue RAS (Dzau 1984, 1987; Campbell 1987).

In addition to plasma membrane receptors, specific ANG II receptors are localized in hepatocyte nuclei (Booz et al. 1992; Tang et al. 1992). ANG II is known to be taken up by adrenocortical cells (Bianchi et al. 1986; Crozat et al. 1986) and accumulated in the nuclei of various tissues (Robertson and Khairallah 1971; Jimenez et al. 1994). These receptors may function to alter RNA synthesis by changing the chromatin conformation (Re and Parab 1984), resulting in the stimulation of cell growth (Geisterfer et al. 1988). ANG II binding proteins have also been identified in the cytosolic fraction of mammalian liver (Hagiwara et al. 1989; Kiron and Soffer 1989). These binding proteins are distinct from the AT_1 and AT_2 receptors of the plasma membrane as well as from nuclear receptors because of the lack of affinity to specific receptor antagonists for each receptor. Recently, two different cDNA clones coding for porcine soluble ANG II binding protein were isolated and their primary structures determined (Sugiura et al. 1992; Kato et al. 1994). Sequence comparisons with other proteins revealed that these proteins are ANG II degrading enzymes that belong to thimet oligopeptidase and microsomal endopeptidase, respectively.

The initial part of this chapter will be devoted to describing the cloning and classification of ANG II receptors, which is one of the recent topics in the field of ANG research. Then a brief account of ANG II receptors in each tissue will

Table 7.1. Angiotensin II receptors in various mammalian tissues

| Tissue | Effect | Angiotensin II receptors | | | | | |
		Animal	Class	Kd (nM)	Number (fmol/mg)	Affinity (AII : AIII)	References
Aorta	Contraction	Rabbit	2	6.2, > 500	–	1 : 0.1	Devynck and Meyer (1976)
Mesenteric artery	Contraction	Rat	1	0.9	54	1 : 0.2	Gunther et al. (1980)
Uterus	Contraction	Rat	1	0.5	194	1 : 0.2	Douglas et al. (1982)
Heart	Inotropic chronotropic	Rabbit	1	4.5	54	1 : 0.6	Baker et al. (1984)
Adrenal cortex	Aldosterone release	Rat	1	1.1	1804	1 : 0.9	Douglas et al. (1980)
Adrenal medulla	Catecholamine release	Rat	1	0.7	169	–	Healy et al. (1985)
Liver	Glycogenolysis angiotensinogen	Rat	2	0.2, 2.9	229, 1820	1 : 0.3	Campanile et al. (1982)
Renal glomerulus	GFR	Rat	1	1.2	1127	1 : 0.8	G.P. Brown et al. (1980)
Renal tubule	Na transport	Rat	1	9.5	162	1 : 0.05	Brown and Douglas (1982)
Pituitary	ACTH release	Rat	1	0.9	293	1 : 0.2	Hauger et al. (1982)
Intestine	Na transport	Rat	1	0.6	22	1 : 0.01	Cox et al. (1986)
Brain	Drinking Na appetite etc	Calf	1	0.2	1600	1 : 3.0	Bennett and Snyder (1976)
Spinal cord	Ionic conductance	Mouse	2	0.4, 25.6	13, 220	1 : 0.1	Laribi et al. (1985)

follow in relation not only to blood-borne ANG II but also to ANG II generated locally by the tissue RAS (see Chap 4). Finally, a series of events that occur inside the cell following hormone–receptor interaction will be reviewed.

7.1 Structure of ANG II Receptors

ANG II receptors have a history of research of more than 20 years during which efforts have been made to isolate and sequence the receptor protein. The initial biochemical characterization of the receptor protein revealed that it is susceptible to neuraminidase (Devynck and Meyer 1976) and to reducing agents (Gunther et al. 1980a), indicating that it is a glycoprotein whose tertiary structure is determined by disulfide bonds. In fact, the receptor protein was later isolated and its amino acid sequence was partially determined (Capponi and Catt 1980; Cauraud 1987; Rondeau et al. 1990; Marie et atl. 1990). Recently, several non-peptide receptor antagonists of ANG II were discovered which facilitated characterization of ANG II receptors at subcellular levels (Timmermans et al. 1991; Smith et al. 1992). ANG II receptors are now divided into two types according to their affinity for different antagonists (Chiu et al. 1989; Bumpus et al. 1991); type 1 receptors (AT_1) which have affinity to Dup753 (losartan) and type 2 receptors (AT_2) which have high affinity to PD123177 and CGP42112A (Figs. 7.1; 7.2). The AT_1 receptors are susceptible to reducing agents such as dithiothreitol, and utilize inositol triphosphate (IP_3) or inhibition of adenylate cyclase for signal transduction (Table 7.2). Therefore, AT_1 receptors may be divided further into two subtypes according to signal transduction systems; one coupled with G_q-type GTP-binding protein and the other coupled with G_i-type. The inhibition of adenylate cyclase by ANG II is reported in the renal tubules, liver, adrenal cortex and pituitary, but its functional role is not yet known (Catt and Abbott 1991). The AT_2 receptors do not seem to be coupled with GTP-binding protein inspite of seven transmembrane structure, and thus their subcellular mechanism is unkown (Table 7.2). The functional role of AT_2 is also undetermined yet, but it may be related to inhibition of basal cGMP level (Sumners et al. 1991). The AT_1 receptors are

Table 7.2. Characteristics of angiotensin II receptor subtypes

	AT_1 receptor	AT_2 receptor
Other names	AII-1, AII-B, AII_α	AII-2, AII-A, AII_β
Potency order	ANGII > ANGIII	ANGII = ANGIII
Selective antagonists	Dup753 (losartan)	PD123177, CGP42112A
Second messenger	IP_3/DG, cAMP	Unknown
Sensitivity to reducing agent	High	Low
Structure (rat)	7- Transmembrane 359 amino acids	7- Transmembrane 363 amino acids

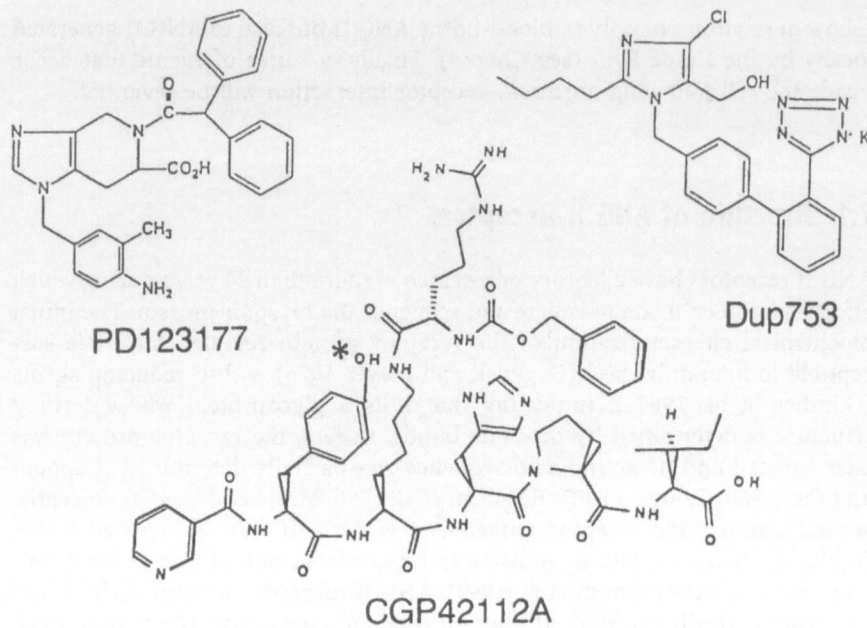

Fig. 7.1. Molecular structures of nonpeptide ANG II receptor antagonists

localized predominantly in the vascular smooth muscle, whereas the AT_2 receptors are concentrated in the adrenal medulla and uterus. In the adrenal cortex, kidney and brain, both types are found in various proportions (Timmermans et al. 1991).

The cDNAs for AT_1 receptors have been cloned from the cDNA library of bovine adrenal cells (Sasaki et al. 1991) and rat vascular smooth muscle cells (Murphy et al. 1991), and their primary structures deduced from the cDNA sequences. The AT_1 receptor of rat consists of 359 amino acid residues and its

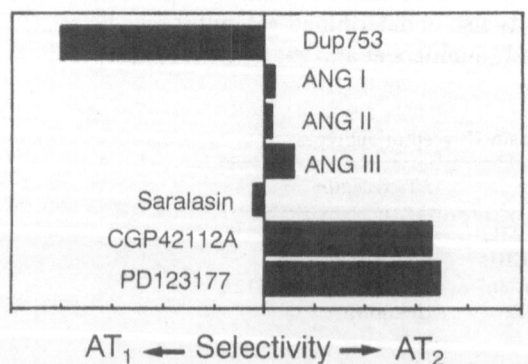

Fig. 7.2. Selectivity of peptide and nonpeptide ligands to angiotensin type 1 (AT_1) and type 2 (AT_2) receptors. (Modified from Timmermans et al. 1992)

94

hydrophobicity profile reveals seven putative transmembrane domains, which is typical of receptors coupled with GTP-binding proteins (Table 7.2). The receptor protein contains three potential N-glycosylation sites; one in the hydrophobic N-terminal extracellular region and the other two in the third extracellular loop. Cysteine residues are present in each of the four extra-cellular domains and are probably important in the formation of disulfide bridges necessary for the ligand-binding conformation. Several potential phosphorylation sites are present in the second intracellular loop and in the C-terminal intracellular domain. Two subtypes were identified for the AT_1 receptors at the cDNA and genomic DNA levels, and were named the AT_{1a} and AT_{1b} receptors (Elton et al. 1992; Iwai and Inagmi 1992; Sasanuma et al. 1992). These two subtypes have more than 90% sequence identity at the nucleotide level. The AT_{1a} receptor is localized on rat chromosome 17 and AT_{1b} on rat chromosome 2 (Lewis et al. 1993). These subtypes are expressed in various target tissues with different proportions (Kakar et al. 1992; Ye and Hearly 1992). It is yet unknown whether these subtypes utilize different signal transduction systems, IP_3/Ca or inhibition of adenylate cyclase/cAMP. The binding of ANG II to recombinant AT_{1a} receptors did not inhibit adenylate cyclase activity. The history of cloning of the AT_1 receptors has been reviewed recently (Bernstein and Alexander 1992; Inagami et al. 1992b).

The AT_2 receptor protein has recently been purified to near homogeneity from neonatal rat kidney using an affinity column of CGP42112 (Ciuffo et al. 1993). Dissociation of labeled CGP42112 from the purified protein occurs with PDl23177, ANG II and [Sar[1]] ANG II but not with Dup753. The cDNA for the AT_2 receptor has also been expression-cloned from the cDNA library of the rat phenochromocytoma cell line (Kambayashi et al. 1993) and of whole rat fetuses (Mukouyama et al. 1993). The cDNA encodes 363-amino acid protein which has approximately 30% sequence identity with that of the AT_1 receptor. Northern blot analysis showed that the receptor is expressed in the adrenal medulla, inferior olive of the brain and uterus, and is more abundantly ex-pressed in the fetuses than in adults. The recombinant receptor protein se-lectively binds PD123319 and CGP42112A but not Dup753. Although the AT_2 receptor has a seven-transmembrane domain topology, it does not seem to be coupled to the GTP-binding protein (Table 7.2). The binding of recombinant AT_2 receptor protein with ANG II does not decrease intracellular cGMP but modulates protein phosphotyrosine levels by a protein kinase C (C-kinase) independent pathway in rat glomerular mesangial cells (Force et al. 1991). Judging from an abundant expression in the fetus, AT_2 receptors seem to play a role in growth and development.

It is known that mineralocorticoid augments vasopressor and dipsogenic responses to centrally and peripherally administered ANG II in the rat (Wilson et al. 1986). The augmented response is accounted for by an increased number of ANG II receptors in the thalamus, hypothalamus and brain stem regions and not by an increased affinity of the receptors. Gutkind et al. (1988) also observed an increase in the receptor density in the medial preoptic area, subfornical organ SFO and paraventricular nucleus in DOCA-or salt-loaded rats, and in

95

DOCA–salt hypertensive rats; the receptor density was also increased in the nucleus solitarius and area postrema. It is further shown that promoter regions of the rat AT_{1a} gene contain putative glucocorticoid responsive elements and the transcriptional activity increased in rat aortic smooth muscle cells in culture after activation of the promoter with glucocorticoid (Uno et al. 1994).

The presence of receptors specific to ANG II-(3-8), named ANG IV, has been suggested in various organs of several mammalian species (Swanson et al. 1992). The definitive function of this ligand and its receptor has not been assigned yet, but interaction with the endothelium for vascular relaxation is a possible function. The presence of specific receptors for ANG II-(1-7) is also suggested in neurons of the paraventricular nucleus as a result of experiments using an electro physiological technique (Felix et al. 1991) and using a selective antagonist to ANG II-(1-7) (Santos et al. 1994).

7.2 ANG II Receptors in Target Tissues

7.2.1 Peripheral Tissues

7.2.1.1 Muscular Tissues

ANG II is one of the most potent naturally occurring vasoconstrictors known to date (Peach 1977). The effect is demonstrated in the rat portal vessel (Voth et al. 1971), rabbit aorta (Shibata and Briggs 1966), rabbit mesenteric artery (Cuthbert and Sutter 1965), rabbit ear vessel (Blair-West et al. 1968), guinea pig aorta (Palaic and Le Morvan 1971), sheep carotid artery (Keatinge 1966), dog mesenteric artery (Somlyo and Somlyo 1968) and in isolated vascular smooth muscle cells (Chamley-Campbell et al. 1979; Gunther et al. 1982; Penit et al. 1983). ANG II is also known to induce protooncogene expression and growth promotion in vascular smooth muscle cells (Geisterfer et al. 1988), which indicates its role in angiogenesis (Heagerty 1991).

ANG II receptors have been characterized in various vascular tissues by radioligand binding studies (Table 7.1). Although ^{3}H-ANG II retains almost full biological activity, ^{125}I-ANG II exhibits only 33% of vasopressor activity in rats (Lin and Goodfriend 1970). In the rabbit aorta (Devynck and Meyer 1976), high affinity binding to ^{3}H-ANG II has a K_d of 6.2 nM, which is similar to an ED_{50} for the contractile effect of ANG II in the rabbit aortic strip (7.8 nM; Regoli et al. 1974). The binding is displaced by ANG II and [Sar^1, Ile^8] ANG II, and to much lesser extents by ANG I and III. Similar results are obtained in the guinea pig aorta (Le Morvan and Palaic 1975) in which the K_d is similar to the ED_{50} for the contractile effect (Le Morvan et al. 1974). ANG II receptors are also identified in rat aortic smooth muscle cells in culture (Penit et al. 1983). Specific binding to ^{3}H-ANG II has a K_d of 2.3 nM and N_0 of 50 000 sites/cell.

ANG II receptors in resistance-type vessels are more important than those of conductance-type aorta with respect to blood pressure regulation (Table 7.1). A single class of binding site was identified in the rat mesenteric artery (Gunther et al. 1980a). The binding is displaced by [Sar^1, Ile^8] ANG II, ANG II, [Sar^1, Ala^8] ANG II, ANG III and ANG I in this order of potency. McQueen et al. (1984) identified two classes of binding sites in the same tissue with a K_d for the high affinity site of 19–74 pM, which is as low as plasma levels after injection of ED_{50} for the vasopressor effect. The binding is stimulated by divalent and monovalent cations and inhibited by guanine nucleotides (Wright et al. 1982). Scatchard analysis showed that stimulation of binding by Na and Mg ions is due to increased K_d with N_0 unchanged, whereas infusion of ANG II decreases N_0 without changes in K_d (Schiffrin et al. 1984). ANG II receptors were also identifed in cultured smooth muscle cells from rat mesenteric artery. The cells have a single class of binding site with a K_d of 2.8 nM and N_0 of 45 000 sites/cell (Gunther et al. 1982). These cultured cells exhibit contractile responses when ANG II is added to the culture medium.

ANG II receptors were detected in the rat portal vein by photoaffinity labelling (Kwok and Moore 1984), but their K_d is as high as 300 nM and more than 60% of the population of receptors appear to be spare. ANG II receptors were found in rat brain microvessels; their K_d of approximately 1 nM is much smaller than those of surrounding brain parenchymal tissues (Speth and Harik 1985). The N_0 of brain microvessel ANG II receptors from spontaneously hypertensive rats is twice as high as that of normotensive Wistar–Kyoto rats (Sugiura et al. 1982).

It is known that sensitivity to the vasopressor action of ANG II is modulated by Na balance. Brunner et al. (1972) found that the amount of ANG II antisera required to block the vasopressor action of exogenous ANG II increases eightfold in Na-loaded rats than in Na-depleted animals. They ascribed the increase to an increase in the number of ANG II receptors within the circulatory system. The notion of 'spare receptors' was not introduced at that time. More direct evidence is obtained by ligand binding studies in which dietary Na restriction for 1 week increases plasma renin activity and decreases the number of ANG II receptors in the rat mesenteric artery (Gunther et al. 1980b). The treatment of captopril further increases plasma renin activity but decreases the number of receptors. Thus, ANG II seems to directly regulate the number of its receptors. Aguirela and Catt (1981) showed that dietary Na restriction increases plasma ANG II levels and decreases the number of mesenteric ANG II receptors, while dietary Na loading has opposite effects. The affinity of ANG II receptors is not altered by changes in Na balance.

ANG II induces contraction of uterine smooth muscles of the rabbit (Regoli and Vane 1964) and rat (Khairallah et al. 1965). Specific ANG II receptors were found in uteri of the rat, rabbit, dog and African green monkey (Table 7.1; Rouzaire-Dubois et al. 1975; Capponi and Catt 1979; Schirar et al. 1980; Petersen et al. 1985b). Their K_ds varied from 0.6 nM to 80 nM. In the rat uterus, monovalent cations increase the affinity and divalent cations increase the number of receptors (Douglas et al. 1982). The effect of ANG II on its receptor is somewhat controversial; nephrectomy increases the number without changing the affinity (Devynck et al. 1976), whereas chronic low-dose infusion of

ANG II decreases the affinity without changing the number (Douglas and Brown 1982). In the rat and rabbit uteri, the number of ANG II receptors varies during the estrous cycle, with a maximum at proestrous and minimum at diestrous II (Schirar et al. 1980). Ovariectomy is followed by a progressive decrease in uterine ANG II receptors, and injection of estradiol causes dose-dependent increases in their number. The structural requirement of ANG II analogs for displacing ANG II binding in the uterus is similar to that in the adrenal cortex of the dog (Capponi and Catt 1979) and rat (Douglas et al. 1980).

A single class of specific ANG II receptors with a K_d of 0.77 nM was identified in the smooth muscle of rat urinary bladder (Aguilera and Catt 1981). Only the number of receptors is altered after changes in Na intake and plasma ANG II levels. ANG II causes dose-dependent increases in tension of the lower eso-phageal sphincter of the opossum, and this effect is completely blocked by [Sar[1], Ala[8]] ANG II (saralasin) (Mukhopadhyay and Leavitt 1978). This result indicates the presence of specific ANG II receptors in the esophageal smooth muscle.

ANG II elicits a positive inotropic response in the dog, cat and rabbit (Fowler and Holmes 1964; Koch-Weser 1965; Illanes et al. 1967; Dempsey et al. 1971). ANG II also causes a contractile responses in rabbit atrial tissues in vitro (Baker et al. 1984). Specific ANG II receptors were identified in the cardiac ventricle of rabbit and calf (Table 7.1; Mukherjee et al. 1982a; Baker et al. 1984; Rogers 1984). The K_ds vary between 0.42 and 104 nM. As with the ANG II receptors in the mesenteric artery, the binding is stimulated by divalent cations and inhibited by guanine nucleotides (Baker et al. 1984). The affinity order of ANG II analogs correlates well with their potency order for the inotropic effect in vivo and in vitro. The calf sarcolemmal membranes have two classes of receptors with K_ds of 1.1 and 52 nM (Rogers 1984). Monovalent and divalent cations inhibit binding. The molecular weight of the receptor is estimated to be 116 kDa. Two classes of receptors were identified in cultured myocytes from the heart of neonatal rat (K_ds 0.7 and 5.6 nM), and ANG II applied to these cells causes chronotropic effects (Rogers et al. 1986).

7.2.1.2 Nonmuscular Tissues

ANG II receptors were identified in the adrenal cortex of the calf and rat (Glossmann et al. 1974), dog (Saltman et al. 1975), and rabbit (Gurchinoff et al. 1976; Table 7.1). Receptors are concentrated in the zona glomerulosa region in the rat (Healy et al. 1985). Two classes of receptors were identified in the rat and bovine adrenal glomerulosa cells (Glossmann et al. 1974), while a single class was detected in other studies (Forget and Heisler 1976; Devynck et al. 1977; Aguilera et al. 1978; Douglas et al. 1980). The K_d of high affinity sites ranges between 0.2 and 2 nM. The K_d is 0.56 nM in dispersed rat adrenal glomerulosa cells (Douglas et al. 1978). Divalent and monovalent cations change the number and affinity of receptors, respectively (Douglas et al. 1982). The affinity of ANG II analogs to the receptors is positively correlated with the potency of their steroidogenic action (Douglas et al. 1980).

The adrenal ANG II receptors are strongly influenced by dietary Na and K intake. Chronic Na restriction increases the number of receptors by 74%, while K loading increases it by 170% (Douglas and Catt 1976). The affinity of receptors is not altered in either case. Short-term Na restriction for 36 h increases the affinity of receptors in dispersed glomerulosa cells by 80% and the number by 25%, while more prolonged Na restriction for 4 days causes further increases in the number but decreases the affinity to normal (Aguilera et al. 1978). Opposite results are obtained during Na loading. These changes in the number of ANG II receptors are correlated well with changes in the potency of steroidogenic action of ANG II. Since blockade of ANG II formation by captopril prevents the increase in the number of adrenal ANG II receptors and the potentiation of ANG II effect on aldosterone secretion after short-term Na restriction (Aguilera et al. 1980), plasma ANG II concentration is a primary factor that regulates expression of ANG II receptors responsible for steroidogenic action. Consistently, low-dose infusion of ANG II increases the number and decreases the affinity of ANG II receptors in rat adrenal glomerulosa cells (Douglas and Brown 1982). Recently, ANG II receptor subtypes were examined in cultured bovine adrenal cells (Ouali et al. 1993). The AT_1 type occupies 80% of total receptors and has an Mr of 50 000, and a small proportion of AT_2 type has an Mr of 70 000.

ANG III, another important circulating component of the RAS, is as potent as ANG II in stimulating aldosterone synthesis and secretion from the adrenal of rat, calf, dog, sheep, rabbit and man. Furthermore, ANG III is suggested as a mediator for the steroidogenic action of ANG II (Blair-West et al. 1971b). This notion is based on the fact that the ANG II effect is blocked by a selective antagonist of ANG III, and that specific ANG II antagonists do not block the ANG III effect completely (Bravo et al. 1975). The presence of ANG III in the adrenal gland also favors this role (Semple and Norton 1976). Devynck et al. (1977) showed that ANG III displays a higher affinity ($K_d = 0.2$ nM) than ANG II ($K_d = 3.3$–5.2 nM) to rat adrenocortical receptors. However, formation of ANG III does not seem to be an obligatory step for the steroidogenic action of ANG II, since ANG peptides bound to dispersed dog adrenal glomerulosa cells after incubation with [^{125}I]-ANG II are mostly ANG II with only 15% of ANG III (Douglas et al. 1978). ANG II and III seem to bind to the same receptors; their binding and steroidogenic action are prevented similarly by competitive antagonists, and the number of adrenal receptors for each peptide is similar (Douglas et al. 1985).

ANG II stimulates catecholamine release from the adrenal medulla (Feuerstein et al. 1978). Specific ANG II receptors were identified in the adrenal medulla of rats by autoradiography (Table 7.1; Healy et al. 1985). The number of receptors decreases after water deprivation (Hwang et al. 1986). High affinity receptors for ANG II were also identified in the particulate fraction from the rat adrenal medulla by a radioligand binding study (Singh et al. 1986). In isolated, perfused adrenal gland, ANG I is as potent as ANG II in stimulating catecholamine release even when converting enzyme activity is inhibited (Peach et al. 1971). ANG I also facilitates norepinephrine release from peripheral sympathetic neurons (F.M. Johnson et al. 1974). Furthermore, the centrally mediated pressor response may be caused by a direct action of ANG I on the area postrema (Buckley 1972). Since ANG I is the first hormonal message from the RAS, the notion of interaction of ANG I with evolutionary primitive structures such as the adrenal medulla, sympathetic nervous system and area postrema, is quite attractive (Peach 1977). A radioligand binding study using [^{125}I]-ANG I identified specific binding sites for ANG I with low affinity in several target organs of ANG II (Goodfriend et al. 1972), but it is unknown whether ANG I and II bind to the same receptor or not.

ANG II is implicated in regulation of gluconeogenesis and glycogenolysis (Hems 1977) and in stimulation of angiotensinogen production (Freeman and Rostorfer 1972) in the liver (Table 7.1). A single class (LaFontaine et al. 1979) or two classes of ANG II receptors were identified in the rat liver (Campanile et al. 1982). The number of receptors is increased by divalent cations and high affinity sites disappear after chelating divalent cations from incubation medium. Interestingly, saralasin binds only to low affinity sites, and thus saralasin binding is not influenced by divalent cations. A positive correlation is established between potency of ANG II analogs to stimulate glycogen phosphorylase activity and their affinity to ANG II receptors in hepatocytes. Guanine nucleotides decrease the number of high affinity receptors but have little effects on the affinity (Crane et al. 1982). Gunther (1984) and Sen (1985) suggested that high affinity receptors are involved in phosphorylase stimulation while low affinity receptors are responsible for inhibition of adenylate cyclase activity. Thus, there seem to be ANG II receptor subtypes which have different biochemical features and mechanisms of action. Sernia et al. (1985) observed that low dietary Na intake or low-dose infusion of ANG II causes similar changes in ANG II receptors in rat hepatocytes; initial decrease and increase, respectively, in the number and affinity of receptors for 1–2 days followed by delayed changes in the opposite direction. More recently, specific ANG II receptors were identified in the nuclear membrane of rat hepatocytes (Booz et al. 1992; Tang et al. 1992). The nuclear ANG II receptor seems to be coupled to GTP-binding proteins because of the sensitivity to GTP, and its structure may be similar to AT_1 receptor because of the susceptibility to Dup 753. However, some properties such as pH sensitivity and affinity to angiotensin analogs are distinct from plasma membrane receptors.

ANG II is shown to affect glomerular microcirculation (Meyers et al. 1975; Blantz et al. 1976), glomerular ultrafiltration coefficient (Davalos et al. 1978; Kon and Ichikawa 1985), renal cell growth (Normal 1991), and renin release from JG cells (Table 7.1; Chap. 5). A single class of ANG II receptors was identified in isolated rat glomeruli with a K_d of 45 pM, which is similar to the ED_{50} (60 pM) for the contractile action of ANG II in isolated glomeruli (Sraer et al. 1974). A single class of ANG II receptors was identified in isolated rat glomeruli (G.P. Brown et al. 1980; Slorecki et al. 1983), while another study identified three classes (Beaufils et al. 1976). In isolated human glomeruli, two classes of receptors were detected with K_ds of 0.1 and 2 nM (Chansel et al. 1982). The occurence of two classes may result from negative cooperativity.

The binding is increased by divalent cations (Douglas et al. 1982) and inhibited by guanine nucleotides (Chansel et al. 1982). The renal ANG II receptors are influenced by Na balance; the number alone decreases (Skorecki et al. 1983) or both number and affinity decrease (Beaufils et al. 1976) after Na loading. Since downregulation occurs after ANG II infusion, the effect of Na manipulation may be mediated by its effect on renin secretion (Kitamura et al. 1986). ANG II receptors are concentrated in the basement membrane of rat glomeruli (Sraer et al. 1977), whereas dense autoradiographic grains accumulate in mesangial cells of rat glomeruli after intraaortic administration of ^{3}H- ANG II (Osborne et al. 1975). Recently, most ANG II receptors in cultured rat renal mesangial cells were identified as AT_1 type (Ernsberger et al. 1992). ANG I seems to act directly on renal vasculature (Itskovitz and McGiff 1974); the effect of ANG I is not blocked by SQ20881, and the site of action of ANG I is more limited than that of ANG II.

ANG II stimulates Na and water reabsorption by rat renal tubules (Munday et al. 1972; Harris and Navar 1985), and bicarbonate reabsorption in the SI proximal convoluted tubule (Liu and Cogan 1987; Table 7.1). Specific ANG II receptors were identified in rabbit renomedullary cells in culture (C.A. Brown et al. 1980). ANG II receptors were also detected in the semipurified plasma membrane fraction of renal tubular epithelial cells (Brown and Douglas 1982; Cox et al. 1983). The tubular receptors are located in both luminal brush border membranes and basolateral membranes, with the greater number and affinity in the basolateral portion (Brown and Douglas 1983; Cox et al. 1984). The largest number of receptors are found in the proximal convoluted tubule of rat, which is in good agreement with its biological action on this segment (Mujais et al. 1986). In another study, autoradiographic grains were con-centrated in the glomerulus of the renal cortex and in the medullary vasa recta and tubular regions, whereas moderate density of grains were found in the MD and proximal tubule of the cortex (Gehlert et al. 1984). Intrarenal ANG II receptors are mostly of the AT_1 type, but the AT_2 type is also expressed in large numbers in the fetal kidney (de Gasparo and Levens 1994).

ANG II modulates the release of various hormones from the anterior pi-tuitary gland (Table 7.1.; Steele et al. 1981; Aguilera et al. 1982; Paglin et al. 1984; see Sect 8.6). A single class of specific ANG II receptors was identified in the anterior pituitary of rat (Mukherjee et al. 1982b) and anterior pituitary cells of rat, dog and rabbit (Hauger et al. 1982), but no ANG II binding was detected in the posterior pituitary (Hauger et al. 1982). Favrod-Coune et al. (1982) also reported a single class of ANG II receptors in the dispersed anterior pituitary cells, and their K_d is similar to ED_{50} for the effect of ANG II on stimulation of ACTH release in vitro. ANG II binds specifically to mammotrophs, cortico-trophs, and presumably thyrotrophs in dispersed anterior pituitary cells of male rats (Paglin et al. 1984). Using quantitative autoradiography, Israel et al. (1985) showed that water deprivation increases the number and decreases the affinity of ANG II receptors in the rat anterior pituitary.

ANG II increases the motility of the intestine, and increases water and Na transport across the intestinal epithelia of rat (Table 7.1; Davies et al. 1970; see

Levens 1985). Specific ANG II receptors were identified in epithelial membranes from the rat jejunum and descending colon (Cox et al. 1986). The affinity of the jejunal ANG II receptors is similar to that of renal tubules, another absorptive epithelium (Cox et al. 1983), but the number is much smaller than that of renal tubules. Specific binding is localized in the mucosa and muscularis of the rat intestine and absent in the submucosa and serosa layer (Sechi et al. 1993). Nearly 90% of the intestinal receptors seem to be of the AT_1 type.

The presence of high affinity binding sites for ANG II was reported in mononuclear cells in the blood (Shimada and Yazaki 1978), but the apprarent binding was later found to be the result of free fluid endocytosis of labeled ligand (Neyses et al. 1984). Platelets have high affinity binding sites for ANG II (Moore and Williams 1982) which appear to be true receptors (Ding et al. 1984). Moore et al. (1984) showed that the number of platelet receptors are regulated inversely by plasma Na and ANG II concentrations. Interestingly, human platelets have an ANG I-processing system which converts ANG I to [des-Leu^{10}] ANG I (Snyder et al. 1985). [des-Leu^{10}] ANG I inhibits converting enzyme activity, thereby possibly regulating plasma ANG II levels.

7.2.2 Central Nervous System

ANG II is known to have a number of central actions such as induction of thirst, Sodium appetite, hypertension, and stimulation of vasopressin release (Table 7.1; Ganong 1984; Phillips 1987b; Ganten et al. 1988). The site of action of ANG II has been examined by topical injection of ANG II, lesioning or by electrophysiological studies. In fact, high affinity ANG II receptors ($K_d = 0.2$ nM) were identified in the brain of calf and rat (Bennett and Snyder 1976). Sodium ions increase the number and affinity of ANG II receptors but do not alter the binding of [Sar^1, Ile^8] ANG II (Bennett and Snyder 1980). ANG II receptors are also found in cultured neurons from the fetal rat brain whose K_d and N_0 are 1 nM and 6000 sites/cell, respectively (Raizada et al. 1981). Na ions increase the number of receptors. Mann et al. (1981) found a close correlation between biological actions of ANG peptides and their binding to brain membranes in the rat, suggesting the presence of ANG II receptors for these central effects.

While ANG II receptors are localized in the cerebellum in calf (Bennett and Snyder 1976), they are most abundant in the thalamus–hypothalamus, midbrain and brainstem areas in rat (Sirett et al. 1977). Among these areas, ANG II receptors are particularly dense in the superior colliculi of midbrain and lateral hypothalamus (Sirett et al. 1979a). Harding et al. (1981) observed highest levels of binding in the area postrema, spetum (including the SFO) and superior colluculi of rats and mice. In the dog, ANG II receptors are most dense in the anterior pituitary followed by the area postrema, neural lobe, organum vasculosum of the lamina terminalis (OVLT), SFO and hypothalamus (Speth et al. 1983). The area postrema, posterior pituitary, OVLT and SFO are so-called circumventricular organs to which peripheral ANG II is accessible (van Houten et al. 1980; Landas et al. 1980). An autoradiographic study using $[^{125}I]$-$[Sar^1]$ ANG II identified high density ANG II receptors in the SFO, paraventricular and periventricular nuclei of the hypothalamus, nucleus tractus solitarius, and area postrema of rat (Mendlsohn et al. 1984). This ligand has similar affinity and higher stability compared with ANG II. The OVLT and the medial preoptic nucleus, inferred dipsogenic receptor sites for ANG II, exhibit moderate density. Similarly, ANG II receptors are dense in the anteior pituitary, OVLT, and SFO of rat (Isarel et al. 1984) and dog (Speth et al. 1985). These ANG II receptors appear as early as 2 days after birth in the rat (Millan et al. 1991). Using antagonists specific to each ANG II receptor, AT_1 receptors were localized in the OVLT, SFO, area postrema, medial preoptic area, paraventricular nucleus and nucleus solitarius of rat, whereas AT_2 receptors were predominant in the locus coeruleus, lateral spetum, superior colliculus and subthalamic nucleus of rat (Song et al. 1992) of rabbit (Aldred et al. 1993), and hamster (Saylor et al. 1992).

The number and affinity of ANG II receptors in the brain were not altered by intracerebro-ventricular infusion of ANG II (Singh et al. 1984) or by dietary Na handling (Speth et al. 1984). However, ANG II receptors were influenced by plasma ANG II and Na levels to different degrees in various regions of the rat brain irrespective of the presence of the blood–brain barrier (Thomas and Sernia 1985). Quantitative autoradiography revealed that ANG II receptors increased after 5 days of water deprivation in the rat (Hwang et al. 1986). The primary baroreceptor afferent fibers terminate in the nucleus tractus solitarius (Palkovits and Zaborsky 1977) where ANG II-containing neurons (Lind et al. 1984) were identified. A topical injection of ANG II into the nucleus produces dose-dependent changes in arterial blood pressure and heart rate in the rat (Catro and Phillips 1985), and vagotomy reduces the number of receptors in the nucleus (Lewis et al. 1986).

ANG III is as potent a dipsogen as ANG II in gerbils, *Meriones unguiculatus* (Wright et al. 1984), nonhuman primates (Simonnet et al. 1979; Lotter et al. 1980), rabbits (Wright et al. 1985), and pigs (Mutter et al. 1984). An autoradiographic study revealed that $[^{125}I]$-ANG III binds with higher affinity than $[^{125}I]$- ANG II to cerebral target sites for ANG II in gerbils (Harding et al. 1981), African green monkey (Petersen et al. 1985a) and rabbits (Wright et al. 1985). In the monkey, however, bound radioligands were shown to be mostly metabolites of ANG III. ANG III binding sites coincided well with putative target sites for dipsogenic action of ANG II in the rat (Phillips 1978). It is known that AT_2 receptors have higher affinity to ANG III than AT_1 receptors (Table 7.2).

In New Zealand strain genetically hypertensive rats (Sirett et al. 1979b) and spontaneously hypertensive rats (Moore and Khairallah 1977; Mizuno and Fukuchi 1981), the number of ANG II receptors in the thalamus–hypothalamus–midbrain area is smaller than that of normotensive controls. However, when compared more locally at the OVLT by radioligand binding assay (Stamler et al. 1980) or at the SFO by quantitative autoradiography (Saavedra et al. 1986b), the number is greater in spontaneously hypertensive rats than in controls. Furthermore, the number of receptors in the SFO for artrial na-triuretic peptide, which is known as a natural antagonist of ANG II, is smaller in hypertensive rats than in controls (Saavedra et al. 1986a). The affinity of ANG II receptors is greater in the nucleus tractus solitarius but not in the area of postrema of spontaneously hypertensive rats than in those sites of normo-tensive rats (Plunkett and Saavedra 1985). The number of receptors does not differ between the two strains. However, only the number of ANG II receptors is greater in neuronal cultures from hypertensive rats than in those from normotensive controls (Raizada et al. 1984a). Cole et al. (1980) reported that the affinity of ANG II receptors in the hypothalamic–thalamic–septal–midbrain area does not differ between two groups of spontaneously hypertensive rats fed with low and high Na diet. The number of receptors is higher in Na-deficient rats than in Na-sufficient ones during the first 60–90 days, but the relationship is reversed thereafter. The number of receptors in neuronal cells from 1-day-old spontaneously hypertensive rats is increased by Na ions to a greater extent than the number from age-matched normotensive controls (Feldstein et al. 1986). The number of ANG II receptors in the rat brain increases after birth, reaches a maximum in 2 weeks, and decreases gradually to adult levels (Baxter et al. 1980). Since the blood–brain barrier is incomplete in rats of 2 weeks of age, and since circulating ANG II levels are high in newborns (Broughton Pipkin et al. 1974), peripherally generated ANG II may play a significant role in the regulation of ANG II receptors in the brains of newborn rats.

ANG II has various concentration-dependent effects on ionic conductance in cultured mouse spinal cord neurons (Table 7.1; Legendre et al. 1984). In cultured spinal cord neurons, two classes of binding sites are present with K_ds of 0.43 nM and 25.6 nM (Laribi et al. 1985). The affinity of ANG II analogs correlates well with their agonistic and antagonistic potencies.

7.3 Signal Transduction of ANG II Receptors

After binding of a hormone to its receptor, information of the first messenger is transmitted intracellularly via so-called second messengers which initiate a series of intracellular reactions that lead to biological actions (Fig. 7.3). Since most ANG II actions are related to muscle contraction or hormone secretion, and since both effects are regulated by intracellular free Ca concentrations, Ca ions may play a crucial role in these ANG II effects. Intracellular mechanisms that follow ANG II binding to its receptor have been reviewed by Peach (1981) and Smith (1986). More recently, Bottani et al. (1993) reviewed a signal transduction system of AT_1 and AT_2 receptors in relation to their possible physiological functions.

In cultured smooth muscle cells from rabbit aorta, a contractile response to ANG II consists of two phases; an initial fast phase independent of Ca ions in the media, and a delayed tonic phase dependent on extracellular Ca con-centrations (Deth and van Breeman 1974, 1977). The first direct evidence for the involvement of Ca ions was provided by experiments using a fluorescent Ca

Fig. 7.3. Schematic drawing of subcellular mechanisms after ANG II binding to its vascular receptor. After binding, phospholipase C (*PLC*) is activated or adenylate cyclase (*AC*) is inhibited via GTP binding proteins, *Gq* and *Gi*, respectively. The activation of *PLC* results in intracellular formation of inositol triphosphate (*IP₃*) and diacylglycerol (*DG*); the former stimulates Ca release from the sarcoplasmic reticulum (*SR*). Intracellular Ca ions are also increased by an influx through receptor-coupled Ca channels. Increased Ca ions coupled with calmodulin (*Cam*) stimulate a protein kinase (*Cam-kinase*) which leads to contraction of vascular smooth muscles. Ca ions also stimulate Na/Ca antiporters in plasma membrane which increases Ca efflux. In renal tubules and intestinal epithelia, increased *DG* stimulates Na/H antiporters and increases Na absorption

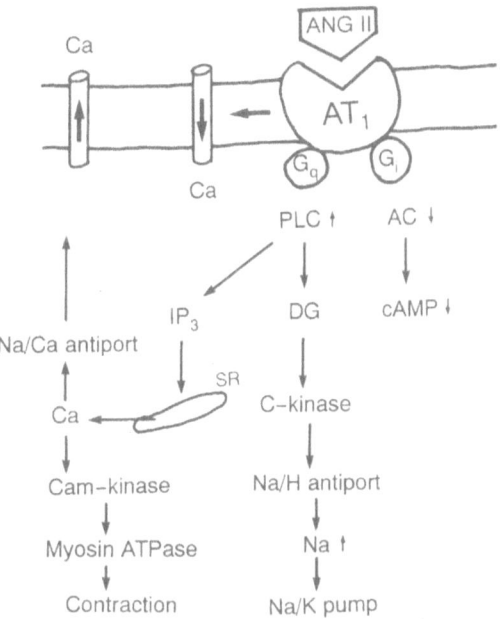

indicator, quin 2 (Tsien et al. 1982). Smith (1986) added ANG II to rat aortic smooth muscle cells in culture and observed a rapid, time-dependent increase in intracellular quin 2 fluorescence. [Sar[1], Ile[8]] ANG II prevents the ANG II-induced Ca accumulation. Quantitative analyses showed that 20 nM ANG II increases cytoplasmic free Ca concentration from 150 to 420 nM. The initial rise in fluorescence is followed by a delayed, sustained increase for 10 min, which is dependent on the concentration of ANG II and extracellular Ca ions. These results are consistent with those of other in vitro experiments (Alexander et al. 1985; Brock et al. 1985; Capponi et al. 1985; Nabika et al,. 1985). Brock et al. (1985) suggested that Ca ions sequestered from intracellular pools are important for the initial rapid increase, because chelating external Ca ions does not abolish the increase (Fig. 7.3). However, the delayed, sustained increase is completely abolished by the chelation. Nabika et al. (1985) showed that both intracellular and extracellular Ca ions are responsible for the rapid increase in cytoplasmic Ca ions evoked by ANG II at concentrations higher than 10 nM, but ANG II at lower than 1 nM induces only a small and slow increase which is exclusively dependent on extracellular Ca ions. In hemisected rat aortic tissues, ANG II induces a rapid and long-lasting increase in Ca uptake from the media, and enhances Ca release from intracellular pools for a short period (Schleiffer and Gairard 1985). Since both saralasin and verapamil inhibit ANG II-induced depolarization and Ca uptake, these ANG II effects may be mediated by receptor-operated, voltage-dependent Ca channels (Fig. 7.3).

Smith et al. (1984) observed an increase in Ca efflux from smooth cells of rat thoracic aorta after addition of ANG II. The ANG II-stimulated Ca efflux consists of fast and slow components. The fast and large efflux occurs in 10 sec, and the slow and smaller efflux continues for an extended period. Both effluxes are abolished by saralasin. The fast component may be due to a rapid initial increase in intracellular Ca ions, which may be caused by two factors. One is by stimulating Ca release from intracellular pools, and the other by stimulating Na/Ca antiporter in the cell membrane (Fig. 7.3). The involvement of Na/Ca antiporter is evidenced because replacement of external Na ions with other monovalent cations greatly inhibits the rapid Ca efflux, and amiloride derivatives, which are potent inhibitors of Na/Ca antiporter, inhibit the efflux (Smith 1986). The role of intracellular pools is likely because the Ca efflux is usually preceded by a Ca influx which occurs after addition of ANG II (Smith et al. 1984). Thus, it is hypothesized that ANG II first stimulates the release of Ca ions from intracellular pools, which then increases Ca efflux via the Na/Ca exchange system. Then the membrane potential changes, which possibly activates voltage-sensitive Ca channels, resulting in an increase in Ca influx (Zelcer and Sperelakis 1981). A similar increase in cytosolic Ca concentration is noted in beating neonatal rat myocardiocytes which is blocked by the AT_1 receptor antagonist Dup753 but not by the AT_2 receptor antagonist PD123319 (Kem et al. 1991).

Various hormones and neurotransmitters that mobilize intracellular Ca ions rapidly increase polyphosphoinositide hydrolysis (Berridge and Irvine 1984). ANG II activates phospholipase C (Fig. 7.3), a phosphodiesterase that selectively hydrolyzes phosphatidylinositol monophosphate (PIP_1), bisphosphate (PIP_2) or triphosphate (PIP_3) to produce inositol triphosphate (IP_3), in rat hepatocytes (Creba et al. 1983), rat adrenal glomerulosa cells (Alexander et al. 1985; Nabika et al. 1985; Smith et al. 1985). The generation of IP_3 occurs maximally within 15 s after addition of ANG II to cultured aortic smooth muscle cells of rat, which is rapid enough to mobilize Ca ions from intracellular Ca pools (Smith 1986). The GTP-binding protein may mediate the activation of phospholipase C (Fig 7.3). Intracellular receptors for IP_3 were identified in particulate fractions of ANG II target tissues which were free from contamination by mitochondrial membranes (Guillemette et al. 1987). The binding is saturable, specific to IP_3, and of high affinity ($K_d = 0.9$–1.7 nM) and low capacity (15–124 fmol/mg protein). These receptors may be localized in the endoplasmic reticulum and responsible for mobilization of Ca ions from the organelle (Fig. 7.3). Concomitant with the stimulation of hydrolysis, ANG II stimulates de novo synthesis of polyphosphoinositides in cultured arterial smooth muscle cells of rats (Smith et al. 1984). Another hydrolysis product of PIP_2, diacylglycerol, also has a similar time coruse of change in intracellular concentration; an immediate, transient increase followed by a sustained increase. Diacylglycerol acts as a second messenger by activating (Nishizuka 1984). In the sustained increase, C-kinase seems to activate phospholipase D thereby stimulating phosphatidylcholine hydrolysis (Lassegue et al. 1991). Possible intracellular Ca reservoirs have been pursured in saponin-treated rat aortic smooth muscle cells. Saponin makes cholesterol-rich plasma membrane permeable to small molecules without damaging cholesterol-poor membranes of intracellular organelles (Endo 1975). The endoplasmic (sarcoplasmic) reticulum and mitochondria in permeabilized cells actively accumulate Ca ions. Endoplasmic reticulum has a higher affinity and a lower capacity than mitochondria for Ca binding. Therefore, Ca ions accumulate selectively in the endoplasmic reticulum by the ATP–dependent Ca pump when cytosolic free Ca levels are low (Smith 1986). In saponin-treated rat aortic smooth muscle cells, IP_3 neither alters Ca fluxes across the mithochondrial membrane nor affects an influx of Ca ions by nonmitochondrial organelles. However, 5 μM of IP_3 causes fourfold increases in Ca efflux from nonmitochondrial organelles in 10 s (Smith et al. 1985). ATP is required for activation of the efflux, although ATP itself is ineffective. Lowering incubation temperature from 37 $°C$ to 4 $°C$ inhibits IP_3-stimulated Ca efflux by only 33%. Thus, IP_3 may activate ion channels rather than ion carriers by a ligand binding reaction rather than by a metabolic reaction.

Similar subcellular mechanisms may function in nonvascular cells to regulate cytosolic free Ca levels after ANG II action. In bovine adrenal glomerulosa cells in culture, ANG II regulates secretion of aldosterone by Ca-dependent mechanisms which involve both uptake of Ca from media and release of Ca from intracellular pools (Kojima et al. 1984). Saponin-treated glomerulosa cells accumulate Ca ions into non-mitochondrial pools by a cyclic AMP–dependent route. ANG II-stimulated Ca influx is rapid, sustained, reversed by ANG II antagonists, and dependent on extracellular Ca ions (Kojima et al. 1985a). The rapid increase in aldosterone release after ANG II stimulation is not blocked by a Ca channel blocker, nitrendipine, but the sustained increase is blocked by nitrendipine (Kojima et al. 1985b). Thus, as shown in the aortic smooth muscle cells, the initial rapid increase is caused largely by the release of Ca ions from intracellular nonmitochondrial organelles, and the sustained increase is due to increased Ca influxes across the plasma membrane. The rapid increase in cytosolic Ca ions is induced by the hydrolysis of PIP_1 and PIP_2 in adrenal cells (Kojima et al. 1984; Farese et al. 1984). ANG II induces a hydrolysis of PIPs rapidly enough to initiate the initial mobilization of Ca ions. The hydrolysis leads to a sustained increase in IP_3 and diacylglycerol, which stimulate aldosterone secretion in a manner similar to that demonstrated in contraction of vascular smooth muscle; a calmodulin branch is activated by a transient increase in cytosolic IP_3 and Ca, and a C-kinase branch is activated by both sustained increase in cytosolic Ca and diacylglycerol (Kojima et al. 1984). Similar mechanisms seem to operate in ANG II action on hepatocytes (Billah and Michell 1979; Garrison et al. 1979; Mauger et al. 1984; Charest et al. 1985). In fact, a long-term activation of C-kinase is demonstrated in bovine adreno-medullary cells in culture which may be responsible for the expression of genes encoding proenkephalin and a catecholamine-producing enzyme (Tuominen et al. 1991). In human astrocytes, two distinct ANG II receptor subtypes were identified; one reacts with ANG II-(1–7) and release prostaglandin E_2 and prostacyclin, and the other responds to ANG II-(2–8) and utilizes IP_3 for intracellular Ca mobilization (Tallant et al. 1991).

As mentioned above, ANG II stimulates Ca efflux across the plasma membrane via Na/Ca antiporter in cultured rat arterial smooth muscle cells (Smith et al. 1984). Furthermore, C-kinase activated by diacylglycerol also stimulates Na/H antiporter (Fig. 7.3). Thus, after addition of ANG II to culture media, considerable amounts of Na ions are accumulated in the cells. ANG II increases exchange of Na and K ions in the cultured rat aortic smooth muscle cells; this increase is accomplished by an increased activity of the Na/K pump (Brock et al. 1982). However, the activation of the Na/K pump is not a direct action of ANG II but is mediated by an increase in cytosolic Na ions (Fig. 7.3), since ANG II no longer activates the pump when the accumulation of intracellular Na ions is inhibited (Smith and Brock 1983).

Another possible second messenger for ANG II action is cyclic AMP, the evidence for this is much less convincing than for Ca. ANG II had no effect on cyclic AMP levels in cultured vascular smooth muscle cells of rats (Penit et al. 1983; Smith and Brock 1983), but inhibited the basal and catecholamine-stimulated activity of adenylate cyclase in a concentration-dependent manner in the rat aorta (Anad–Srivastava 1983). ANG II stimulated guanylate cyclase activity in plasma membranes from rat aorta, heart, and kidney (Vesey 1981). ANG II also increased cyclic GMP accumulation, but not cyclic AMP accumulation, in cultured rat endothelial cells (Buonassisi and Venter 1976).

7.4 Nonmammalian Vertebrates

The presence of specific ANG II receptors has been suggested in non-mammalian species using competitive antagonists. Whereas vasopressor and depressor actions of human ANG II in the chicken are similarly attenuated by both [Sar[1], Ala[8]] and [Sar[1], Ile[8]] ANG II (Moore et al. 1981b), the effects of chicken ANG II are blocked by [Sar[1], Ile[8]] ANG II but not by [Sar[1], Thr[8]] ANG II in the same species (Nakamura et al. 1982). In the quail, [Sar[1], Ile[8]] ANG II is a potent antagonist but [Sar[1], Ala[8]] and [Sar[1], Thr[8]] ANG II are only weak antagonists (Takei and Hasegawa 1990). Both [Sar[1], Ile[8]] and [Sar[1], Ala[8]] ANG II inhibited the vasopressor effect of fish ANG II in the turtle, *Pseudemys scripta elegans* (Stephens 1981), and the latter ANG II decreased the vasopressor effect of human ANG II to one-third in the rat snake, *Ptyas korros* (Ho et al. 1984). In the American alligator, however, only [Sar[1], Ile[8]] ANG II inhibited the vasopressor effect, and [Sar[1], Ala[8]] and [Sar[1], Thr[8]] ANG II were without antagonistic effect (Silldorff and Stephens 1992). The contractile response of the aortic strip to ANG II was blocked by [Sar[1],Ile[8]] and [Sar[1], Ala[8]] ANG II in *Pseudemys* (Stephens 1984) and the cobra, *Naja naja* (Yung and Chiu 1985). The Vasopressor effect of homologous ANG II was blocked in the bullforg by [Sar[1], Ile[8]] ANG II but not by [Sar[1], Ala[8]] ANG II and [Sar[1], The[8]] ANG II (Harper and Stephens 1985). The hydrosmotic effect of ANG II in the skin of the toad, *Bufo arenarum*, was inhibited by [Leu[8]] ANG II (Coviello et al. 1974). [Sar[1], Thr[8]] and [Sar[1,] Ile[8]] AGN II had only weak antagonistic activities in the eel, and higher doses displayed agonistic effects (Nishimura et al. 1978). Since all analogs used above completely block ANG II effects in mammals, vascular ANG II receptors of nonmammalian species may be slightly different from those of mammals.

In addition to a unique vasodepressor effect in the chicken, ANG II induced relaxation of the precontracted aortic ring of chicken in vitro (Yamaguchi and Nishimura 1988). Thus, ANG II may act on the aortic smooth muscle to cause relaxation. Two classes of ANG II receptors with K_ds of 0.15 nM and 29 nM were identified in endothelium-removed aortic smooth muscles of the chicken (Takei et al. 1988d). However, ANG II receptors were also present in endothelial cells (Stallone et al. 1990), and the removal of endothelium abolished the vasorelaxant effect of ANG II in the chicken aortic ring (Yamaguchi and Nishimura 1988; Hasegawa et al. 1993). It seems therefore that ANG II receptors are present in endothelial cells which may be responsible for production of the endothelium-derived relaxing factor. Unlike mammalian vascular receptors, divalent cations did not change the affinity of vascular ANG II receptors in the chicken (Stallone et al. 1988). Furthermore, ANG II receptors in chicken endothelial cells and smooth muscles were almost insensitive to selective AT_1 and AT_2 receptor antagonists as demonstrated by bioassay and radioligand binding assay (Nishimura et al. 1994). Therefore, it is likely that a new type of ANG II receptors which are pharmacologically distinct from mammalian receptors are present in the chicken vasculature. Two classes of ANG II receptors were found in the heart of 10-day-old chicks (Baker and Aceto 1989). The cardiac ANG II receptors may be related to cardiac development, because ANG II increases protein synthesis and cell growth in embryonic chick myocytes in culture (Baker and Aceto 1990). A cDNA coding for the ANG II receptor was cloned recently from the turkey adrenal gland (Murphy et al. 1993). The predicted protein had 359 amino acids with approximately 75% sequence identity with the mammalian AT_1 receptor. However, the receptor protein expressed in COS cells had a different affinity for ANG II peptides and had little affinity for Dup753 compared with mammalian AT_1 receptors, although ANG II stimulated inositol phosphate production in the expressed COS cells as observed in mammalian AT_1 receptors.

ANG II induced drinking in the quail and chicken by acting on the SFO and OVLT (Takei 1977b; Massi et al. 1986; Takei and Kobayashi 1990). ANG II binding was localized in the SFO and OVLT of the duck by autoradiography (Fig. 7.4; Gerstberger et al. 1992). A K_d for avian [Asp[1], Val[5]] ANG II in these sites is 1.2 nM (Gerstberger et al. 1987; 1992). The affinities of [Sar[1], Ala[8]] ANG II and ANG III in these sites were much lower than those of mammals. Similar binding characteristics were reported in adrenal ANG II binding sites of the duck (Gray et al. 1989). ANG II actions in the central nervous system in birds have recently been reviewed by Simon et al. (1992) including cerebral ANG II receptors.

ANG II induces antidiuresis in *Bufo arenarum* (Reboreda and Segura 1989) and stimulates adrenal corticosteriod secretion in *Rana ridibunda* (Perroteau et al. 1984) and *Rana temporaria* (Hanke and Maser 1985). Kloas and Hanke (1992a) identified specific, saturable ANG II binding to renal glomeruli and interrenal tissue of *Rana temporaria* by autoradiography. The K_d was 0.5 nM

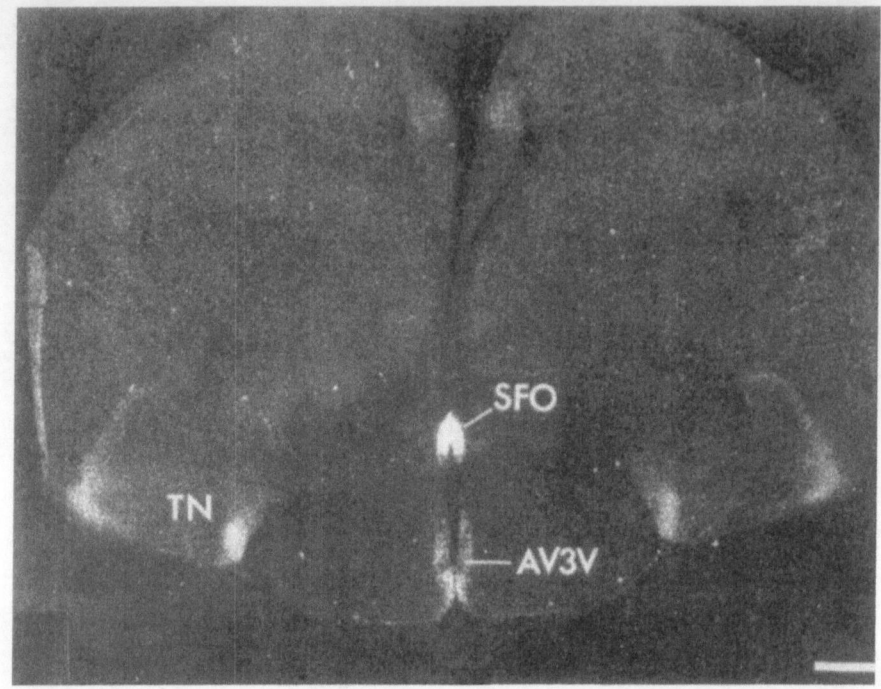

Fig. 7.4. Autoradiographic localization of ANG II-specific binding sites in coronal sections of the adult duck brain using ^{125}I-[Sar^1, Ile^5] ANG II. Specific ANG II binding sites are labeled in the subfornical organ (*SFO*), anteroventral third ventricle (*AV3V*), and amygdala-analogous structure of the nucleus taeniae (*TN*). Bar 1mm. (Simon et al. 1992)

for the kidney and 0.6 nM for the interrenal gland. In *Xenopus*, however, no ANG II binding sites were found in interrenal tissues, and ANG II does not stimulate corticosteroid secretion in this frog (Kloas and Hanke 1992b).

ANG II inhibited glomerular filtration rate by decreasing the number of filtering glomeruli in freshwater-acclimated rainbow trout, and by decreasing the single nephron filtration rate in seawater-acclimated fish (Brown et al. 1980; Gray and Brown 1985). Specific ANG II binding was demonstrated in the glomerulus of rainbow trout by autoradiography (J.A. Brown et al. 1990). The affinity of the glomerular ANG II receptor appears to be very low, and the ANG II binding can be displaced by Dup753, but not by saralasin and PD123177 (Cobb and Brown 1993). Other binding sites are aorta, urinary bladder, renal proximal tubules, gills, digestive tract, liver, heart, adrenocortical tissue and brain (Cobb and Brown 1992).

ANG II receptors were localized in the subscapsular zone of the adrenal gland by autoradiographic and radioligand binding studies in the Pekin duck (Gray et al. 1991). Atrial natriuretic peptide also binds to the subscapsular region, but ANG II and atrial natriuretic peptide bind to different receptors since the former binding was not displaced by the latter. ANG II receptors were also identified in the isolated adrenal steroidogenic cells of the turkey (Kocsis et al. 1994a, b). The turkey adrenal ANG II receptors discriminated poorly between Dup753 and PD123177, indicating some difference from mammalian AT_1 and AT_2 receptors.

Recently, ANG II receptors were identified in *Xenopus* ovarian follicles (Lacy et al. 1989, 1992; Sandberg et al. 1990; Woodward and Miledi 1991). These receptors induced a rapid increase in cytosolic Ca concentrations via PIP_2 breakdown, resulting in depolarization. The signal received by the follicular cells seems to be transmitted to oocytes via intercellular communication by IP_3(Sandberg et al. 1990), although ANG II receptors were also present on the oocytes (Sakuta et al. 1991). The *Xenopus* heart had two classes of ANG II receptors with K_ds of 1.6 nM and 22 nM (Sandlberg et al. 1991). These receptors were susceptible to dithiothreitol and, unlike mammalian myocaridal receptors, had rather high affinity to ANG I and low affinity to $[Sar^1, Ala^8]$ ANG II. Since these *Xenopus* receptors exhibited low affinity to Dup753 and PD123177, they may be pharmacologically different from the AT_1 and AT_2 receptors identified in mammals (Ji et al. 1991). More recently, a cDNA clone encoding an ANG II receptor was isolated from the *Xenopus* heart (Bergsma et al. 1993). The *Xenopus* receptor consists of 363 amino acid residues and shares 63% sequence identity with human AT_1 receptor. Although this *Xenopus* receptor displays pharmacological properties distinct from mammalian AT_1 and AT_2 receptors, it is evident that the receptor is an amphibian counterpart of mammalian AT_1 receptor, based on the conservation of structural features and similarity of the signal transduction mechanism. It seems that ANG II receptors have been subjected to greater divergence than the ligand during vertebrate evolution.

Chapter 8

Biological Actions of ANG II

ANG II generated in blood has a wide spectrum of biological actions on a variety of tissues. Included are effects on smooth and cardiac muscles, modification of other hormone release, and brain actions such as drinking behavior, neurogenic hypertension and sodium appetite. All are fast-acting with short durations, suggesting that ANG II is a hormone that responds to emergent changes in the internal or external environment and initiates physiological reactions to restore the status *quo ante*. In the case of hemorrhage, for example, renin is secreted immediately in response to hypovolemia/hypotension and sympathetic activation; the resulting elevation of plasma ANG II induces drinking and sodium appetite, and increases intestinal absorption of water and Na^+ to restore their loss. ANG II acts also on the kidney to decrease the urinary loss of water and Na^+. Furthermore, ANG II stimulates the secretion of vasopressin and aldosterone to decrease the loss of water and Na^+. In this way, blood volume is restored. The circulating RAS seems to be the first humoral messenger responsible for the initiation of an emergency reaction (Kobayashi and Takei 1982; Kobayashi et al. 1983; Sects. 8.2.5, 8.3.1, 9.4.3).

The RAS was found in the brain, kidney, adrenal, ovary, testis, pituitary, heart, vascular tissues and other areas in vertebrates (Phillips et al. 1993; Table 4.1). ANG II generated within these tissues or organs may act locally through paracrine or autocrine functions. For example, blood vessel ANG II seems to be involved in vascular remodelling (see Phillips et al. 1993) and stimulates angiogenesis in the chick embryo (Le Noble et al. 1991). The RAS is present in the pituitary gonadotrophs. LHRH stimulates gonadotrophs to secrete both LH and ANG II, which coexist in secretory granules in the gonadotrophs. In this case, ANG II may modulate LH release from gonadotrophs through autocrine function. Further, ANG II released from gonadotrophs seems to stimulate lactotrophs containing ANG II receptors to secrete prolactin through paracrine regulation (see Schwartz and Cherny 1992). Thus, it is certain that there are two ways for the RAS to function: one, systemically, as circulating RAS and the other, locally, as tissue RAS.

In this chapter, we shall describe several biological actions of circulating ANG II, since the physiological significance of tissue RAS has not been fully elucidated. Additional information on ANG II can be found in Chapter 7 where ANG II receptors are discussed. Several reviews have already dealt with the biological actions of ANG II from the comparative point of view (Nishimura 1980a,b, 1987; Henderson et al. 1981; Wilson 1984a, b; Simon et al. 1992; Takei et al. 1989; Takei 1992a, b; Henderson and Deacon 1993).

8.1 Renal Dipsogen – Historical Background

The presence of a renal factor causing the sensation of thirst was first suggested by Linazasoro et al. (1954). They observed water intake for 24 h in normal and nephrectomized rats after 48 h of fasting and found that nephrectomized rats drank significantly less water than did normal rats. An injection of kidney extracts from pig prevented a decrease in drinking in nephrectomized rats. Linazasoro and his colleagues confirmed the presence of a renal thirst factor (Jiménez-Díaz et al. 1959); however, they concluded that this factor is not renin, since they could not demonstrate the dipsogenic nature of renin. However, this failure may have been due to the fact that the animals were given only physiological saline to drink. Renin increased water intake but had little effect on the intake of saline in rats (Fitzsimons 1979, p. 266). Nairn et al. (1956) found that renal extracts and preparations of renin caused serious effusions and increased drinking in nephrectomized rabbits, guinea pigs and rats. Masson et al. (1956) injected renin into nephrectomized rats and observed drinking and effusions. Masson and his colleagues argued that the formation of serious effusions depleted plasma volume and stimulated thirst: drinking was not due to the primary stimulation of thirst centers by renin. Asscher and Anson (1963) found that nephrectomized rats injected with saline extracts of kidney drank more water than nephrectomized controls. Treated rats showed generalized edema, ascites, pleural effusions, a rise in hematocrit and a fall in plasma volume. The authors concluded that thirst was induced by a fall in plasma volume owing to leakage of plasma from the vascular compartment into the tissues.

Fitzsimons (1964) showed that caval ligation resulted in a sharp diminution in venous return to the heart and a fall in cardiac output. This ligation was an effective stimulus to drinking in rats with intact kidneys, but ineffective in nephrectomized rats. He suggested that the kidney plays a role in an increase in drinking, possibly by release of renin. Further, Fitzsimons (1966) and Gutman et al. (1967) constricted the renal pedicles and caused increased drinking in the rat. Considering his results and others, Fitzsimons (1966) reached the conclusion that renin might exert its effect on drinking through angiotensin. In a preliminary experiment, he observed that [Asn^1, Val^5]ANG II amide (Hypertensin, CIBA) stimulated drinking when the dose was large. Fitzsimons (1967, 1969) also showed that partial obstruction of the abdominal aorta above the level of the renal arteries caused thirst. The effects of caval ligation and aortic constriction were considerably less in bilaterally nephrectomized rats, and aortic constriction was ineffective when the constriction was placed below the level of the renal arteries where it did not interfere with renal circulation. These data indicated that the kidney plays an important role in thirst mechanisms. Fitzsimons (1969, 1970) observed further that (1) the renal dipsogenic factor was found in saline extracts of the renal cortex, but not in the renal medulla and liver; (2) thirst and pressor activities were found only in the renal cortex; (3) these two activities were always present to the same degree in a particular fraction, and it was impossible to separate the two factors during the fractionation procedure; (4) both the extractable dipsogenic factor and the extractable pressor activity were reduced by pretreatment of the rat with saline and DOCA for several weeks before extraction to deplete the kidney of renin; and (5) intravenous injection of pig renin caused drinking in normal and nephrectomized rats. Thus, the kidney dipsogenic factor was postulated to be identical with renin in the rat. Thus, arguments that the renal dipsogen is renin, suggested as early as 1956, were convincing (Nairn et al. 1956; Masson et al. 1956). As briefly mentioned above, Fitzsimons (1966) and Fitzsimons and Simons (1968, 1969) demonstrated that injection or infusion of ANG II induced copious drinking in normal rats. They concluded that renin exerts its effect on drinking through ANG II.

8.2 Exogenous ANG II and Drinking

Since the findings of Fitzsimons and Simons (1968, 1969), many investigators have studied the dipsogenic action of ANG II. Nishimura (1980b) and Rolls and Rolls (1982) catalogued and referenced the various experimental animals used, and the routes and effects of ANG II administration. Recent data obtained by other investigators have been added to the list of Rolls and Rolls (Table 8.1); our detailed data have been transferred to separate lists (Tables 8.2–8.7).

8.2.1 Mammals

Drinking was induced in rats by intravenous infusions of ANG II (25–40 pmol/kg/min, i.e. 25–40 ng/kg/min), which probably produced plasma concentrations of ANG II within the physiological range (Epstein and Hsaio 1975; Hsaio et al. 1977). In the dog, Trippodo et al. (1976), Fitzsimons et al. (1978b) and H. Suzuki et al. (1987) made similar observations with 8.3–40 pmol/kg/min infusions. Man, monkey and cat also responded to ANG II by drinking (Table 8.1).

In contrast, goat and sheep were relatively unresponsive to ANG II and needed pharmacological doses to induce drinking. Wild and New Zealand rabbits were also insensitive to ANG II as a stimulus for drinking (Table 8.1, 8.2). Another insensitive species was the dasyurid marsupial, *Antechinus stuartii*, a scansorial insectivore in which drinking is not required to maintain water balance since their food (insects) provides their water supply. This species did not respond to intraperitoneal injections of 1–30 μg of ANG II (Blair-West et al. 1983). Although the Australian possum, *Trichosurus vulpecula*, (Young and McDonald 1978) and the North American opossum, *Didelphis virginiana*, (Elfont et al. 1980; Findlay et al. 1980) were responsive to

Table 8.1. Drinking responses of vertebrate species to ANG II. Recent data, excluding ours, were added to the list compiled by Rolls and Rolls (1982)[a]

Species[b]	Route of administration	Investigators
Man	Intravenous	Rolls et al. (1986)
Rat	Intravenous	Fitzsimons and Simons (1968, 1969); Epstein and Hsaio (1975); Hsaio et al. (1977); Mann et al. (1980)
	Intracranial	Booth (1968); Epstein et al. (1969); Epstein et al. (1970); Severs et al. (1970); White et al. (1972); Gronan and York (1979); Wright et al. (1984); Harland et al. (1988)
	Subcutaneous	Fregly et al. (1991)
Mouse[*]	Subcutaneous	Rowland and Fregly (1988)
Golden hamster[*]	Intracranial	Miceli and Malsbury (1983)
Cat	Intravenous	Cooling and Day (1975)
	Intracranial	Sturgeon et al. (1973); Brophy and Levitt (1974)
Dog	Intravenous	Trippodo et al. (1976); Suzuki et al. (1987)
	Intracarotid	Fitzsimons et al. (1978b)
	Intracranial	Rolls and Ramsay (1975); Ramsay and Reid (1975); Fitzsimons and Kucharczyk (1978); Ramsay et al. (1979)
Rhesus monkey	Intracranial	Setler (1971); Myers et al. (1973); Sharpe and Swanson (1974); Simonnet et al. (1979)
Cebus monkey	Intracranial	Block, cited in Schwob and Johnson (1977)
Baboon	Intracranial	Lotter et al. (1980)
Goat[*]	Intracranial	Andersson and Westbye (1970); Andersson and Eriksson (1971)
	Intracranial, intracarotid	Thornton and Baldwin (1985)
Sheep[*]	Intracranial	Breuhaus and Chimoskey (1990)
	Intracarotid, intracranial	Abraham et al. (1975)
	Intracarotid	Abraham (1976); Weisinger et al. (1977)
Pig[*]	Intracranial	Baldwin and Thornton (1986)
Mongolian gerbil[*]	Intracranial	Block et al. (1974); Wright et al. (1984)
Rabbit[*]	Intracranial	Denton et al. (1985); Tarjan et al. (1988)
Dasyurid marsupial[*]	Intraperitoneal	Blair-West et al. (1983)
Marsupial possum[*] (Australian)	Intravenous	Young and McDonald (1978)
Marsupial opossum[*] (North American)	Intracranial, intravenous	Elfont et al. (1980); Findlay et al. (1980)
Pigeon	Intraperitoneal, intravenous, intracranial	Evered and Fitzsimons (1976, 1981)
	Intravenous	Kaufman and Peters (1980)
	Intracranial	deCaro et al. (1982)
Chick	Subcutaneous, intracranial	Volmert and Firman (1991)
Chicken	Intravenous, intracranial	Snapir et al. (1976)
	Intramuscular, intracranial	Schwob and Johnson (1977)
Peking duck	Intracranial	deCaro et al. (1980)

116

Table 8.1. *Contd.*

Species[b]	Route of administration	Investigators
Turkey	Intracranial	Denbow (1985)
Iguana	Intraperitoneal	Fitzsimons and Kaufman (1977)
	Intracranial	Fitzsimons (1979)
Euryhaline killifish	Intraperitoneal	Malvin et al (1980)
Japanese eel	Intravenous	Hirano and Hasegawa (1984)
Euryhaline flounder (*Platichthys flesus*)	Intravenous	Carrick and Balment (1983); Balment and Carrick (1985)
Marine flounder (*Pseudopleuronectes americanus*)	Intramuscular	Beasley et al. (1986)
Long-horned sculpin (marine)	Intramuscular	Beasley et al. (1986)
Goldfish	Intraperitoneal	Beasley et al. (1986)
Mottled sculpin	Intramuscular	Beasley et al. (1986)
Common shiner	Intramuscular	Beasley et al. (1986)
Carp	Intramuscular	Perrott and Balment (1985)
Several species of SW and FW teleosts	Intramuscular	Perrott et al. (1992)
Dogfish (*Scyliorhinus canicula*)	Intramuscular	Hazon et al. (1989)

[a]Our experimental data are listed in Tables 8.2, 8.4, 8.5, 8.6, 8.7.

[b]*,These animals were insensitive to ANG II in drinking (see Sect. 8.2).

intravenous infusion of ANG II, they required large doses (0.2–3.5 nmol/kg/min) compared with those needed by rats and dogs. Young and McDonald (1978) reported that drinking of free water by *Trichosurus vulpecula* in the wild had never been observed. These findings have raised questions concerning the physiological role of ANG II in thirst. The goat, sheep and rabbit are herbivorous, depending on grass for most of their ingested water. The dasyurid marsupial takes water from insects. None of these animals drink much free water. The Australian possum and North American opossum drink little water in nature. All these animal species seem to be insensitive to ANG II in drinking or require high doses of ANG II to induce drinking. In such species, an angiotensin-thirst mechanism might have been of little use or attenuated through adaptive evolution to their habits or habitats. Therefore, data obtained from such species do not preclude the possibility of the physiological involvement of ANG II in normal drinking in other species.

We used ten mammalian species to examine the dipsogenic effect of intraperitoneal (i.p.) injection of ANG II (Hypertensin CIBA; Table 8.2), and found that flying fox, rat and cat (Nos. 1–3) responded to 5 to 10 μg/100 g of ANG II by drinking. The chipmunk, *Tamias sibiricus* (No. 4), needed relatively high doses (50 μg/100 g) to induce drinking, compared with the other three species. Since chipmunks generally eat dry nuts, water conservation mechanisms may have become well developed, instead of the development of the ANG

Table 8.2. Drinking responses of mammals to a single intraperitoneal injection of ANG II

Species[a]	No of animals	Response	Minimum effective dose (μ/100 g)	Latency (min)	Remarks
1. *Pteropus gigantius* (flying fox)	3	+	<5	About 10	Studied in India
2. *Rattus norvegicus albus* (Wistar rat)	9	+	5	2–5	
3. *Felis catus* (cat)	3	+	<10	3–10	Eat dry nuts
4. *Tamias sibiricus* (chipmunk)*	5	+	50	15–50	Originated in arid areas in or near Mongolia
5. *Meriones unguiculatus* (Mongolian gerbil)	17	–			Both strains originated in dry areas in Central Asia
6a. *Mus musculus* (mouse, C57BL)	7	–			
6b. *Mus musculus* (mouse, BALB/C)	12	–			
7. *Mesocricetus auratus* (golden hamster)**	5	–			Found in the Syrian Desert
8. *Lepus brachyurus* (hare)	4	–			Herbivorous
9. *Oryctolagus cuniculus* var. *domesticus* (rabbit)**	5	–			Herbivorous
10. *Cavia cobaya* (guinea pig)**	5	–			Herbivorous

[a], This species lapped water only momentarily each time they approached water (3 to 16 times). **, These species did not respond to ANG II at doses from 1 to 100 μg/100 g in experiments carried out in a quiet animal room with constant temperature (20 °C). The results differed from experiments done earlier (Kobayashi et al. 1979).

118

II-induced drinking mechanism. They may not need to drink much water in nature. The Mongolian gerbil, *Meriones unguiculatus* (No. 5) and the mouse, *Mus musculus* (Nos. 6a, 6b), were insensitive to ANG II (Table 8.2). Rowland and Fregly (1988) made a similar observation in mice. The Mongolian gerbil originated in arid areas in or near Mongolia (Robinson 1975). The mouse used in our experiments originated in dry areas in Central Asia (Schwarz and Schwarz 1943). The hamster, *Mesocricetus auratus* (No. 7), which is a Syrian desert species, was insensitive. Thus, those species which are native to arid areas and drink little water in nature, are insensitive to ANG II in drinking. The hare (No. 8) and the rabbit (No. 9), which are also listed in Table 8.1, and the guinea pig (No. 10) did not respond to ANG II by drinking. These species are herbivorous, like the sheep, and do not drink much water. It can be concluded again here that in those animal species which drink little water in nature, ANG II-induced drinking mechanisms have been attenuated or lost during their adaptive evolutionary process. In contrast, in animals which drink much water the ANG II-induced drinking mechanism has been preserved as part of their water-acquiring behavior during evolutionary process.

One possibility for the unresponsiveness is that the receptor for ANG II has become lost or changed in its characters in these species. For example, the possum, *Trichosurus vulpecula*, has no receptor in its brain for ANG II and ANG III in drinking, although its liver, adrenal and vascular tissues have the receptor (Sernia et al. 1990).

It is worth noting that intracerebroventricular injection of ANG II (100 ng) dissolved in water or dextrose was ineffective, but ANG II in 0.15 M NaCl induced vigorous drinking in pigs; the drinking response to central ANG II required the presence of sodium ions (Baldwin and Thornton 1986). In rats exposed chronically to cold, responsiveness to acute subcutaneous administration of ANG II increased significantly as assessed by the drinking response (Fregly et al. 1991). These data suggest that internal or external environmental circumstances should be carefully considered when such experiments are performed.

8.2.2 Birds

The aerial life of birds results in metabolic burdens reflected in increased respiratory gas exchange and in high body temperature, leading to evaporation through the respiratory surfaces and the skin with accompanying water loss (Willoughby and Peaker 1979). Evaporative water loss in birds varied from 5 to 35% of body weight per day at about 25 °C (Bartholomew and Cade 1963). In addition, water is lost through excreta. The rate of total water loss varies depending on the species and environmental conditions. Birds compensate for water loss primarily by drinking. Birds, especially if small, are good subjects for testing drinking behavior, since they tend to drink frequently and the effects of drug administration can usually be observed within 1 h (Fig. 8.1; Table 8.3).

Table 8.3. Drinking response of Japanese quail to a single intraperitoneal injection of ANG II

Dose of AII (μg/100 g)	Number of birds	Latent period (min)	Water intake[a]	
			(ml/100 g/30 min)	(ml/100 g/h)
0 (saline)	16	–	0.31±0.05[b]	0.74±0.07
1	5	–	0.42±0.07	0.67±0.12
3	9	7.9±1.8	0.88±0.20*	0.95±0.18
5	4	4.5±1.2	1.47±0.14*	1.56±0.13*
10	5	4.8±0.5	2.23±0.36*	2.31±0.32*
20	5	8.3±0.9	2.98±0.41*	3.14±0.47*
50	9	6.6±0.9	3.40±0.57*	3.88±0.36*
100	8	>20	0.17±0.07	0.39±0.16

[a]*,Highly significant ($p<0.01$) compared with saline group.
[b]Mean ± SE.

The dipsogenic action of intracranial and i.p. injections of ANG II was initially studied in the white-crowned sparrow (Wada et al. 1975), but the effective dose was huge, because the ANG II used had deteriorated. Later, 5μg/ 100 g (i.p.) was found to be sufficient to induce drinking (Kobayashi et al. 1979). A dipsogenic action of ANG II was also reported by other investigators

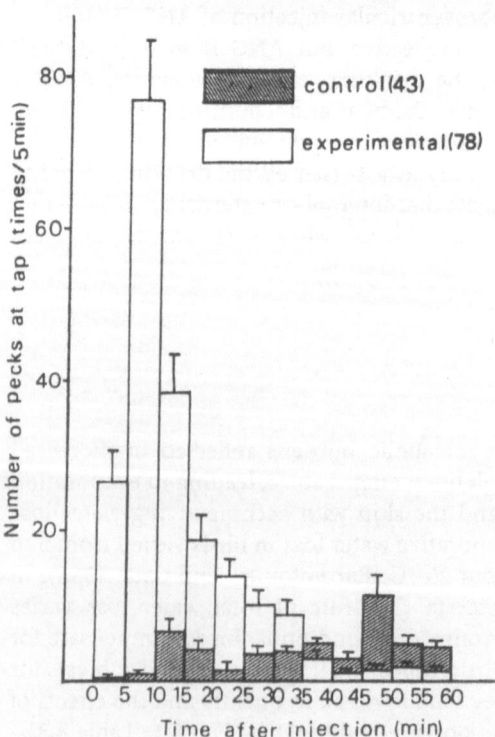

Fig. 8.1. Number of pecks (per 5-min interval) after intravenous injection of ANG II (50 μg) in the Japanese quail. *Open columns* show the means of pecking number for the experimental group, *shaded columns* for the control group. Number of observations was 78 in the experimental group, and 43 in the control group. *Vertical lines* indicate standard error (SE). (Takei 1977a)

in the pigeon, chick, domestic chicken, turkey and duck (Table 8.1). We have tested the dipsogenic effect of ANG II (Hypertensin, CIBA), which is almost as potent as avian ANG II with respect to cardiovascular effects (Takei and Hasegawa 1990), in 40 avian species (Table 8.4). Water-searching behavior was assessed by counting the number of pecks at the tap of a water bottle (Fig. 8.1), or the number of swallows of water from a container after a single i.p. injection. Observations were usually limited to 30 or 60 min, since drinking induced by minimum effective doses of i.p. injected ANG II terminated within 30 min in most species (Fig. 8.1; Table 8.3). In some species, the amount of ingested water was measured with a graduated cylinder as described by Takei (1977a). As shown in Table 8.4, most granivorous and omnivorous birds responded to ANG II (1-10 µg/100 g) with latent periods from 1 to 20 min (shorter times predominate; Table 8.3). In the Japanese quail, water intake was around 0.4 ml/ h in the laboratory (Fig. 8.2). The dipsogenic action of ANG II varied with different routes of ANG II administration. Water intake after a subcutaneous injection of 50 µg of ANG II (5.24 ± 0.44 ml/h, $n = 37$) was significantly ($P < 0.01$) greater than that induced after an intravenous injection at the same dose (3.45 ± 0.33 ml/h, $n = 50$). Vigorous pecking was observed from 5 to 15 min after intravenous injection (Fig. 8.1); it continued from 5 to 25 min after subcutaneous injection. The latent period following subcutaneous injection (329.8 ± 31.2 s, $n = 37$) was significantly ($P < 0.05$) longer than that following intravenous injection (207.2 ± 29.1 s, $n = 17$; Takei 1977a).

Carnivorous birds (Nos. 14, 15, 16, 17, 35 in Table 8.4), on the other hand, did not respond to ANG II administration (Kobayashi et al. 1979; Kobayashi and Takei 1982). Carnivorous birds ingest water mostly from meat and rarely drink free water (Skadhauge 1981). The white-breasted kingfisher, *Halcyon smyrnensis* (No. 13), did not respond until dose levels of ANG II reached 50 µg.

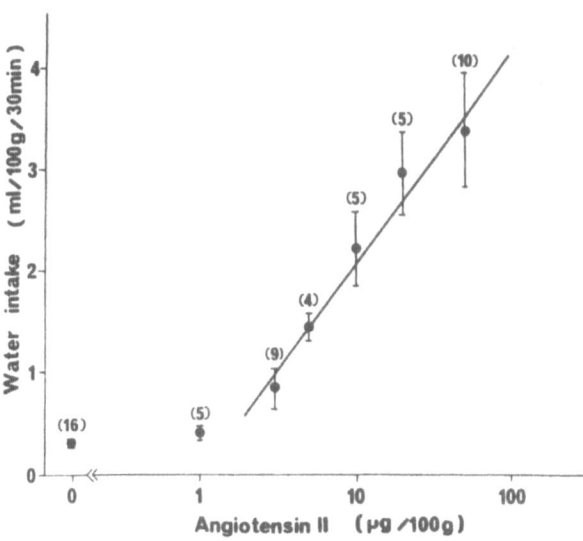

Fig. 8.2. Dose-response relationship for water intake during the 30 min following i.p. injection of ANG II (1–100 µg/100 g body wt.) in the Japanese quail. The *vertical lines* indicate SE. Number of birds is shown in *parentheses*

This species is a fish eater. They may take water from fish flesh and drink only a little free water in nature.

A group of birds originating in dry regions were also insensitive to ANG II. Among 3 species of munia, *Lonchura malabarica* (No. 7) was fairly insensitive to ANG II, but the related species, *L. punctulata* (No. 6) and *L. malacca* (No. 5), responded to 1 μg/100 g. Unlike the other two, *L. malabarica* lives in arid areas of India and can survive without drinking water for extended periods (Ghosh and Ghosh 1972). The Port Lincoln parrot, *Barnardius zonarius zonarius* (No. 26), and the twenty-eight parrot, *Barnardius zonarius semitorquatus* (No. 27), responded to about 10 μg/100 g of ANG II, but the mallee ringneck parrot, *Barnardius barnardi barnardi* (No. 28), required about 100 μg (Kobayashi 1981). The Port Lincoln parrot and the twenty-eight parrot have apparently lived in the wetter southwest of Australia for a long period of time (more than 5000 years), while the mallee ringneck parrot is from the dry areas of south-eastern Australia (C.D. Fisher 1979). The red-capped parrot, *Purpureicephalus squrius* (No. 29), was fairly insensitive to ANG II and needed 100 μg/100 g to start drinking (Kobayashi 1981). This species although now ubiquitous in

Table 8.4. Drinking responses of birds to a single intraperitoneal ionjection of ANG II

Species	Minimum effective dose of ANG II (μg/100 g)	Remarks
1. *Dicrurus adsimilis* (fork-tailed drongo)	1	
2. *Columba livia domestica* (domestic pigeon)	1	
3. *Dendrocitta vagabunda* (rufous treepie)	1	
4. *Estrilda amandava* (red munia)	1	
5. *Lonchura malacca* (chestnut munia)	1	
6. *Lonchura punctulata* (scaly-breasted munia)	1	
7. *Lonchura malabarica* (white-throated munia)	100	
8. *Dinopium benghalense* (black-rumped goldenback)	10	
9. *Sturnus malabaricus* (chestnut-tailed starling)	10	
10. *Padda oryzivora* (Java sparrow)	10	
11. *Corvus splendens* (house crow)	10	
12. *Petronia xanthocollis* (yellow-throated sparrow)	10	
13. *Halcyon smyrnensis* (white-breasted kingfisher)	50	
14. *Otus bakkamoena* (collared scops-owl)	100	
15. *Centropus sinensis* (common crow-pheasant)	–	No response to 1–100 μg/100 g
16. *Accipiter badius* (shikra)	–	No response to 30–1000 μg/100 g
17. *Athene brama* (spotted owlet)	–	No response to 100–1000 μg/100 g
18. *Vanellus indicus* (red-wattled lapwing)	5	
19. *Psittacula krameri* (rose-ringed parakeet)	10	

Table 8.4. *Contd.*

Species	Minimum effective dose of ANG II (µg/100 g)	Remarks
20. *Psittacula cyanocephala* (plum-headed parakeet)	10	
21. *Molpostes cafer* (red-vented bulbul)	10	
22. *Emberiza bruniceps* (red-headed bunting)	5	
23. *Acridotheres tristis* (common mynah)	5	
24. *Uroloncha striata* var. *domestica* (Bengales)	10	
25. *Melopsittacus undulatus* (green budgerigar)	10[a]	
(yellow budgerigar)	10	
26. *Barnardius zonarius zonarius* (Port Lincoln parrot)	10	
27 *Barnardius zonarius semitorquatus* (twenty-eight parrot)	10	
28. *Barnardius barnardi barnardi* (mallee ringneck parrot)	100	
29. *Purpureicephalus spurius* (red-capped parrot)	100	
30. *Nymphicus hollandicus* (cockatiel)	–	No response to 1–100 µg/100 g
31. *Domicella lorius* (lory)	5	
32. *Agapornis fischeri* var. *blew* (blue lovebird)	10	
33. *Agapornis roseicollis* (rose-faced lovebird)	1	
34. *Zonotrichia leucophrys gambelii* (white-crowned sparrow)	5[b]	
35. *Falco tinnuculus* (kestrel)	1000	
36. *Streptopelia risoria* (ring dove)	1	
37. *Emberiza aureola ornata* (Ussurian yellow-breasted bunting)	30	
38. *Passer montanus* (tree sparrow)	5	
39. *Gallus gallus* var. *domesticus* (domestic fowl)	10	
(one-day-old)	10	
40. *Coturnix coturnix japonica* (Japanese quail)	5	

[a]Kasuya et al. (1987a) found that both green and yellow budgerigars responded similarly to 5 µ g of ANG II, unlike former data (Kobayashi et al. 1979).
[b]Kobayashi et al. (1979) found that 5 µg/100 g of ANG II could induce drinking in this species (No.34). In previous studies (Wada et al. 1975), ANG II seemed to have deteriorated before use.

Australia, was once confined to the arid eastern areas for a long period of time (Serventy 1938, cited in Forshaw 1969). Thus, it is likely that birds that originated in dry areas or have lived in such areas for a long time have become relatively insensitive to ANG II with respect to drinking. Examination of angiotensin receptors in the brain of ANG II-insensitive birds, similar to that performed in the Australian possum (Sernia et al. 1990), is necessary for clarification of the cause of the insensitivity.

The budgerigar, *Melopsittacus undulatus*, appears to be an exception to the above. This bird originated in arid parts of Australia, drinks little water in nature (Bartholomew and Cade 1963) and can survive for extended periods without water (Cade and Dybas 1962; Uemura 1964; Krag and Skadhauge 1972; Kasuya et al. 1987a). However, both the yellow (Kobayashi et al. 1979) and green (Kasuya et al. 1987b) budgerigars responded to 10 µg/100 g of ANG II by drinking. It is not known why the budgerigars behaved differently from *Lonchura malabarica* and the mallee ringneck parrot in this regard. In budgerigars, plasma ANG II concentrations did not change in parallel with the daily drinking rhythm. Further, captopril (SQ 14225), a converting enzyme inhibitor, did not affect natural drinking (Kasuya et al. 1985). Therefore, in budgerigars, plasma ANG II may not be involved in natural physiological drinking, even though their responsiveness to exogenous ANG II has apparently not been lost. Although we indicated initially that the green budgerigar (wild type) needed more ANG II than the yellow to respond, we later found that both types of budgerigars reacted similarly to ANG II (Kasuya et al. 1987a). It should be noted, however, that young budgerigars are more sensitive to ANG II than older ones (Kasuya et al. 1987a).

8.2.3 Reptiles

Although the body surface of reptiles is covered with scales, they lose water by evaporation through the skin, with the amount varying among species. Water also evaporates through the lungs and pulmonary passages in terrestrial environments, although the rate of water loss is relatively small compared with birds and mammals (see Bentley 1971, p. 138). In order to compensate for the water loss, ANG II-induced drinking mechanisms, which are not possessed by amphibians (see below), have evolved in the brain in reptiles. They search for drinking water actively, and as described below, have a well-developed ANG II-induced drinking mechanisms as part of their water-acquiring behaviour. We tested sixteen reptilian species (Table 8.5). After injection of ANG II (Hypertensin, CIBA), searching behavior for water and approach to water were observed, and the number of swallowing movements of the throat was counted for 30 min. Animals responding to ANG II by drinking showed latent periods ranging from 1 to 30 min.

In Chelonia, three of four species (Nos. 1–3 in Table 8.5) responded readily to ANG II. In these, the ANG II-induced drinking mechanism functions as water-acquiring behavior. However, the Mediterranean terrestrial tortoise, *Testudo graeca*, was relatively insensitive. This species lives in arid desert regions of the Mediterranean and does not have free access to water in nature (Pritchard 1967). As reported in mammals and birds originating in arid areas, an angiotensin-thirst mechanism may have become attenuated in this species during its evolution.

Table 8.5. Drinking responses of reptiles to a single intraperitoneal injection of ANG II

	Species	Minimum effective dose (μg/100 g)[a]	Remarks
Chelonia	1. *Kinosternon subrubrum* (eastern mud turtle)	<10	
	2. *Chrysemys scripta elegans* (red-eared turtle)	<1	
	3. *Geoclemys reevesii* (Chinese turtle)	<10	
	4. *Testudo graeca* (European tortoise)	1000	No response to 10–500 g/100 g
Lacertilia	5. *Hemidactylus flaviviridis* (house gecko)	5	
	6. *Calotes vesicolor* (Indian garden lizard)	–	Hibernating, no response to 5–500 μg/100 g
	7. *Anolis carolinensis* (green anole)	–	No response to 10–100 μg/100 g
	8. *Eumeces latiscutatus* (Japanese skink)	<5	
	9. *Eumeces okadae* (Okada's skink)	5	
	10. *Takydromus tachydromoides* (Jqapanese grass lizard)	<5	
	11. *Takydromoides septentrionalis* (northern grasss lizard)	1	
Ophidia	12. *Elaphe conspicillata* (Japanese burrowing rat snake)	<30	
	13. *Elaphe quadrivirgata* (four-lined rat snake)	16.5[b]	
	14. *Elaphe climacophora* (Japanese rat snake)	16.5[b]	
	15. *Natrix piscator* (fisherman water snake)	–	Hibernating, no response to 10–1000 μg/100 g
	16. *Rhabdophis tigrinus* (tiger water snake)	50	

[a]Latency to smallest effective dose used was 1 to 30 min.
[b]In later experiments, they did not respond to 20–200 μg of ANG II.

In Lacertilia, five of seven species (Nos. 5, 8–11) responded to ANG II (Table 8.5), suggesting that in most lacertilians ANG II is involved in the thirst mechanism. Two species, in the genera *Anolis* and *Calotes*, behaved differently. *Anolis carolinensis* (No. 7), weighing 3.5 to 8.0 g, did not drink after injection of ANG II. This species lives in subtropical humid areas in North America and eats small insects. It would be interesting to examine angiotensin receptors in the brain of *Anolis*. They drank little water (0–0.01 ml/10 g/h between 10:00 and 18:00) in the laboratory. In contrast, *Eumeces okadae* (No. 9), which weighs 7.5 to 12.5 g and lives in the bush of Miyake Island, one of the Izu Islands of Japan, in the laboratory drank 0.04–0.07 ml/10 g/h between 10:00 and 18:00. They responded to 10 μg/100 g of ANG II by drinking (0.12 ml/10 g/ h). When *Anolis carolinensis* and *Eumeces okadae* were deprived of drinking water for 50 h, the former lost no body weight, but the latter lost about 50% of their body weight. These observations suggest that *Anolis*, has well-developed water conservation mechanisms. Hibernating Indian garden lizards, *Calotes vesicolor* (No. 6), did not drink after injection of ANG II at about 15 °C in February in India. Even after keeping them at 35 °C for 5 days to break hibernation, they did not respond to a dose as high as 500 μg/100 g. Experiments during nonhibernating periods may clarify the responsiveness in *Calotes*. The common iguana drank in response to i.p. injection of mammalian ANG II, after injection of the optimal dose of 2×10^{-8} mol/100 g, the mean latency was 10.5 + 4.68 min ($n = 8$; Fitzsimons and Kaufman 1977; Table 8.1).

In Ophidia, four of five species, *Elaphe conspicillata* (No. 12), *Elaphe quadrivirgata* (No. 13), *Elaphe climacophora* (No. 14) and *Rhabdophis tigrinus* (No. 16), reacted to ANG II (16.5–50 μg/100 g) initially. However, *E. climacophora* and *E. quadrivirgata* did not respond to ANG II (30–200 μg) in later experiments. Further studies are needed to clarify this discrepancy. Ophidians seem to require more ANG II to respond than chelonians and lacertilians. Indian water snakes, *Natrix piscator* (No. 15), which were hibernating in air, failed to respond to ANG II at about 15 °C in February in India. Since this species is aquatic when active, we transferred them to water at 29 °C for 2 weeks to break hibernation. They became active, but did not respond to ANG II. Injections of ANG II into *Natrix* should be done when they are not hibernating in nature.

Although crocodilian species were not examined, we may conclude that reptiles, the first truly terrestrial vertebrates, developed an angiotensin-drinking mechanism as part of their water-acquiring behavior during their evolution in the early Mesozoic period, since none of the tested amphibians responded to ANG II by drinking (see next section).

8.2.4 Amphibians

It is generally believed that amphibians do not normally drink, but rehydrate by absorbing water across their permeable skin (Bentley and Yorio 1979). It is interesting to examine whether they respond to ANG II by drinking. For drinking tests, animals were transferred after intraperitoneal injection of ANG II to water containing phenol red at 0.004%; 30 or 60 min later phenol red in the gut was extracted, measured with a spectrophotometer, and the amount of ingested water thus estimated (Kobayashi et al. 1979).

ANG II did not induce drinking in any species of amphibians examined. Doses of ANG II (i.p.) were 0.1 to 1000 μg/100 g. Urodeles used were the Japanese clawed salamander (*Onychodactylus japonicus*), the sword-tailed newt (*Cynops ensicauda*), and the fire-bellied newt (*Cynops pyrrhogaster*). Anurans used were the Western-Japanese common toad (*Bufo bufo japonicus*), the Japanese tree frog (*Hyla arborea japonica*), the Japanese brown frog (*Rana japonica japonica*), the Tokyo daruma pond frog (*Rana brevipoda porosa*), the bullfrog (*Rana catesbeiana*), the kajika frog (*Rhacophorus buergeri*), the common platanna (*Xenopus laevis*) and the tropical platanna (*Xenopus mülleri*). Hirano et al. (1978) showed that systemic injection of ANG II did not induce drinking in *Rana brevipoda porosa*. Water intake of tadpoles of *Rana catesbeiana* was not enhanced by ANG II (10 and 200 µg/100 g; Kobayashi et al. 1979), although they ingest about 20 µl/10 g/h throughout the tadpole stage. Thus, amphibians did not develop a mechanism for oral drinking induced by ANG II. Instead, they have developed a mechanism for water absorption through the skin. It might be of interest to inject *Crinia* angiotensin II, Ala-Pro-Gly-Asp-Arg-Ile-Tyr-Val-His-Pro-Phe, which was isolated from the skin of the frog *Crinia georgiana* (Erspamer et al. 1979) and induced drinking in pigeons and rats (Cantalamessa et al. 1982).

It is worth noting that the sodium isotope flux and short-circuit current in isolated skin of *Rana pipiens* were stimulated by ANG II (McAfee and Locke 1967). Further, ANG II stimulated sodium and water transport in isolated skin of the toads, *Bufo arenarum* and *Bufo paracnemis*, (Coviello and Brauckmann 1973; Coviello et al. 1974, 1975). These effects were inhibited by [Leu[8]] ANG II, a competitive antagonist of ANG II (Coviello et al. 1974). Osmotic water permeability, short-circuit current and cyclic AMP levels were also increased by ANG II in isolated frog skin (Coviello et al. 1976). Tokuda et al. (1995) found that in the tree-frog, *Hyla arborea japonica*, two different water absorption systems are present in the isolated ventral skin: a rapid enhanced flow, which was observed in dehydrated frogs or those stimulated by adrenaline ß-agonists or vasotocin, and a slow basal flow, which was seen in normally hydrated frogs during the nonbreeding season. Ouabain completely blocked the rapid flow, but not the slow basal flow. ANG II totally inhibited the basal water absorption, but not the rapid one. Thus, ANG II affects the function of the frog skin.

During in vivo experiments, Reboreda and Segura (1989) observed that systemic and intracerebroventricular injections of ANG II and systemic injection of 100 µg/kg of [Sar[1], Val[5], Ala[8]] Ang II (saralasin, an ANG II antagonist) had no effect on water intake across the skin of *Bufo arenarum*. However, Hoff and Hillyard (1991) demonstrated in the toad, *Bufo punctatus*, that i.p. injections of ANG II (1, 5, 200 µg/100 g) increased both the duration of cutaneous water-absorption behavior and water weight gain, and that these effects were eliminated by i.p. injection of saralasin (100 µg/100 g) 30 min before ANG II (5 µg) injection. Therefore, the effects of ANG II may not be due to its release of neurohypophysial hormone (Sect. 8.6.1). Tran et al. (1992) observed that *Bufo woodhousei* responded to ANG II (i.p. 100 µg/100 g) by increases in the duration of water-absorption behavior and in the amount of water absorbed, and that in toads with empty bladders the responses were greater. Hoff and Hillyard (1993) reported that cutaneous water-absorption behavior following dehydration was inhibited by saralasin, suggesting that ANG II was involved in the regulation of water reabsorption behavior in *Bufo punctatus*. Propper and Johnson (1994) using two species of desert anurans, *Scaphiopus couchii* and *Bufo cognatus*, observed that dehydration induced water-absorption behavior in both species. Fully hydrated toads injected i.p. with ANG II exhibited significant water-absorption behavior. The minimum effective dose for inducing this behavior was 10 µg/100 g for *S. couchii* and 100 µg/100 g for *B. cognatus*. Further, they noted that when dehydrated toads were treated with Thr^8-saralasin, *S. couchii* showed an increase in the behavior, while *B. cognatus* did not. Captopril did not affect the behavior in dehydrated toads of either species. From these findings, they concluded (1) that ANG II might be involved in water absorption behavior in amphibians; (2) that the failure of saralasin or captopril to inhibit water-absorption behavior in dehydrated toads suggests that the receptor mechanisms involved in thirst regulation in toads might be different from those in mammals; and (3) that the RAS might not be the only potential mediator of water-absorption behavior in these species (Propper and Johnson 1994). From the findings mentioned above, it is evident that the frog skin has receptors for ANG II as seen in the in vitro experiments using isolated skin, and also it seems likely that the amphibian brain has receptive sites for ANG II involved in cutaneous water absorption behavior.

8.2.5 Fishes

To measure the amount of ingested water in fishes, the phenol red method was used (Kobayashi et al. 1983). As shown in Table 8.6, ANG II, (Hypertensin CIBA) did not induce drinking in species (Nos. 1–10) found exclusively in fresh

Table 8.6. Drinking responses of freshwater fishes to a single intraperitoneal injection of ANG II

Unresponsive species	Responsive species	Minimum effective dose (µg/100 g)
1. *Lampetra japonica japonica* (arctic lamprey)	11. *Carassius auratus* (goldfish)	10
2. *Pseudorasbora parva* (topmouth gudgeon)	12. *Leuciscus hakonensis* (Japanese dace)	1
3. *Rhodeus ocellatus* (rose bitterling)	13. *Carassius carassius* (crucian carp)	10
4. *Rhodeus lanceolatus* (slender bitterling)	14. *Oryzias latipes* (Asiatic ricefish, medaka)	1
5. *Cyprinus carpio* (carp)[a]	15. *Gambusia affinis* (common gambusia)	10
6. *Ctenopharyngodon idella* (grass carp)	16. *Gyrinocheilus anymonieri* (algae eater)	1
7. *Cobitis anguillicaudatus* (Asian pond loach)	17. *Parasilurus asotus* (Japanese catfish)	50
8. *Salvelinus leucomaenis* (Japanese char)	18. *Chaenogobius annularis* (floating goby)	20
9. *Sarotherodon mossambicus* (Mozambique tilapia)	19. *Tridentiger obscurus* (Japanese trident goby)	10
10. *Pungitius sinensis* (Chinese stickleback)	20. *Anguilla japonica* (Japanese eel)	1

[a]Perrott and Balment (1985) and Perrott et al. (1992) observed that ANG II induced drinking in the carp (see Table 8.1)

water where environmental osmosis is constant. Doses of ANG II used were from 0.1 to 100 μg per 100 g of body weight. Perrott and Balment (1985) and Perrott et al. (1992) observed that carp responded to ANG II by drinking, in contrast to our data in Japanese carp. Those freshwater fishes (Nos. 12, 13, 16, 17, 18, 19) which can survive in estuarine brackish water did respond to ANG II. Furthermore, some freshwater fishes which can survive in hypertonic water (Nos. 11, 15) or in seawater (Nos. 14, 20) also responded to ANG II. Drinking induced by ANG II was reported earlier by Hirano et al. (1978), Takei et al. (1979b), Hirano and Hasegawa (1984) and Perrott et al. (1992) in the eel.

Of 17 seawater fishes examined (Table 8.7), the cyclostome, *Eptatretus burgeri* (No. 1), and the elasmobranchs, *Triakis scyllia* (No. 2) and *Heterodontus japonicus* (No. 3), did not respond to ANG II by drinking. However, Hazon et al. (1989) demonstrated that intramuscular injection of ANG II induced drinking and vasopressor responses in the elasmobranch, *Scyliorhinus canicula*. The pressor activity of ANG I and II (Opdyke and Holcomb 1976) and the presence of radioimmunoreactive ANG II in plasma and other organs (Galli-Phillips 1991) have been demonstrated in elasmobranchs. The amino acid sequence of ANG I has recently been determined in *Triakis scyllia* (Takei

Table 8.7. Drinking responses of seawater fishes to a single intraperitoneal injection of ANG II

Unresponsive species	Responsive species	Minimum effective dose (μg/100 g)[a]
1. *Eptatretus burgeri* (inshore hagfish)	12. *Glossogobius giuris fasciatopunctatus* (spottyband goby)	5
2. *Triakis scyllia* (banded dogfish)	13. *Callionymus richardsoni* (Richardson's dragonet)	0.1, 10[*]
3. *Heterodontus japonicus* (bull-head shark)	14. *Hypodytes rubripinnis* (redfin velvetfish)	0.5[*], 10[*]
4. *Platichthys bicoloratus* (stone flounder)	15. *Chasmichthys gulosus* (gluttonous goby)	2
5. *Sardinops melanosticata* (Japanese pilchard)	16. *Sillago japonica* (Japanese whiting)	10
6. *Trachurus japonicus* (Japanese horse mackerel)	17. *Mugil cephalus* (black mullet)	50
7. *Acanthopagrus schlegeli* (black porgy)		
8. *Takifugu niphobles* (grass puffer)		
9. *Parapristipoma trilineatum* (chicken grunt)		
10. *Sebastes inermis* (black rockfish)		
11. *Rudarius ercodes* (Japanese file fish)		

[a]*, Inhibitory effect on drinking.

129

et al. 1993b; (Table 4.2). Unfortunately the drinking response to ANG II was not examined in these studies.

ANG II did not induce drinking in teleostean fishes (Nos. 4–11) living exclusively in sea water where environmental osmosis is usually constant. However, seawater fishes which can survive in the brackish water of estuaries or inlets (Nos. 12, 14, 15) or in tide pools (No. 13) responded to ANG II. The doses used were 1–100 μg per 100 g body weight. However, *Platichthys* (No. 4), *Acanthopagrus* (No. 7) and *Takifugu* (No. 8) failed to respond to ANG II, although they can survive in brackish water. In *Callionymus* (No. 13), 0.1 μg of ANG II stimulated, but 10 μg suppressed drinking. In *Hypodytes* (No. 14), 0.5 and 10 μg inhibited drinking. It is not known whether both these doses were too high. Although some exceptions occurred among the present experimental data, the drinking response to ANG II appeared to be characteristic of brackish water fishes which encounter hypertonic and hypotonic estuarine water every day. Accordingly, a drinking mechanism induced by ANG II could be considered a compensatory emergency reaction to dehydration stress (Kobayashi et al. 1983). Malvin et al. (1980) observed that ANG II induced drinking in the euryhaline killifish, *Fundulus heteroclitus*, caught in brackish water (Table 8.1). Carrick and Balment (1983) and Balment and Carrick (1985) obtained similar results in the euryhaline flounder, *Platichthys flesus* (Table 8.1). These observations in two species of the euryhaline fishes concur with our data.

Beasley et al. (1986) reported that three stenohaline freshwater species, the goldfish (*Carassius auratus*), common shiner (*Natropis cornutus*) and mottled sculpin (*Cottus bairdi*) did not respond to ANG II. Japanese goldfish did drink water in response to ANG II (Kobayashi et al. 1983; Fig. 8.3). The discrepancy could be due to differences between strains of goldfish (Beasley et al. 1986). Although goldfish are called stenohaline, Japanese goldfish can survive in diluted seawater. Beasley et al. (1986) did find that the longhorn sculpin and the

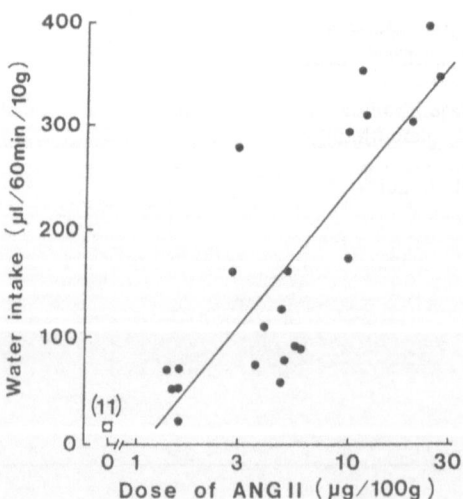

Fig. 8.3. Dose-response relationship for water intake during the 60 min following i.p. injection of ANG II (1–30 μg/10 g body wt.) in the goldfish. Number of experimental fish (•) was 22; number of control fish (□) was 11. (Okawara and Kobayashi 1988)

winter flounder, which live exclusively in seawater, responded to ANG II by drinking. Recently, Perrott et al. (1992) has shown that intramuscular injection of ANG I induced drinking in fishes (plaice, dab and whiting), generally thought to be stenohaline, placed in sea water containing ^{125}I-polyvinyl pyrrolidone. However, Kobayashi et al. (1983) reported that fishes found exclusively in sea water did not drink in response to ANG II. It is difficult to explain the difference between our results and those by Beasley et al. (1986) and Perrott et al. (1992). In drinking experiments using fish, the definition of euryhaline or stenohaline may not be important, but the native habitats, behavior and tolerance for changes in osmosis of water surrounding the fishes should be carefully considered. Further, the presence or absence of ANG II receptors in the brains of fishes with different habitats and behavior should be examined.

It is generally thought that freshwater fish drink little or no water, and that seawater fish drink copiously. In our drinking experiments (Kobayashi et al. 1983), we found that most freshwater fishes drank less than 20 μl (1–19 μl)/10 g/h and most seawater fishes drank more than 20 μl (20–127 μl). However, some freshwater species (Nos. 2, 3, 7, 11, 14, 15, 16; Table 8.6) drank as much water (21.1–63.8 μl) as seawater fishes, and some seawater fishes (Nos. 1–4, 6, 12; Table 8.7) drank as little (0.2–7.5 μl) as freshwater fishes. The physiological significance of these exceptions deserves attention particularly in respect to microhabitats and body structures such as bladders.

8.2.6 Summary

Most of the mammals responded to intraperitoneal injection of ANG II by drinking, but species from arid habitat origins did not. Marsupials and herbivores seemed relatively insensitive to ANG II. However, investigations on more species are needed. Most birds responded to ANG II, but carnivorous birds and those from dry areas were insensitive to ANG II. In reptiles, most species responded to ANG II, though some did not, perhaps due to their special habits or habitats. Amphibians did not respond to ANG II by drinking, suggesting that amphibians do not possess the angiotensin-drinking mechanism. Therefore, it was concluded that this mechanism first arose in reptiles during their evolutionary process. However amphibians respond to ANG II by water absorption behavior and by absorption of water through the ventral skin.

ANG II did not induce drinking in teleosts living exclusively in fresh water or sea water where environmental osmosis is usually constant. However, ANG II did induce drinking in brackish water fishes and freshwater fishes which can survive in seawater or hypertonic water. These findings suggest that in these fishes, endogenous ANG II may regulate drinking depending upon osmotic changes in the surrounding water. Thus, the drinking response to exogenous ANG II in teleosts seems to depend on the habitats and habits of the animal species. The RAS appeared to be present in the shark. Exogenous ANG II induced drinking in one species, though not in two other species of shark. Further studies are necessary to determine whether ANG II participates in water ingestion through the mouth in elasmobranchs. In cyclostomes, the lamprey and the hagfish did not respond to injection of ANG II by drinking.

During the course of studies on ANG II-induced drinking, we found that some freshwater fishes drank as much water as seawater fishes and some seawater fishes drank as little water as most freshwater fishes.

8.3 Endogenous ANG II and Drinking

It is generally accepted that the sense of thirst is generated in the central nervous system, which receives information primarily through osmoreceptors, baroreceptors and volume receptors. In the process causing thirst some humoral factors, such as hormones and/or biologically active peptides, are involved (see Uemura et al. 1983: Szczepanska-Sadowska 1991 for reviews). Among these factors, ANG II has been shown to be a strong dipsogen when injected into animals. As discussed previously, exogenous administration of ANG II induced drinking in a wide variety of animal species. However, questions exist about the role of endogenous ANG II in physiological thirst (Fitzsimons 1979, pp. 276–282; Rolls and Rolls 1982, pp. 52–55; Johnson et al. 1986, pp. 161–180).

8.3.1 Plasma ANG II and Drinking

Fitzsimons et al. (1978b) showed that ANG II infused at a rate calculated to produce increases in arterial concentration within the physiological range, induced drinking in a water-replete dog. Trippodo et al. (1976) and Hsiao et al. (1977) observed stimulation of drinking with levels of ANG II that could occur physiologically in the dog and rat. With the development of a radioimmunoassay technique for ANG II, the relationship between plasma ANG II concentrations and drinking has become more understandable. Semple (1980) showed that the plasma concentration of ANG II was around 30 fmol/ml and increased to 220 fmol/ml 30 min after hemorrhage; deprivation of dietary sodium for 10 days caused an increase of ANG II from 25 to 136 fmol/ml in rats. Unfortunately the author did not measure water intake. Mann et al. (1980) determined dipsogenic threshold values by measuring plasma ANG II concentrations 15 and 60 min after infusing different doses of ANG II (0, 1, 25, 50 and 100 pmol/kg/min)in nephrectomized (unanesthetized) rats (see Johnson et al. 1986 for review). Control nephrectomized rats had around 30 fmol/ml concentrations 15 and 60 min after saline infusion. Infusion of ANG II at 25 pmol/kg/min produced approximately 200 fmol/ml 15 and 60 min following infusion. Infusion of 50 and 100 pmol/kg/min elevated plasma levels to around

380 and 800 fmol/ml, respectively, 60 min after infusion. Conscious rats with intact kidneys receiving 25 and 100 pmol/kg/min showed almost the same plasma levels after 60 min. Since Hsiao et al. (1977) reliably induced drinking in rats with infusion of 25 pmol/kg/min of Hypertensin, Mann et al. (1980) concluded that the intravenous dipsogenic threshold of ANG II was approximately 200 fmol/ml plasma.

Mann et al. (1980) and Johnson et al. (1981) measured plasma ANG II concentrations in normal rats and in rats treated with various manipulations (see Jounson et al. 1986 for review). They showed that ANG II concentrations in plasma ranged from 50 to 100 fmol/ml in control rats. Dipsogenic treatments by water deprivation, caval ligation, isoproterenol and polyethylene glycol administrations caused elevations of plasma ANG II concentration beyond the dipsogenic threshold. For instance, ANG II plasma levels rose to approximately 400 (48 h water deprivation), 900 (1 h caval ligation) and 1200 fmol/ml (isoproterenol 300 µg/kg).

Di Nicolantonio and Mendelsohn (1986) reported that rats dehydrated for 48 h had significantly elevated renin (from 1.5 to 4.5 ng/ml) and plasma ANG II (from 50 to 310 fmol/ml) concentrations. Copious drinking was observed during the first hour after water was returned. Although drinking rates of the dehydrated rats returned to normal by 2 to 4 h after rehydration, the levels of plasma renin and ANG II exhibited a further increase after rehydration, and remained significantly above dehydration levels (renin: 7 to 11 ng/ml, ANG II: 450 to 500 fmol/ml) for 2 to 4 h after rehydration. The levels then decreased gradually, but remained above the dipsogenic threshold levels for circulating ANG II for 8 h after rehydration. The investigators concluded that termination of drinking by 2 to 4 h after rehydration was caused by mechanisms, such as distention of the stomach or hypoosmotic plasma, etc., which overrode the dipsogenic action of circulating ANG II.

To determine the plasma ANG II levels at the moment of drinking in rats, van Eekelen et al. (1988) infused intravenously 10 to 200 pmol/kg/min of ANG II in rats; they observed that 25, 50 and 75 pmol/kg/min induced variable water intake after a relatively long latency following infusion (45–65 min). The control levels were between 30 and 120 fmol/ml. The average ANG II level in blood samples taken at the moment of drinking was 458 ± 58.1 fmol/ml. This value was comparable to plasma ANG II levels of approximately 323 fmol/ml observed by Yamaguchi (1981) after 46 h of water deprivation, and approximately 400 fmol/ml reported by Mann et al. (1980) after 48 h of water deprivation. Endogenous ANG II increase may be one of several mechanisms required to bring body fluid volume back into balance in emergency situations, such as dehydration where ANG II levels reach the dipsogenic threshold. The idea that ANG II-induced drinking is an emergency reaction has been proposed by Fitzsimons (1979, p. 380) and by Kobayashi et al. (1980, 1983; Kobayashi and Takei 1982).

In normal Japanese quail, plasma ANG II concentrations were between 20 and 50 fmol/ml. After dehydration for 4, 10, 25 and 48 h, plasma ANG II concentrations increased from 50 to 70, 90, 170 and 370 fmol/ml, respectively.

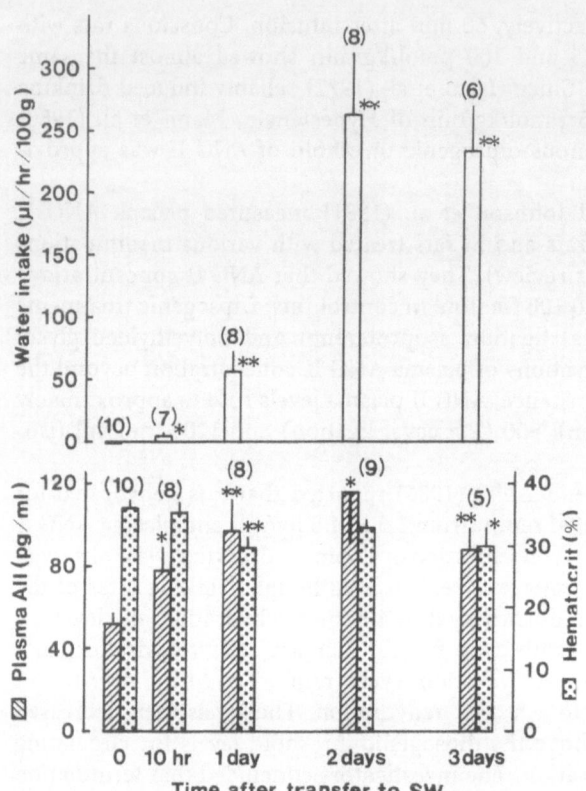

Fig. 8.4. Changes in drinking rate, plasma ANG II and hematocrit following transfer of freshwater eels to seawater. The amount of water ingested was measured for the last hour of 10 h, 1, 2 and 3 days after immersion in seawater. Each *column* represents the mean with SE. Number of eels is indicated in *parentheses.* ∗ P<0.05; ∗∗ P<0.01 compared with control (0 h). (Okawara et al. 1987)

Marked drinking occurred 4 h after dehydration, indicating that the dipsogenic threshold was about 70 fmol/ml and that the quail was sensitive to changes in plasma ANG II concentration. Fifteen minutes after hemorrhage of 1.5 ml blood, the birds showed about 130 fmol/ml levels; vigorous drinking started 30 min after hemorrhage (Kobayashi and Takei 1982; Takei et al. 1988a).

Plasma ANG II concentrations were approximately 68 fmol/ml in Pekin ducks adapted to freshwater (FW ducks) and approximately 260 fmol/ml in ducks adapted to 2% NaCl solution for drinking (SW ducks; Gray and Simon 1987). After 48-h dehydration, plasma ANG II increased to approximately 250 fmol/ml in FW ducks. Plasma ANG II did not decrease within 150 min on rehydration in FW ducks or on hydration in SW ducks. In the ostrich (*Struthio camelus*), water deprivation for 5 days elevated plasma arginine vasotocin (AVT) from 10.2 to 32.3 pg/ml and ANG II from 44.3 to 143.1 pg/ml (fmol/ml). Rehydration by drinking reduced plasma AVT to basal levels after 2.5 h, but plasma ANG II remained elevated (Gray et al. 1988). As mentioned above, this phenomenon was also seen in rats (Di Nicolantonio and Mendelsohn 1986). In the Japanese quail, birds dehydrated for 48 h drank vigorously for about 1.5 h after rehydration and then drinking became normal. Plasma ANG II con-

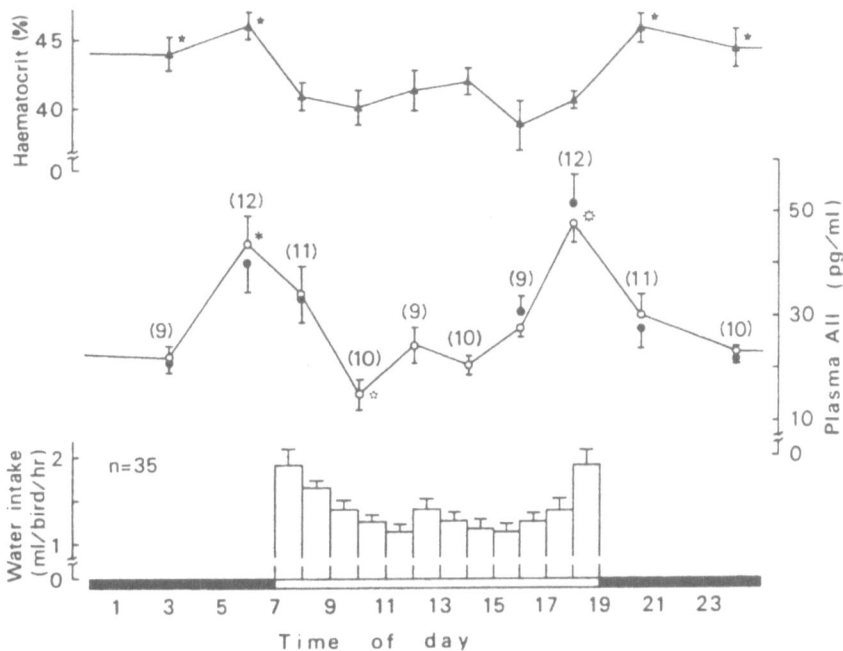

Fig. 8.5. Changes in hematocrit (▲), plasma AII absolute value (o) and values corrected by hematocrit (●), and hourly water intake in the Japanese quail. The *black abscissa bar* shows a dark period, and the *white bar* a light period. Each value is the mean with SE. The number of birds used for estimation of plasma AII and hematocrit appears in *parentheses*. Hematocrit: ★ , $P<0.05$, compared with the mean of all other values obtained for a light period. Plasma AII: * , $P<0.05$, compared with values found at 03:00, 10:00, 14:00, and 20:30 h. ✪, $P<0.05$ compared with all other values. ✿, $P<0.05$, compared with all other values except that at 06:00 h. Water intake: two peaks appeared from 07:00 to 08:00 and from 18:00 to 19:00 h and no significant difference in drinking rate was noted from 08:00 to 18:00 h. (Okawara et al. 1985)

centrations decreased from 450 fmol/ml to 280 and 200 fmol/ml 40 and 70 min after rehydration, respectively. These values were still above the dipsogenic threshold. Distention of the crop sac might have depressed vigorous drinking (Takei et al. 1988a).

When the Japanese eel, *Anguilla japonica*, was transferred from fresh water to sea water, water intake increased significantly at 10 h following transfer and further increased into day 2. Plasma ANG II gradually increased from 53 fmol/ml to 77.6, 96.9, and 116.6 fmol/ml at 10 h, day 1 and day 3, respectively, following the transfer (Okawara et al. 1987; Fig. 8.4). In this study copious drinking immediately after transfer from FW to SW was not observed. Hirano (1974) observed copious drinking immediately after transfer of FW eels to SW using esophagus-cannulated eels. The reason for the discrepancy between the two experiments is not clear.

It seems, therefore, that endogenous plasma ANG II contributes to drinking induced by dipsogenic treatments in the rat, Japanese quail, and Japanese eel. Since the RAS seems to function as an emergency reaction to stressful conditions such as dehydration and hemorrhage, we have concluded that ANG II may work as a trigger of various phenomena induced by such stress. ANG II stimulated secretion of arginine vasotocin, corticosterone and aldosterone, just as water deprivation stimulated secretion of these hormones in the Japanese quail (Kobayashi and Takei 1982).

The Japanese quail had a diurnal rhythm of water intake: one peak 1 h after the light was on (06:00) and another peak 1 h before the light was off (18:00) under 12L:12D. Plasma ANG II concentration peaked twice also, 1 h before each of the peaks in water intake (Fig. 8.5) (Okawara et al. 1985). The quail seems to anticipate the times when the light is to be turned on and off. The concentrations of plasma ANG II were approximately 50 fmol/ml during both peaks, higher by approximately 30 fmol/ml than during other times of the day. These observations indicate that endogenous ANG II functions as a tool regulating daily physiological phenomena such as natural drinking. Kraly and Corneilson (1990) reported that endogenous ANG II mediates drinking elicited by eating in the rat.

8.3.2 Drinking and ANG II Antagonists, Renin Inhibitors and Converting Enzyme Inhibitors

In earlier studies to assess the involvement of the renin–angiotensin system in thirst, the kidney was removed; however, nephrectomy induced pathological conditions in the experimental animals (Johnson et al. 1986). Another assessment method was the administration of ANG II competitive antagonists, renin inhibitors or converting enzyme inhibitors. Vaughan et al. (1973) were the first to show that saralasin [Sar[1], Val[5], Ala[8]] ANG II (P-113), a competitive inhibitor of ANG II, inhibited drinking induced by intraperitoneal injection of crude renal extracts in nephrectomized rats. Several investigators observed that intravenous or intracranial injections of saralasin inhibited ANG II- or renin-induced drinking in mammals (Summy-Long and Severs 1974; Cooling and Day 1975; Fitzsimons et al. 1978a; see Fitzsimons 1979). Spontaneous drinking by Japanese quail was reduced after a single injection of saralasin (100 μg/bird; Kobayashi and Takei 1982). These observations indicate that saralasin inhibits the dipsogenic action of exogenous and endogenous ANG II. deCaro et al. (1982) observed that ANG II analogs, [Sar[1], Ile[8]]-, [Sar[1], Leu[8]]-, [Sar[1], Ala[8]]- and [Sar[1], Gly[8]]-ANG II, inhibited ANG II-induced drinking in the pigeon, but did not significantly affect drinking induced by eledoisin and bombesin. They suggested that the dipsogenic response elicited by these peptides was mediated by the activation of central receptors different from those of ANG II-induced drinking.

Saralasin, however, did not affect drinking induced by water deprivation in mammals (Ramsay and Reid 1975; Abraham et al. 1976; Phillips and Hoffman 1977; Lee et al. 1981). The dose might not have been sufficient to block the increase in ANG II induced by water deprivation. It was reported that the renal RAS has the capacity to overcome a functional blockade by low doses of antagonists (Johnson et al. 1986). Further, since water deprivation includes both cellular and extracellular depletion, the ANG II antagonist would not be expected to totally abolish drinking (Johnson et al. 1986). Thirst produced by injection of hypertonic saline was not inhibited by saralasin (Summy-Long and Severs 1974). It should be noted that saralasin has an agonistic effect when used at higher doses (Mann et al. 1978).

Recently, ANG II receptor subtypes (AT_1 and AT_2) have been identified in mammals; brain structures contain many AT_1 receptors and few AT_2 receptors. Several AT_1 and AT_2 antagonists, which are structurally dissimilar to angiotensin antagonists, have been synthesized. In rats, the selective AT_1 receptor antagonist Dup753 (losartan) (Fig. 7.1) inhibited drinking induced by subcutaneous injection of ANG II (Fregly and Rowland 1991; Dourish et al. 1992), but the selective AT_2 receptor antagonist WL19 did not (Dourish et al. 1992). Similar findings were reported by Beresford and Fitzsimons (1992) in rats. Kirby et al. (1992) demonstrated that Dup753 inhibited increases in water intake and blood pressure induced by centrally administered ANG II and III. The above observations suggest that ANG II and III act on water intake and blood pressure via the AT_1 receptor subtypes. However, Hogarty et al. (1992), using Dup753 and PD123177 (AT_2 receptor antagonist, formerly EXP655, a close structural analog of WL19; Fig. 7.1) concluded that, in rats, the pressor response to ANG II given centrally is predominantly AT_1 mediated, while the drinking and AVP responses may be mediated by both receptor subtypes. Dourish et al. (1992) observed that neither Dup753 nor WL19 affected drinking in either water-repleted rats or rats subjected to various cellular and extracellular thirst challenges. They suggested that endogenous ANG II plays little or no part in normal drinking that occurs under physiological conditions. Further experiments are necessary to determine the involvement of AT_1 and AT_2 in ANG II-induced drinking.

Pepstatin, a pentapeptide isolated from *Streptomycetes* and an inhibitor of renin, inhibited drinking induced by renin and renin substrate, but did not disturb drinking induced by ANG I or ANG II (Fitzsimons et al. 1978a). Renin antagonists, such as pepstatin, have not been widely used, because of a lack of general availability of such drugs and the relatively low solubility of these inhibitors in biological fluid (Johnson et al. 1986).

In rats, SQ20881 (Pyr-Trp-Pro-Arg-Pro-Gln-Ile-Pro-Pro), a nonapeptide inhibitor of converting enzyme from venom of the snake, *Bothrops jararaca*, possessed an inhibitory effect on drinking induced by renin, renin substrate and ANG I, but enhanced ANG II-induced drinking (Fitzsimons et al. 1978a). Severs et al. (1973) found inhibition of ANG I-induced drinking by SQ20881. However, several investigators reported no inhibition by SQ20881 on ANG I-induced drinking (See Fitzsimons 1979, p 315). The discrepancy seems to be due to the difference in ratios of antagonist to ANG I. The effective blocking ratio was 200 : 1 (Severs et al. 1973). Malvin et al. (1980) observed a partial blockade of drinking after i.p. administration of SQ20881 in seawater-adapted euryhaline killifish, *Fundulus heteroclitus*. In rats, intraventricularly injected SQ20881 inhibited drinking induced by intraventricular ANG I, but peripherally injected SQ20881 increased drinking (Lehr et al. 1973). These authors suggested that the plasma level of ANG I became very high due to peripherally injected SQ20881, and that this ANG I was enzymatically converted locally in the brain into ANG II, resulting in enhancement of drinking. Severs et al. (1973) showed that SQ20881 does not cross the blood–brain barrier. It is interesting that intracranial injection of SQ20881 augmented drinking induced by ANG II in the rat and pigeon (Evered and Fitzsimons 1976; Fitzsimons et al. 1978a).

Captopril (SQ14225, D-3-mercapto-2-methylpropanoyl-L-proline), another more potent converting enzyme inhibitor, attenuated drinking induced by intracerebroventricular ANG I (Evered et al. 1980). Thirst following poly-ethylene glycol treatment, which decreased the extracellular fluid, was inhibited by captopril at 100 mg/kg in rats (Mann et al. 1986). Water intake stimulated by water deprivation was inhibited by captopril in dogs (H. Suzuki et al. 1987), and it suppressed water consumption in genetic polydipsic STR/N mice (Silverstein and Friedland 1990). Captopril inhibited natural drinking in the Japanese quail (Okawara et al. 1985; Kasuya et al. 1985; Figs. 8.6, 8.7). Plasma ANG II was reduced in parallel with the decrease in water intake for 60 min

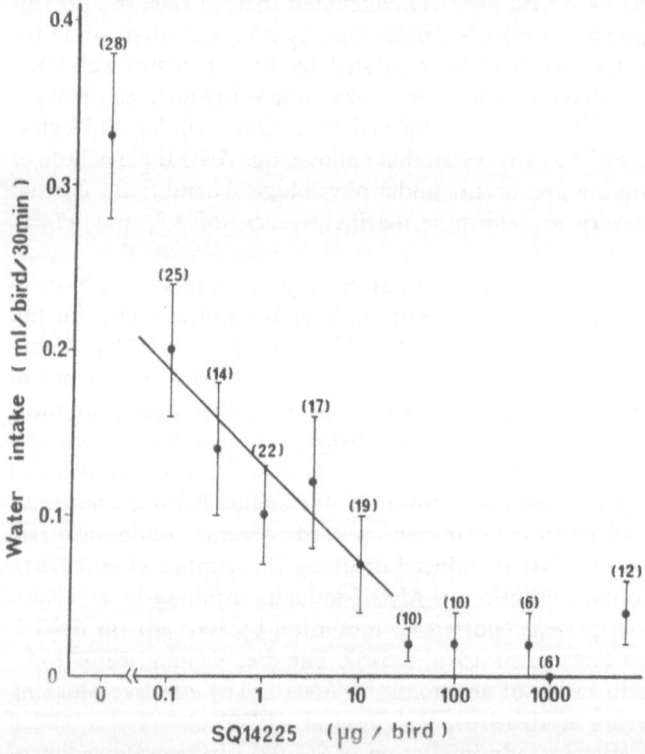

Fig. 8.6. Dose-response relationship for water intake during the 30 min following i.p. injection of captopril in the Japanese quail. Number of birds is indicated in *parentheses*.

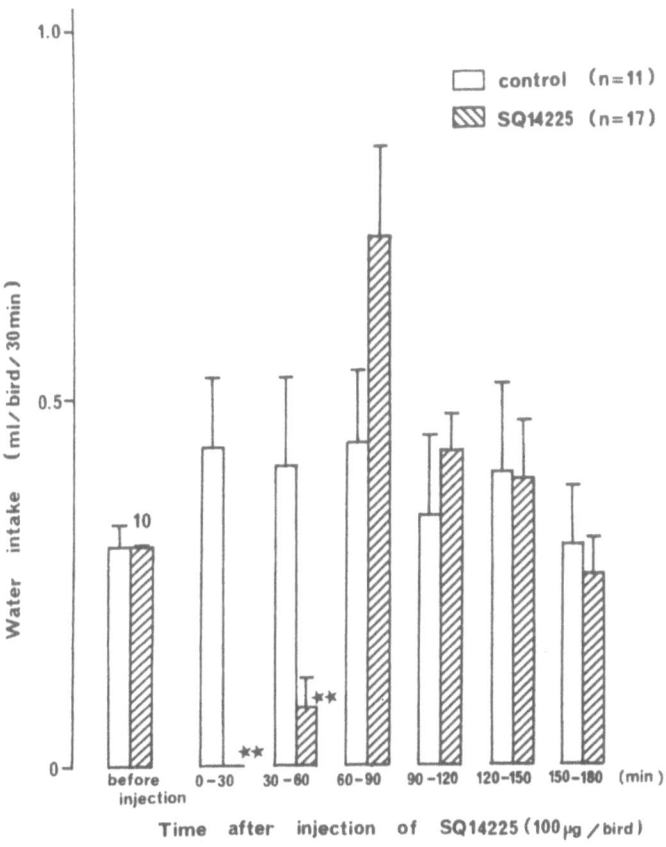

Fig. 8.7. Inhibitory effect of i.p. injection of captopril (100 µg/bird) on water intake in the Japanese quail. The effect lasted for 60 min, after which copious drinking was observed. ★★, $P<0.01$, compared with controls

following the captopril injection (Fig. 8.8). In Japanese eel adapted to seawater for 10 days, water intake was about 57.3 ± 14.3 µl/100 g ($n=7$) for 30 min, but that of eels injected with captopril (1 mg/fish) was 1.7 ± 1.4 µl ($n=8$; Okawara et al. 1987). These observations suggest that endogenous ANG II is involved in drinking induced by ANG I or dipsogenic treatments and that, at least in Japanese quail and eels, endogenous ANG II is associated with natural drinking.

In fish, however, the participation of plasma ANG II in physiological drinking mechanisms seems to differ among species. In goldfish, weighing about 4 g, intraperitoneal injection of captopril (40 and 200 µg/fish) did not reduce water intake; they maintained the basal drinking rate of control fish. On the contrary, 0.04 and 4 µg/fish of captopril stimulated drinking (Okawara and Kobayashi 1988). As hypothesized in rats (Evered and Robinson 1984), lower doses of captopril may block converting enzyme only in the circulation and

139

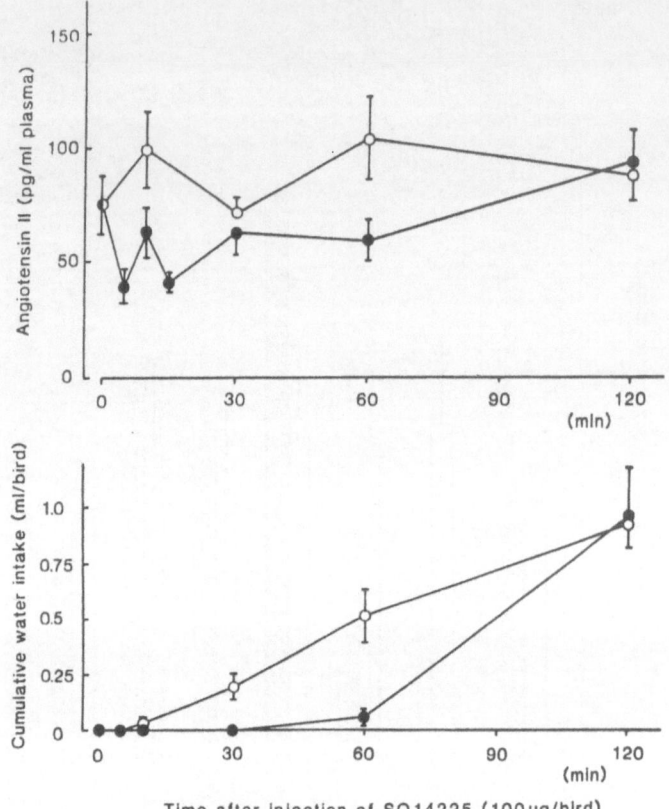

Fig. 8.8. Effects of i.p. injection of SQ14225 (100 µg/bird) on plasma ANG II concentration and water intake in the Japanese quail. ○, Control birds; ●, experimental birds, Each point in both graphs is the mean with SE of 10 birds, except at 30 and 120 min in ANG II measurements where, respectively, 8 and 9 birds were used

accumulated ANG I could enter the brain and be converted to ANG II. However, higher doses might inhibit the enzyme in both circulation and the brain, resulting in insufficient ANG II in the brain to induce drinking. Beasley et al. (1986) found that captopril did not inhibit basal drinking in the marine winter flounder, *Pseudopleuronectes americanus*, and the longhorn sculpin, *Myoxocephalus octodecemspinosus*. Carrick and Balment (1983) reported partial inhibition of the basal drinking rate by captopril in the SW-adapted flounder, *Platichthys flesus*. Both experiments suggested that there was a component involved in drinking that was independent of the RAS. Recently, Perrott et al. (1992) observed that ANG I stimulated drinking, but captopril greatly reduced its stimulatory action in the carp and suggested that ANG II formation was important in the production of the dipsogenic response. In the SW eel, captopril inhibited spontaneous drinking, but not in the FW eel. In whiting held in SW, captopril reduced the normal drinking rate. Further studies using more species of fish living in different habitats are needed to provide a more general understanding of this topic.

As did SQ20881, captopril had agonistic action in some cases. Mann et al. (1986) demonstrated that 100 mg/kg of captopril inhibited drinking induced by polyethylene glycol, but 1 mg/kg enhanced drinking in the rat. Stimulation of water intake by captopril has also been observed in the rat (Schiffrin and Genest 1982). They concluded that ANG I pooled by captopril in the brain was converted into ANG II, resulting in stimulation of drinking. Enhancement of water intake by captopril (0.4 and 4µg/fish) in goldfish was mentioned above (Okawara and Kobayashi 1988). Captopril has no effect on drinking induced by hypertonic saline treatment in rats (Johnson et al. 1986), results similar to those obtained with saralasin (Summy-Long and Severs 1974).

8.3.3 Brain ANG and Drinking

The presence of renin-like activity, angiotensinogen and angiotensin in the brain was first reported by Fischer-Ferraro et al. (1971) in the rat and dog and Ganten et al. (1971) in the dog. Subsequently, all components necessary for an RAS have been found in the brain as in the case of the tissue RAS (See Printz et al. 1982; Unger et al. 1986; Ganten et al. 1988 for reviews). Angiotensin was also found in cerebrospinal fluid (CSF) of humans, sheep, dog and rat (for review, see Schelling et al. 1977, 1980). Bunnemann et al. (1991) have reviewed their own and others' recent studies on the distribution of components of the brain RAS. Angiotensinogen was produced in glia and neurons in some regions. Angiotensinogen may be released from glia in the extracellular space and cleaved to ANG I by renin and then to ANG II by ANG II converting enzyme, or directly to ANG II by cathepsin G or tonin (Demassieux et al. 1976; Fig. 1.2). Thus, to initiate drinking brain ANG II must reach a receptive site such as the median preoptic nucleus through extracellular fluid pathways (Oldfield 1991; Fig. 8.12; see Sect. 8.5). ANG III has strong dipsogenic activity in the brain. In addition to ANG II and ANG III, ANG II fragments such as ANG-(4–8), ANG-(3–8), ANG-(5–8) and ANG-(6–8) have weak dipsogenic functions (see Wright and Harding 1988 for review; Fig. 8.9). It is interesting that ANG-(1–7), which is distributed in the hypothalamus and medulla oblongata, is equipotent with ANG II in its activation of arginine vasopressin release. However, ANG-(1–7) induces neither centrally mediated dipsogenesis nor peripheral vasoconstriction (Schiavone et al. 1990).

ANG II concentrations do not differ between plasma (29.3 ± 2.7 pg/ml) and CSF (30.8 ± 2.8 pg/ml) in euhydrated conscious dogs. Twenty-four h of dehydration with sodium-rich food significantly increased the ANG II concentration in plasma (59.8 ± 16.5 pg/ml) and CSF (71.8 ± 20.1 pg/ml). Following rehydration by drinking, there was no change in plasma ANG II within 90 min, but a decrease occurred in CSF ANG II (Simon-Opperman et al. 1986). Thus, the concentration of CSF ANG II changed independently from that

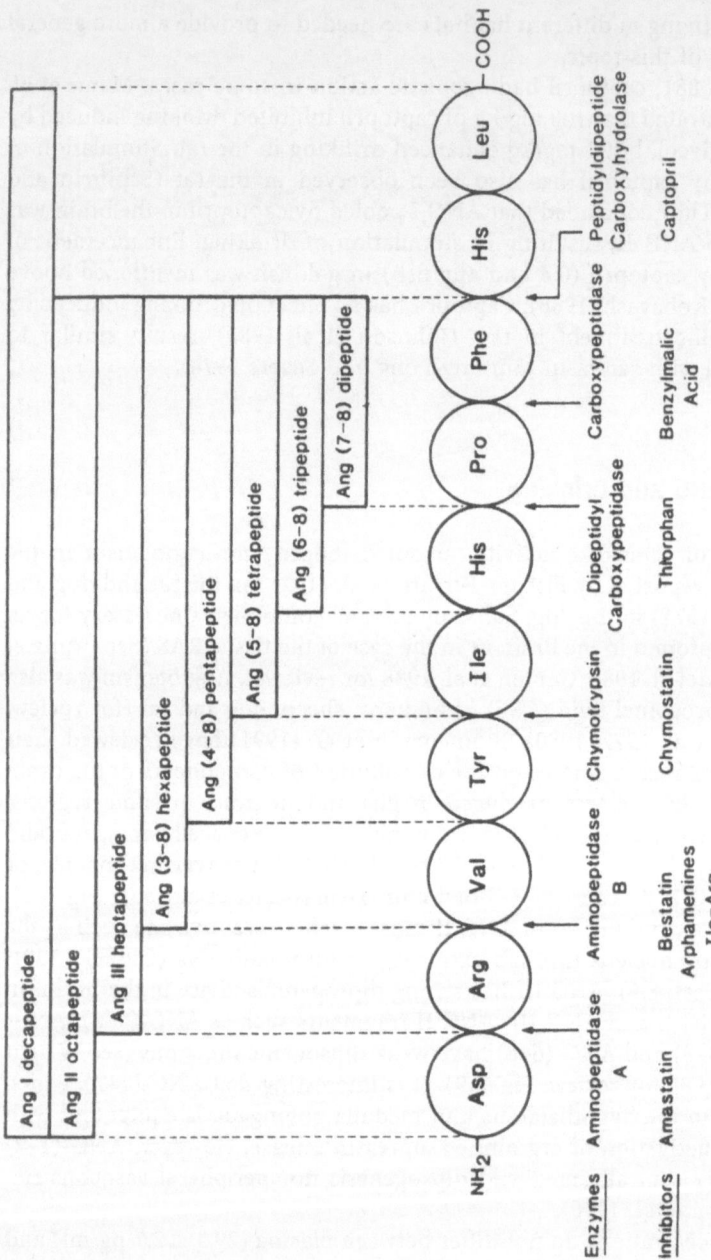

Fig. 8.9. Amino acid composition of the angiotensins in the majority of mammalian species (see Fig. 1.2 and Table 4.2); the specific aminopeptidase enzymes, and their presumed sites of action as indicated by the *arrows*; and the corresponding aminopeptidase inhibitors. (Wright and Harding 1988)

of systemic ANG II. These data suggest that central endogenous ANG II may function as a central osmoregulatory mediator independently of systemic ANG II. Similar experiments were carried out using Pekin ducks which were adapted to either fresh water (FW ducks) or salt water (2% NaCl, SW ducks) for drinking (Gray and Simon 1987). Arginine vasotocin (AVT) and ANG II were measured in CSF and plasma collected simultaneously. In FW ducks, AVT in CSF was 54.4 ± 6.3 pg/ml and in plasma 5.8 ± 0.7 pg/ml; ANG II in CSF was 25.2 ± 4.9 pg/ml and in plasma 67.7 ± 28.7 pg/ml. In SW ducks, AVT increased approximately threefold and ANG II fourfold in both CSF and plasma. Dehydration in FW ducks (24-48 h) caused increases of AVT and ANG II in both CSF and plasma, the relative rise being greater in plasma. Within 150 min after rehydration, plasma AVT decreased but CSF AVT was unchanged, whereas CSF ANG II decreased but plasma ANG II was unchanged. Hydration of SW ducks with freshwater had similar effects. These data indicate separate avenues for release of central and systemic AVT and ANG II, and support the idea of an independent role of central ANG II as a mediator in osmoregulation.

The possibility that ANG III is an active ligand in the central renin–angiotensin system has been discussed. Brain binding sites for ANG III are widely distributed in several mammalian species examined. They are found in the SFO, AV3V and the olfactory bulb. ANG II and ANG III seem to act at a common receptor site in the brain. The equivalent or greater potency of ANG III compared to ANG II in the brain has been reported with respect to drinking and pressor activity (see Wright and Harding 1988). Microiontophoretic application of ANG III induced higher firing rates of SFO neurons in the cat (Felix and Schlegel 1978) and PVN neurons in the rat (Harding and Felix 1987), than did similar application of ANG II. It is not fully understood why peripherally injected ANG III has no dipsogenic action or less than ANG II in dogs and rats (Fitzsimons 1979), pigeons (Evered and Fitzsimons 1981) or budgerigars (Kasuya et al. 1985; Fig. 8.10).

8.3.4 Summary

Plasma ANG II concentration increased following dipsogenic treatments such as dehydration, hemorrhage or polyethylene glycol injection. Thus, endogenous ANG II seems to have a role in drinking under these stressful emergency situations. Investigations using competitive inhibitors of ANG II or converting enzyme inhibitors indicated that, in most cases, they inhibit drinking induced by administration of ANG II, ANG I or other dipsogenic treatments. These observations give further support to the idea that endogenous ANG II has a role in drinking caused by dipsogenic treatments. However, in some cases the inhibitors do not inhibit and even enhance drinking at higher doses. Therefore, it is difficult to draw a definite conclusion from these investigations whether endogenous ANG II is involved in spontaneous natural drinking. However, in the Japanese quail, saralasin inhibited spontaneous drinking and plasma ANG

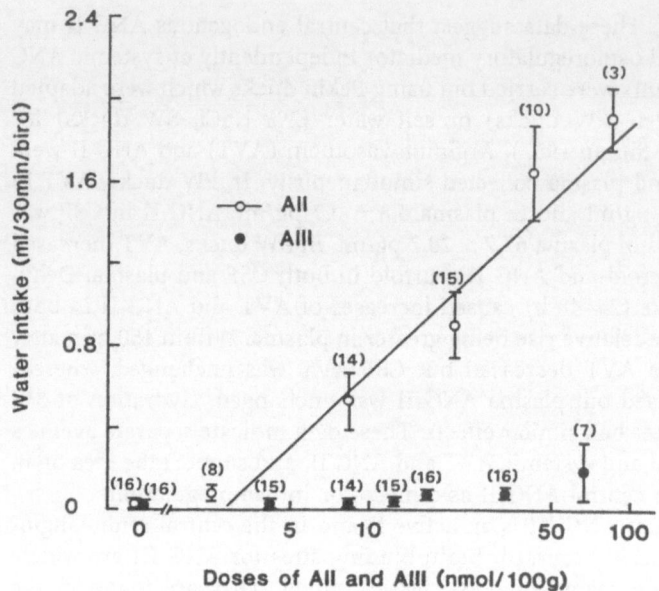

Fig. 8.10. Dose-response relationship for water intake during the 30 min following i.p. injection of ANG II or ANG III in the budgerigar, *Melopsittacus undulatus*. ANG III had no effect when injected i.p. (Kasuya et al. 1985)

II concentrations changed in parallel with the daily drinking pattern. Saralasin and captopril changed drinking rates of fishes. Thus, endogenous ANG II seems to be involved in spontaneous natural drinking. The brain RAS appears to be involved in the ongoing regulation of body fluid homeostasis. The two active ligands, ANG II and III, seem to act at a common brain receptor site.

8.4 Sites of Action of ANG II in Drinking

8.4.1 Subfornical Organ (SFO)

Intravenous infusion of ANG II (Fitzsimons 1966; Fitzsimons and Simons 1968, 1969) and intracranial injections of ANG II (Epstein et al. 1969; Severs et al. 1970) induced drinking in the rat. Epstein et al. (1970) demonstrated that the regions of the brain that were most sensitive to ANG II included the anterior hypothalamus, the septum and especially the preoptic region. These findings were confirmed later by Johnson and Epstein (1975) with the exception of the septal area which was not found to be sensitive to ANG II. Earlier, Booth (1968) had noted, while studying the effect of intracranial injection of noradrenaline on feeding behavior, that injections of large doses of ANG II into the hypothalamus caused rats to drink copiously.

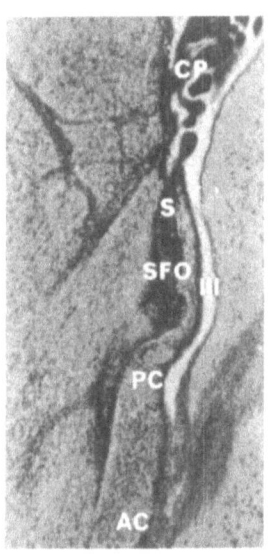

Fig. 8.11. Sagittal section of the subfornical organ (*SFO*) facing the third ventricle (*III*) in the Japanese quail. *AC* Anterior commissure; *CP* choroid plexus; *PC* pallial commissure; *S* sinus. × 84. (Takei et al. 1979a)

Blood-borne ANG II does not normally cross the blood–brain barrier (Volicer and Loew 1971; Schrager et al. 1975). Fitzsimons (1971) suggested that the subfornical organ (SFO), which lies outside the blood–brain barrier (Rohr 1966; Koella and Sutin 1967; Schrager et al. 1975), could serve as the ANG-sensitive receptor for drinking (Figs. 8.11, 8.12). Simpson and Routtenberg (1973) showed that the SFO was sensitive to direct injection of small amounts of ANG II in rats. Later, the same results were obtained by several other investigators, and confirmed also in the bird, *Coturnix coturnix japonica* (Takei 1977b). SFO lesions inhibited the dipsogenic effect of ANG II injected peripherally in the rat (Simpson and Routtenberg 1975), Japanese quail (Takei 1977b), opossum (Findlay et al. 1980), dog (Thrasher et al. 1982) and pigeon

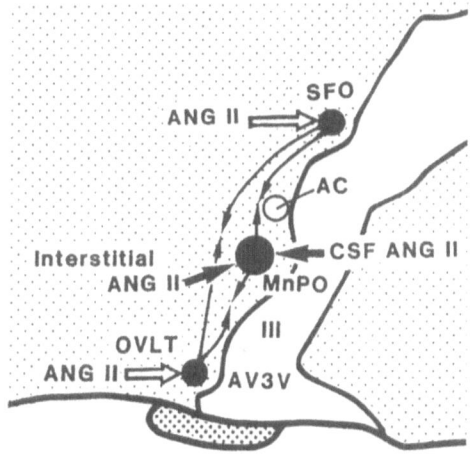

Fig. 8.12. Diagram showing the functional relationship between *SFO, MnPO* and *OVLT* in relation to ANG II-induced drinking, depicted according to the hypothesis of Oldfield (1991). The SFO and OVLT perceive circulating ANG II ($\Rightarrow$) and the MnPO receives information from CSF-borne ANG II ($\leftarrow$) and interstitial ANG II in the brain ($\rightarrow$). Reciprocal projections among the SFO, OVLT and MnPO are denoted by *double arrows* ($\rightarrow\leftarrow$). For explanation, see text. Nerve connections between these three regions and other areas are not depicted

(Massi et al. 1986). In contrast, lesions of the SFO in the sheep did not significantly reduce drinking in response to intracarotid, intravenous or intracerebroventricular infusion of supraphysiological doses of ANG II (McKinley et al. 1986). As mentioned in Section 8.2.1, sheep are herbivorous and do not drink much water in nature (Table 8.1); an angiotensin-thirst mechanism as part of their water-acquiring behavior may have become attenuated during their evolution. It seems possible that the SFO may not be important in sheep thirst responses, and the supraphysiological dosis of ANG II may have stimulated some regions other than the SFO in the brain.

Other data supported the SFO as a site of action of ANG II. Infusion of saralasin into the SFO abolished drinking in response to intravenous ANG II (Simpson et al. 1975). ANG II-sensitive cells were identified electrophysiologically in the cat SFO (Felix and Akert 1974; Phillips and Felix 1976; Felix and Schlegel 1978; Felix et al. 1986), and ANG II-induced excitation was specifically blocked by saralasin. In contrast, ACh-stimulated neurons were not affected by saralasin (Felix et al. 1986). There is no blood–brain barrier in the region of the SFO (Rohr 1966; Koella and Sutin 1967; Schrager et al. 1975). The SFO possesses receptors with high affinity to ANG II (see Chap. 7). It appears certain that the SFO is a receptive site for ANG II in drinking.

However, to interpret the results obtained from lesion experiments on the SFO, the afferent and efferent nerves of the SFO must be taken into consideration. It was reported that intact efferent nerves from the SFO are necessary to allow ANG II-induced drinking. Knife cuts interrupting efferent SFO pathways either just below the SFO, or further ventrally as they enter the median preoptic nucleus (MnPO) of the anteroventral region of the third ventricle (AV3V; Fig. 8.12), eliminated drinking (Lind and Johnson 1982a, b). These findings indicate that the SFO is a receptive site for blood-borne ANG II and the information regarding ANG II is transferred to AV3V.

8.4.2 Preoptic Area

Drinking after intraventricular administration of ANG II was not completely attenuated by lesions of the SFO in rats (Hoffman and Phillips 1976a), opossums (Findlay et al. 1980) and sheep (McKinley et al. 1986). These findings suggested the presence of other site(s) sensitive to ANG II in drinking. The preoptic area (POA) was initially reported to be a site sensitive to intracranial ANG II (Epstein et al. 1970). However, Johnson and Epstein (1975) thought that ANG II injected into the POA diffused into the ventricular fluid and stimulated the SFO. Assaf and Mogenson (1976) observed drinking after they injected ANG II into the rat POA, with no leakage of ANG II into the ventricle. Fitzsimons and Kucharczyk (1978) reported drinking after injection of ANG II into the POA in the dog. The dose–response curves for ANG II injected into the POA and the SFO were similar. Further, pretreatment of the SFO with saralasin (500 ng – 1.0 µg) did not inhibit a drinking response to ANG II injected into the POA (Mogenson et al. 1977). Kucharczyk et al. (1976) showed that lesions of the midlateral hypothalamus and paramedial rostral midbrain interrupted drinking induced by ANG II administered into the POA, but not drinking induced by ANG II injected into the SFO. They concluded that the SFO and the POA are independent ANG II-sensitive sites.

8.4.3 Organum Vasculosum of Lamina Terminalis (OVLT)

It was soon found that the POA, particularly the tissues surrounding the anteroventral region of third ventricle (AV3V), is essential for a drinking response to intraventricular ANG II (Kucharczyk et al. 1976; Hoffman and Phillips 1976b; Buggy 1977; Buggy and Johnson 1977, 1978). ANG II injected into the AV3V caused drinking, and lesions of this portion prevented drinking. The investigators suggested that the organum vasculosum of the lamina terminalis (OVLT; Fig. 8.12), located within the AV3V, may be involved in ANG II-induced thirst, since the OVLT is outside the blood–brain barrier. In the goat (Anderson et al. 1975) and sheep (McKinley et al. 1986), large lesions of the AV3V including the OVLT caused a decrease in ANG II-induced drinking. Thrasher and Ramsey (1986) demonstrated that the selective destruction or ablation of the OVLT in the dog appeared to inhibit drinking after intravenous ANG II administration. The OVLT within AV3V seems to be another receptive site for ANG II in drinking.

8.4.4 Median Preoptic Nucleus (MnPO)

Neuroanatomical studies of the AV3V region have identified several cell groups (Simerly et al. 1984). The AV3V region includes the median preoptic nucleus (nucleus medianus, MnPO), the medial preoptic area (MPO), the OVLT and the anteroventral periventricular nucleus. The MnPO is an unpaired midline structure, which surrounds the dorsal, anterior and ventral aspects of the anterior commissure (Fig. 8.12). Lind and Johnson (1982a) reported that rats with only the OVLT removed responded to ANG II injected peripherally and centrally. This result differs from that obtained in the dog (Thrasher and Ramsey 1986). Lind and Johnson (1982a) observed that destruction of both the OVLT and the MnPO blocked angiotensin-induced thirst in rats. Further, they showed that lesions of the MnPO blocked drinking responses to both central and peripheral ANG II. Mangiapane et al. (1983) and Gardiner and Stricker (1985) found that electrolytic lesions of the MnPO attenuated drinking induced by systemic injection of ANG II. It should be considered that the lesions of the MnPO may have transected a neural pathway important for drinking, which passes through, or terminates in, the MnPO (Miselis et al. 1979; Miselis 1981; Lind and Johnson 1982b).

147

8.4.5 Neural Connections Between SFO, OVLT and MnPO

The MnPO, which contains ANG-binding sites (Mendelsohn et al. 1984), may function as a receptive site for CSF-borne ANG II through the gap junctions of the ependymal lining (Phillips 1987a; Fig. 8.12). The information perceived by the MnPO may be transmitted to the SFO via efferent nerves found by Hernesniemi et al. (1972), Lind et al. (1982) and Lind et al. (1985a, b), or the information may leave the nucleus for motor systems via separate neural systems (Kucharczyk et al. 1976) for eventual expression of the ingestive response. Thus, the MnPO seems to be an important nucleus for drinking induced by ANG II given peripherally or centrally.

On the other hand, the MnPO receives neural input from the SFO (Miselis et al. 1979; Miselis 1981; Lind and Johnson 1982b). Lind et al. (1985a, b) showed that ANG II-like immunoreactive cells and fibers were present in the SFO. Further, Lind and Johnson (1982b) observed that SFO projections, probably ANG II immunoreactive fibers that pass through the MnPO, extend to the OVLT and the suprachiasmatic and supraoptic nuclei. There are bidirectional nerve connections between the SFO and the MnPO (Fig. 8.12). These investigators transected the pre- and postcommissural connections between the SFO and the MnPO, and found that drinking induced by peripheral ANG II was completely inhibited.

Oldfield (1991) summarized the routes of ANG II which affect the SFO, OVLT and MnPO (Fig. 8.12). Neurons in the SFO and OVLT are stimulated by ANG II, which crosses the capillaries that have fenestrations, and the stimulus is transmitted to the MnPO. Further, the neurons in the SFO and OVLT receive information from the MnPO, which is sensitive to ANG II. The SFO and the OVLT may not be stimulated by CSF-borne ANG II, since tight junctions of the ependymal cells covering these organs prevent penetration of ANG II. These tight junctions are not present in other parts of the ventricles except the circumventricular organs. ANG II and ANG III in the interstitial spaces produced by the brain RAS have no access to either organ because of a complex array of tanycyte processes sealed by tight junctions (Fig. 8.12). Neurons in the MnPO are stimulated by CSF-borne ANG II which penetrates the clefts of the ependymal lining, and are also stimulated by brain ANG II in the interstitial spaces of the brain. In addition, the MnPO neurons are stimulated by the efferent nerves of the SFO and the OVLT.

Thus, information from ANG II regarding thirst is integrated within these intracranial compartments of the lamina terminalis. The interaction of these compartments with higher brain centers is far less well understood.

8.4.6 Other Possible Receptive Sites for ANG II

In addition to the SFO and the AV3V region, more than 30 areas or nuclei in the brain have been reported in relation to drinking induced by ANG II. These are areas or nuclei containing ANG II binding sites or terminal fields of ANG II-immunoreactive fibers. Among these areas sensitive to ANG II, the lateral septum, the ventral–basal complex of the thalamus, the nucleus reuniens of the thalamus, the nucleus of the solitary tract, the midbrain central gray, and the bed nucleus of the stria terminalis responded to ANG II by causing drinking, although it is not always clear whether the dipsogenic action of ANG II on these sites is direct or indirect (see Lind 1988 for review). This question should be examined carefully using methods that avoid artifacts of a particular technique.

Recently, a structure related to drinking and body fluid regulation in rats, the organum cavum prelamina terminalis (OCPLT), was described by Nicolaïdis and Ghissassi (1991a; Fig. 8.13). This organ is a hollow, sagittal, forebrain structure consisting of a vertical interhemispheric horn which overlies the anterior aspect of the lamina terminalis, and a horizontal horn which lies in the septum below the corpus callosum. The narrow cavities of the horizontal and vertical horns communicate freely. These horns do not communicate with the cerebral ventricular systems or the subarachnoid space. The posterior wall of the OCPLT has numerous vascular plexi, which may be the source of the fluid that fills the OCPLT and, perhaps, contain peptides that act on it. Neu-

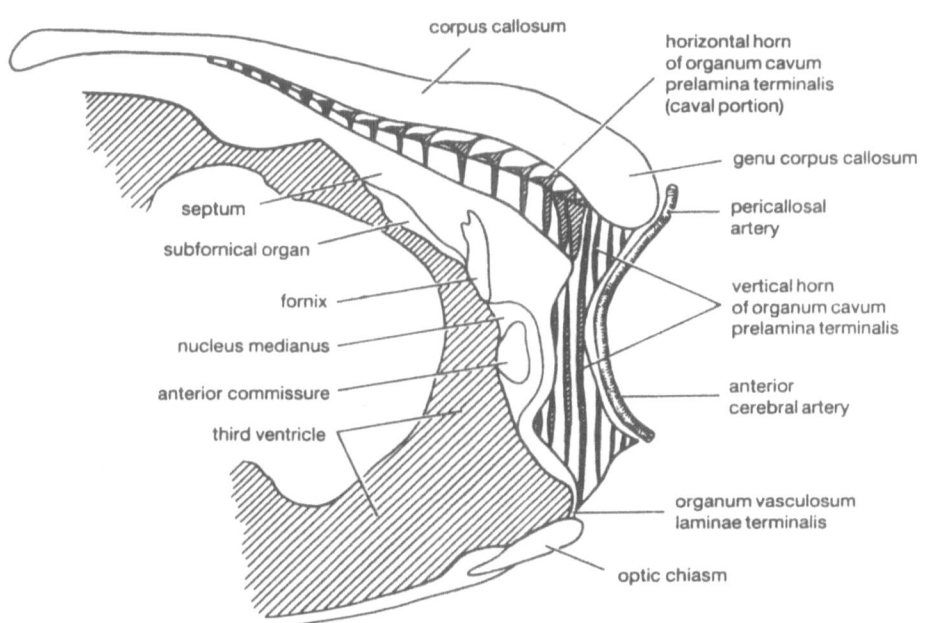

Fig. 8.13. Schematic view of the organum cavum prelamina terminalis (OCPLT) in successive sections showing the sagittal flat irregular pouch of the vertical horn, widening under the genu of the corpus collosum and becoming T-shaped, then a tubular cavity. (Nicholaïdis et al. 1991)

rons lining the OCPLT were found to be electrophysiologically activated by ANG II. ANG II (> 0.2 ng) injected anywhere into the hollow systems (vertical horn and horizontal horn) induced copious drinking. Nicholaïdis and Ghissassi (1991b) found that a specific concentration of NaCl seems to play a role in the manifestation of ANG II's dipsogenic action by acting on the angiotensin-sensitive neurons lining the OCPLT. Nicholaïdis et al. (1991) suggested that systemic and/or regional-borne peptides might be released into the lumen of the OCPLT from the bordering vascular network; from such a peptide "sink", the peptides might affect a vast neural network projecting into the nuclei that control drinking and autonomous regulatory responses. The OCPLT seems to be the place for convergence of chemicals which, in turn, modulate the responsiveness of neurons in vicinities that function as elementary integrators of neural information on the hydrational state of the organism. The functional relationships among the OCPLT, SFO, OVLT and the MnPO are not known.

Apart form neuronal responsiveness to ANG II, local vasoconstriction by ANG II in highly vascularized areas of the SFO and the OVLT was thought to be a cause of drinking (Nicholaidis and Fitzsimons 1975; Fitzsimons 1979). However, not all vasoconstrictors caused drinking and not all vasodilators nullified ANG II-induced drinking (Phillips and Hoffman 1977).

8.4.7 Nonmammalian Vertebrates

Induction of drinking by intracranial injections of ANG II was reported in the white-crowned sparrow (Wada et al. 1975), pigeon (Evered and Fitzsimons 1976; deCaro et al. 1982), chicken (Snapir et al. 1976; Schwob and Johnson 1977), Pekin duck (deCaro et al. 1980), and turkey (Denbow 1985). These reports noted that the POA is one of the receptive sites. In the Japanese quail, it was shown that direct injection of ANG II into the SFO (Takei 1977a, b) and the preoptic area induced drinking (Takei et al. 1979a). However, in the quail with lesions of the SFO, drinking was not induced when ANG II was injected into the POA. Takei et al. (1979a) described green fluorescent monoaminergic, perhaps noradrenergic, fibers proceeding from the POA to the SFO (Fig. 8.14), and Tsuneki et al. (1978) found that there are many synapses of monoaminergic fiber terminals around the subfornical perikarya (Fig. 8.15). Transection of monoaminergic fibers (and perhaps some other fibers) between the two regions attenuated drinking when ANG II was injected into the POA (Takei et al. 1979a; Fig. 8.16). Therefore, it was thought that information perceived by cells in the POA was transmitted first to the SFO via monoaminergic fibers, and then drinking was induced. The role of the MnPO was not studied in these experiments. In the Japanese quail, injections of noradrenaline into the SFO induced vigorous drinking but carbachol did not (Takei et al. 1979a), whereas carbachol did evoke thirst in the rat when injected into the SFO (Simpson and Routtenberg 1972). Thus, the neural mechanisms appear to

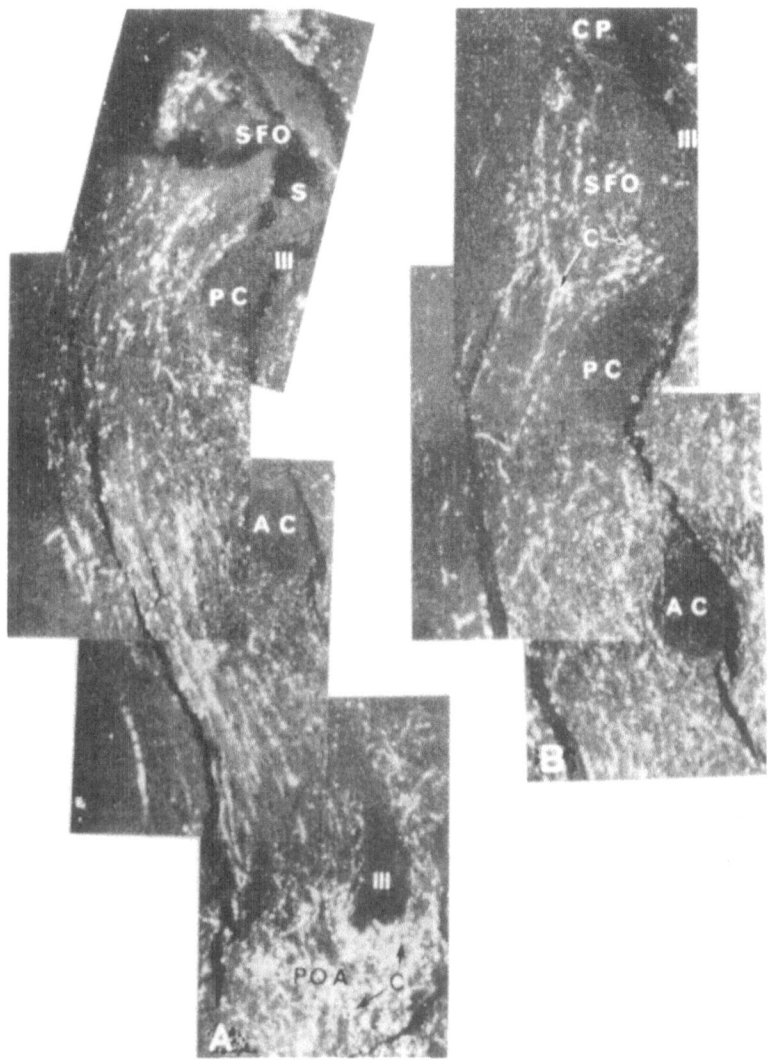

Fig. 8.14. A Midsagittal section of the anterior wall of the third ventricle showing the *SFO* and *POA* regions in the Japanese quail. Fluorescent perikarya (*C*) are found in the POA, but not in the SFO. Strong yellow-green fluorescent fibers run from the POA to the SFO. B Sagittal section 0.1 mm lateral to the medial plane. *AC* Anterior commissure; *C* cells surrounded by fluorescent terminals; *CP* choroid plexus; *PC* pallial commissure; *S* sinus; *III* third ventricle. A, B, × 105 (Takei et al. 1979a)

be somewhat different between the rat and the Japanese quail. The functional anatomy of the POA has not been studied in birds.

The reptiles, *Elaphe climacophora*, *Takydromus takydromoides* (Y. Takei, unpubl. data, 1995) and *Iguana* (Fitzsimons 1979), responded to intracranial injection of ANG II by drinking. Therefore, receptive sites for ANG II in

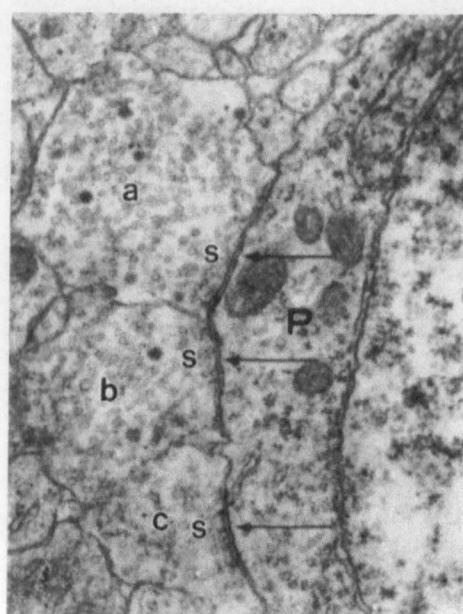

Fig. 8.15. Three axon terminals (*a, b, c*) making synaptic contacts (*S* and *arrows*) on a perikaryon (*P*) of the SFO in the Japanese quail. The terminals contain both dense-cored granules and small vesicles. × 22200

drinking seem to exist in the brain. In amphibians, the ventral skin has receptors for ANG II which are involved in the stimulation of water-absorption and the brain seems to have receptors for ANG II, which are concerned with the cutaneous water-absorption behavior (Sect. Chapter 8.2.4). Distribution of the receptors in the brain has not yet been studied. The fish, *Anguilla japonica*, reacted to intraventricular injections of ANG II. Even decerebrated eels or diencephalon-damaged eels responded to ANG II administered through an arterial catheter (Takei et al. 1979b; Figs 8.17, 8.18). Drinking seems to be elicited by a stimulatory effect of ANG II on the reflex center in the medulla oblongata. Fishes, unlike terrestrial animals, do not need to search for water, and the mechanism inducing drinking by ANG II may well differ between aquatic and terrestrial animals. However, the possibility of the presence of receptive site(s) for ANG II in brain portions other than the medulla oblongata cannot be excluded, since site-specific intracranial injections of ANG II have not yet been performed. JG cells in several species of elasmobranchs (Lacy and Reale 1990) and immunoreactive ANG II in the brain of the nurse shark (Galli-Phillips 1991) have been found. Exogenous ANG II induced drinking in *Scyliorhinus canicula* (Hazon et al. 1989). The sites of action of ANG II in the elasmobranch brain need to be studied.

8.4.8 Summary

The SFO and the OVLT are receptive sites for blood-borne ANG II, but may not be for brain interstitial ANG II and CSF-borne ANG II. The MnPO seems to be

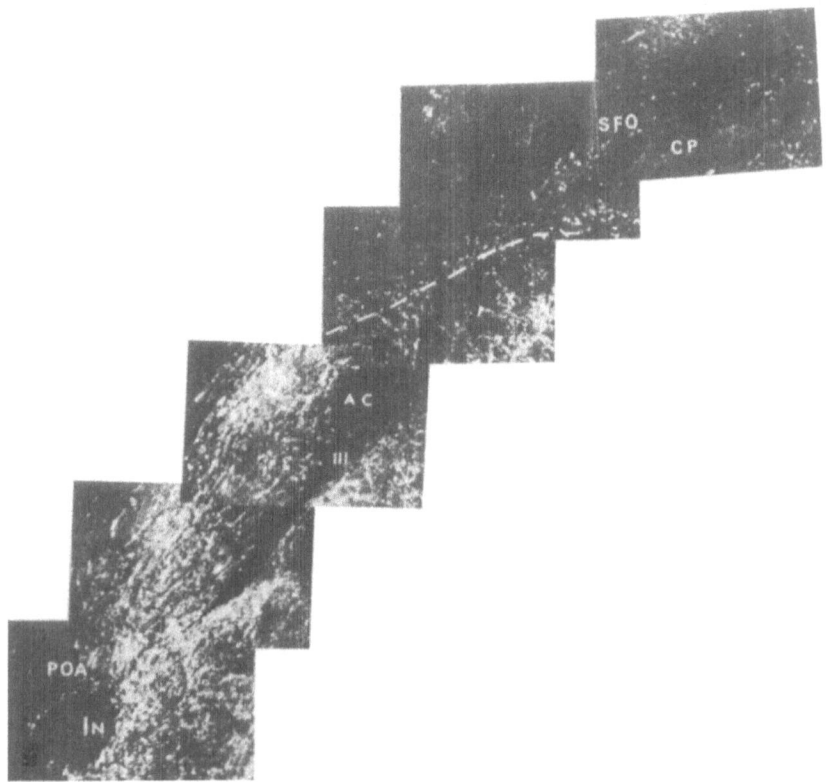

Fig. 8.16. Sagittal section of the anterior wall of the third ventricle (*III*) showing the *SFO* and *POA* regions in the Japanese quail. Yellow-green fluorescence disappeared in the SFO region after transection (*broken line*) of fluorescent fibers between the SFO and the POA. Some fluorescence near the cut and in the SFO is autofluorescence. This bird did not respond to ANG II injected into the POA by drinking. *AC* Anterior commissure; *In* injection site of ANG II; *CP* choroid plexus. × 75 (Takei et al. 1979a)

a receptive site for both brain-interstitial ANG II and CSF-borne ANG II (Oldfield 1991). Among these three sites there are bidirectional neural connections, some of which are angiotensinergic. The functional significance of these connections deserves further investigation. In addition, a number of areas or nuclei related to ANG II-induced drinking have been reported, but it is not fully known whether they are involved in the induction of drinking directly or indirectly. The organum cavum prelamina terminalis (OCPLT) was recently proposed as a site for intracerebral, peptide-mediated communication (Nicolaïdis and Ghissassi 1991a).

In birds, the SFO and POA are important areas of sensitivity to ANG II-induced drinking. The MnPO has not been studied. In reptiles, the brain has receptive areas for ANG II. An angiotensin-induced drinking mechanism as part of water-acquiring behavior developed first in the forebrain in the reptiles

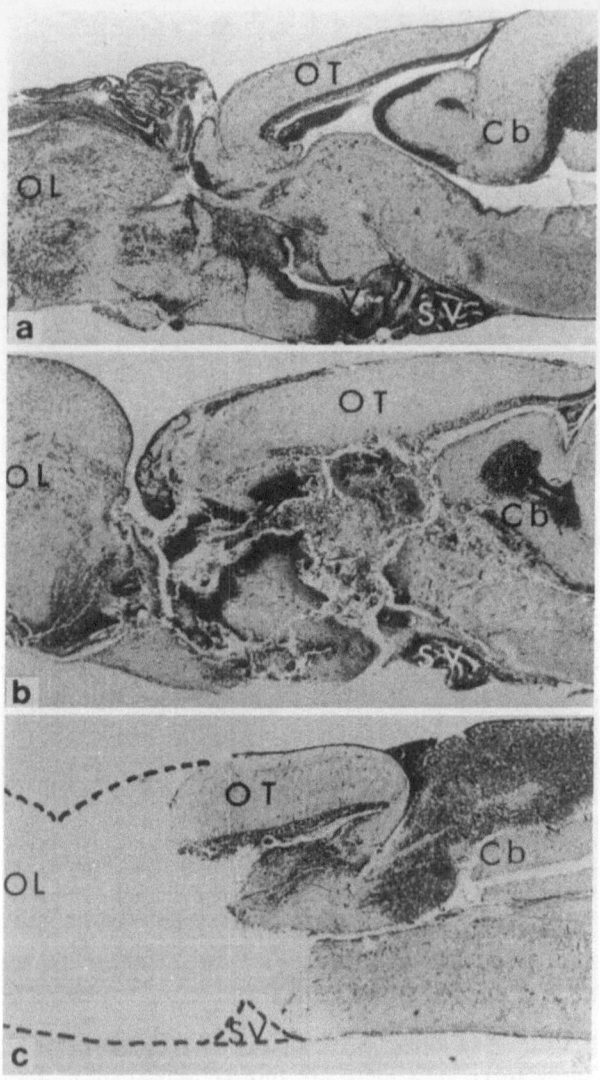

Fig. 8.17a–c. Sagittal section of the brain of the Japanese eel. **a** Intact eel; **b** eel with electric lesion in diencephalic region; **c** eel with decerebrated brain. Water intake of these eels is shown in Fig. 8.18. *Cb* cerebellum; *OL* olfactory lobe; *OT* optic tectum; *SV* succus vasculosus; *V* third ventricle. × 80 (Takei et al. 1979b)

during vertebrate evolution, since ANG II does not induce drinking in amphibians. However, in amphibians, the skin has receptors for ANG II involved in water absorption and also the brain seems to have receptive sites for ANG II, causing cutaneous water-absorption behavior. Fishes are different from tetrapods: decerebrated eels respond to ANG II, suggesting that the swallowing center in the medulla oblongata might be responsible. However, injections of

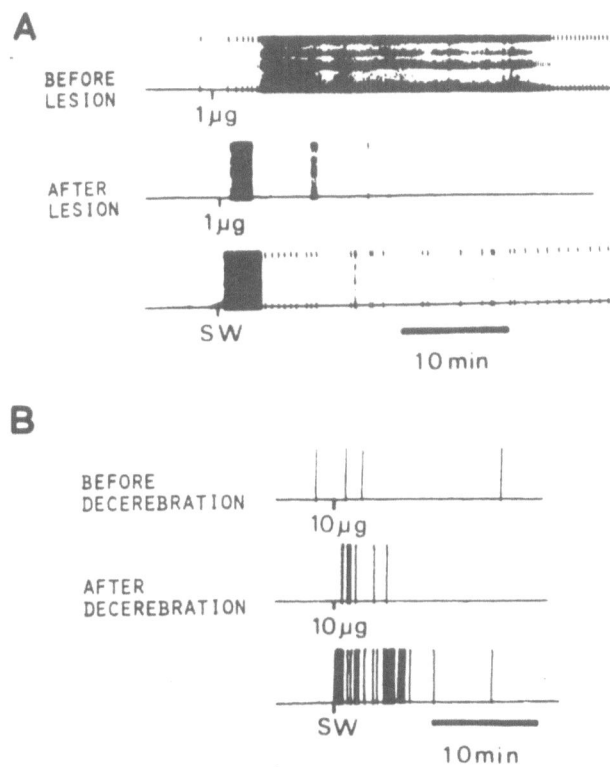

Fig. 8.18. A Induction of drinking after intraarterial injection of ANG II (1 µg) or after seawater (*SW*) exposure in a freshwater eel with electric lesion. Each spike represents 0.03 ml of swallowed water. For the extent of the lesion, see Fig. 8.17b. **B** Induction of drinking after intraarterial injection of ANG II (10 µg) or after exposure to SW in a decerebrated freshwater eel. For the extent of the decerebration, see Fig. 8.17c. (Takei et al. 1979b)

ANG II in various areas of the brain should be attempted to identify other possible receptive sites. To elucidate the evolutionary aspects of drinking mechanisms, more studies are required in nonmammalian vertebrates. Figure 8.19 is a schematic representation of humoral factors relevant to the ANG II-induced drinking mechanism, based on the data from mammalian studies. Interrelations between ANG II and other active substances such as ACTH, VP, ANP and aldosterone are described in Section 8.6.

8.5 ANG II and Sodium Appetite

In addition to elicitation of thirst, ANG II is implicated in arousal of sodium appetite. The essence of this effect in mammals has been summarized in several

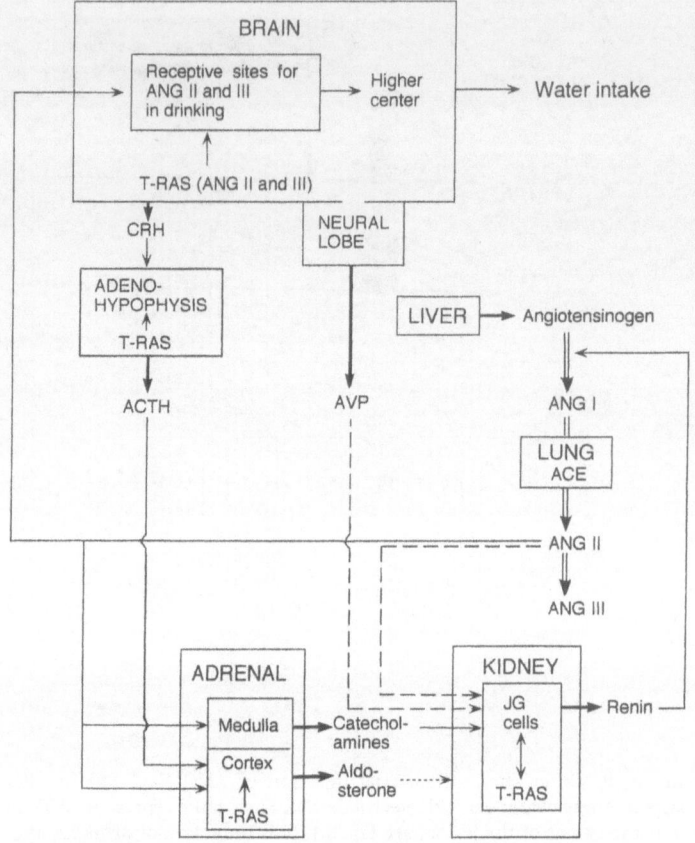

Fig. 8.19. Schematic representation of humoral factors relevant to water intake induced by ANG II following water deprivation in mammals. Water deprivation induces release of renin from JG cells into the blood, and then ANG I, II and III are sequentially produced in the blood (see also Fig. 1.2). Tissue RAS (*T-RAS*) is present in the kidney. Some interactions seem to occur between renin produced by JG cells and ANG II produced by the T-RAS. ANG II produced in the blood and ANG II and III derived from the brain T-RAS reach the receptive sites for ANG II and III (see Fig. 8.12 for details), leading to drinking. Circulating ANG II has a negative feedback action on renin release from JG cells (Sect. 5.1.4.3). ANG II stimulates the adrenal medulla and cortex to secrete catecholamines and aldosterone, respectively. Aldosterone causes a decrease in plasma ANG II (Sect. 5.1.4.4), although possibly not through direct action on JG cells. Catecholamines stimulate JG cells to secrete renin (Sect. 5.1.4.2). ANG II produced by the T-RAS of the adrenal cortex may modulate aldosterone production or secretion, however, no definite evidence has yet been found. The paraventricular and supraoptic nuclei (not shown) are stimulated directly or indirectly by ANG II and/or ANG III, and secrete AVP (Sect. 8.6.1) and CRH (Sect. 8.6.2). CRH stimulates ACTH release, which may be modulated by ANG II produced by adenohypophysial T-RAS and, in turn, secretion of aldosterone from the cortex is enhanced. A negative feedback system between AVP and renin secretion appears to exist (Sects. 5.1.4.3 and 8.6.1). *Solid line* (——➤), stimulation; *broken line* (– – ➤), negative feedback regulation; *dotted line* (······➤), inhibition; *heavy line* (——➤), secretion; *double line* (⟹), pathway from angiotensinogen to ANG III

reviews (Fitzsimons 1979, 1993; Denton 1982, 1991; Grossman 1990). In non-mammalian species, however, the study of sodium appetite is still in its infancy; only in birds have a number of studies been performed. In the initial part of this chapter, therefore, the present state of knowledge about ANG II-induced sodium appetite is reviewed briefly in mammals. The more thorough accounts are made on sodium appetite in birds.

8.5.1 Mammals

In contrast to the situation for animals living in the sea, salt is not freely available to those living in fresh water or on the land. Thus, the latter animals constantly face the need to ingest salt from food or by drinking salt water. Among salts, Na^+ salt is essential for body fluid homeostasis because it is a major constituent ion in the extracellular fluid including plasma. Carnivores are capable of ingesting enough Na^+ from food, but herbivores never satisfy their Na^+ requirements from food because of its low Na^+ content. Therefore, herbivores ingest crystalline salt or salty water in nature (see Fitzsimons 1979 for review). Extensive studies have been performed on sodium appetite in herbivores such as sheep, goat, and rabbit, and in the omnivorous rat. The renin–angiotensin–aldosterone system plays a key role in Na^+ retention. Renin is released in response to Na^+ deficiency or hypovolemia, and ANG II formed in the blood stimulates aldosterone synthesis and release. ANG II and aldosterone act in concert to decrease Na^+ excretion by the kidney (Harris and Navar 1985) and to stimulate Na^+ absorption by the intestine (Levens 1985). It is tempting to speculate, therefore, that the renin–angiotensin–aldosterone system plays an important role in the oral Na^+ intake or arousal of sodium appetite.

The effect of aldosterone or other mineralocorticoids on sodium appetite is controversial. Bilateral adrenalectomy was shown to stimulate sodium appetite in the rat (Richter 1936), and peripheral administration of mineralocorticoid into these animals abolished the increased appetite (Richter 1956). However, the stimulation of sodium appetite after adrenalectomy is apparently indirectly mediated via Na^+ loss after removal of the endogenous source of aldosterone. Other procedures that decrease plasma aldosterone levels such as peripheral administration of captopril (Fregly 1980) or production of hypothyroidism (Fregly and Waters 1965) also stimulate sodium appetite in the rat. Furthermore, a peripheral injection of high doses of desoxycorticosterone acetate (DOCA) stimulates sodium appetite in the normal rat (Rice and Richter 1943); lower doses inhibit the appetite, resulting in a U-shaped dose–response curve. A subcutaneous injection of aldosterone increases the intake of isotonic NaCl solution linearly in the rat, although the rat goes into a positive Na^+ balance with increased intake (Wolf and Handel 1966). The intake was specific for Na^+-containing solutions. However, since isotonic NaCl solution is palatable, as pointed out by Fitzsimons (1979), higher and aversive concentrations of NaCl solution should be used to examine sodium appetite. The results of such an experiment showed that mineralocorticoids have a minor but distinct role in the arousal of sodium appetite in the rat (see Fitzsimons 1979, pp. 471–478, for review).

When administered systemically, renin and ANG II have little influence on sodium appetite. In normal sheep and those made Na^+-deficient by draining saliva from a parotid fistula, administration of ANG II had no effect on sodium appetite or was rather inhibitory on sodium appetite (Bott et al. 1967). Nephrectomy of the Na^+-deficient sheep also fails to eliminate sodium appetite. Similarly, the renal RAS appears to have little influence on sodium appetite in the rat. A peripheral injection of components of the RAS, or activation of the renal RAS by caval ligation or other treatments, often elicits sodium appetite in the rat (Findlay and Epstein 1980), but this effect may be attributable to renal Na^+ loss caused by pharmacologically increased plasma ANG II. The onset of Na^+ intake after injection of ANG II is much slower than that of water intake, and the dose required to cause a reliable response is much higher. Sodium appetite induced by subcutaneous injection of formalin (Fitzsimons and Stricker 1971) or by adrenalectomy (Fitzsimons and Wirth 1978) is abolished by bilateral nephrectomy in rats, but rats made auric by bilateral ureteral ligation also lose the appetite. The loss of sodium appetite by nephrectomy was not restored by administration of renin or ANG II. Furthermore, procedures that decrease plasma ANG II concentration, such as treatment with captopril, induce sodium appetite in the rat (Fregly and Rowland 1985). Bryant et al. (1980) showed that repeated systemic injections of isoproterenol or renin, both of which increase plasma ANG II levels, increase ingestion of an aversive concentration of 2.7% NaCl solution in the rat. The major difference between this experiment and the previous ones with negative results is utilization of prolonged stimulation of the RAS by repeated injections over many days. Collectively, the role of peripheral RAS in eliciting sodium appetite remains undetermined.

Involvement of the central RAS in arousal of sodium appetite is more convincing. Intracerebral administration of each component of the RAS consistently induces an intake of aversive concentration of NaCl solutions in the rat (Avrith and Fitzsimons 1980; Bryant et al. 1980). The ANG II-induced sodium appetite is specific for Na^+ ions, and drinking continues even though rats go into positive Na^+ balance (Avrith and Fitzsimons 1980). Another potent intracranial dipsogen, carbachol, caused only thirst and no sodium appetite (Avrith et al. 1980). The latency to onset of Na^+ intake and the minimal effective dose are somewhat greater than those observed for water intake after cerebral administration of ANG II. However, when ANG II and mineralocorticoid are given together, the effect is more than additive (Epstein 1982; Fluharty and Epstein 1983); intracerebroventricular administration of physiological doses of ANG II in combination with systemic injection of physiological doses of DOCA increases the intake of 3% NaCl in the rat, although separate administration of each hormone requires pharmacological doses to induce the intake. Interaction of ANG II and mineralocorticoid is also noted in sodium appetite caused by Na^+ deficiency. Peripheral injection of high doses of captopril or enalaprilic acid, which inhibits both central and peripheral converting enzyme activity, suppresses deficiency-induced sodium appetite in the rat (Moe et al. 1984; Elfont and Fitzsimons 1985; Weisinger et al. 1988). The high dose of captopril almost eliminates but the low dose rather stimulates sodium appetite in the sheep (Weisinger et al. 1987) and cattle (Blair-West et al. 1988). The low dose of captopril may not have inhibited the conversion of peripherally accumulated ANG I to ANG II in the brain. Blockade of central but not peripheral mineralocorticoid receptors with RU-28318 reduces sodium appetite induced by Na^+ depletion, and the combination of RU-28318 and captopril completely abolished the appetite in the rat (Sakai et al. 1986). It seems that sodium appetite is regulated by a synergistic action of ANG II and mineralocorticoid in omnivorous rats, whereas Na^+ concentration in the cerebrospinal fluid plays a crucial role in herbivorous ruminants. The synergism of ANG II and aldosterone appears in 17-day rat pups, but not in 12-day pups (Thompson and Epstein 1991). The site of action of ANG II for arousing sodium appetite may be located in the AV3V in the rat as is the case for the dipsogenic site of action (de Arruda Camargo et al. 1991). Both the AT_1 receptor antagonist, Dup753, and the AT_2 antagonists, PD123319 and CGP42112A, when injected into the cerebral ventricle, inhibit NaCl intake of rats injected with ANG II via the same route (Rowland et al. 1992). Dup753 is a more effective inhibitor than AT_2 antagonists, and the dipsogenic effect is inhibited only with Dup753.

8.5.2 Birds

Birds are the only nonmammalian species in which ANG II and sodium appetite have been investigated. We previously reviewed hormonal regulation of sodium appetite in birds (Takei and Kobayashi 1990). Epstein and Massi (1987) have shown that a robust sodium appetite is induced in the pigeon after Na^+ depletion by frusemide. The birds drank 21 ml of 3% NaCl solution in 24 h compared to 1-2 ml prior to the depletion. Intramuscular injection of DOCA to the bird at doses of 2-4 mg/day also increased daily intake of 3% NaCl. A pulse injection of 1 ng/bird of purified hog renin into the cerebral ventricle of the pigeon induced immediate water intake followed by increased intake of 3% NaCl. The greatest intake occurred the next morning, and the increased intake continued for 2 days. Thus, the onset of renin-induced Na^+ intake in the pigeon is much slower than in the rat probably because of a heterologous renin–angiotensinogen reaction (Avrith and Fitzsimons 1980). Since infusion of 1 ng/min of ANG II into the pigeon brain does not cause significant changes in Na^+ excretion (Fitzsimons et al. 1982), ANG II-induced sodium appetite is a direct action not mediated via increased Na^+ loss. Both RAS and mineralocorticoid affect sodium appetite in the pigeon just as was observed in the rat (Epstein 1982). Massi and Epstein (1987) showed that an intramuscular injection of captopril reduces sodium appetite induced by Na^+ depletion in the pigeon, whereas the same treatment does not affect DOCA-induced sodium appetite. Intracerebroventricular injection of [Sar^1, Ile^8] ANG II produces marked inhibition of sodium appetite induced by Na$^+$ deprivation in the pigeon. Pretreatment with RU-28318 fails to reduce this sodium appetite, although it does inhibit DOCA-induced appetite. Therefore, the RAS plays a crucial role in deficiency-induced sodium appetite in the pigeon. More recently, Massi and Epstein (1990) showed that concurrent treatment with intramuscular injection of DOCA and intracerebroventricular injection of ANG II induces an excess 3% NaCl intake in the sodium-replete pigeon, but continuous infusion of hypertonic 0.3 M NaCl solution into the cerebral ventricles is without effect. Therefore, pigeon resembles rats more than ruminants in that ANG II and mineralocorticoid have synergism for arousal of sodium appetite.

8.5.3 Summary

It is apparent that sodium appetite is governed by the Na$^+$ concentration in the cerebrospinal fluid in the herbivorous sheep, goat and cows, and Na$^+$ sensors in the brain parenchyma adjacent to the third ventricle seem to regulate the appetite (Weisinger et al. 1982, 1986). However, sodium appetite in the omnivorous rat is regulated by the synergistic action of angiotensin II and aldosterone in the brain (Epstein et al. 1984; Yang and Epstein 1991). Thus, there are species differences in the control of sodium appetite between herbivores and omnivores. In birds, pigeon is the only species in which sodium appetite has been studied. Since synergy of ANG II and aldosterone is apparent in the pigeon, this herbivorous bird is closer to the omnivorous rat than to herbivorous ruminants with respect to hormonal regulation of sodium appetite.

8.6 Release of Hormones by ANG II

8.6.1 Neurohypophysial Hormones

Release of vasopressin by ANG II has been a subject for many investigators since the suggestion (Severs et al. 1970) in rats and the original report (Bonjour and Malvin 1970) in dogs that systemic administration of ANG II increased vasopressin secretion. It is now known in mammals that when administered peripherally, pharmacological doses of ANG II can release vasopressin, but ANG II at physiological levels does not release vasopressin (Reid 1984; Phillips 1987a, b; see Saavedra 1992 for review).

Centrally administered ANG II caused vasopressin release at high doses (Yamaguchi et al. 1980; Phillips 1987a, b; see Saavedra 1992 for review). Supporting this, Ochiai and Nakai (1984) noted that exocytosis occurred in neurosecretory nerve terminals after intraventricular injection of ANG II.

Vasopressin was released by ANG II-(1–7) (Schiavone et al. 1988) and by ANG III (Yamaguchi et al. 1979) to the same extent as by ANG II when given centrally in rats, although ANG II-(1–7) and ANG III possessed minimal pressor activity when administered intravenously. The stimulation of vasopressin release by intraventricular ANG II was inhibited by intraventricular injection of saralasin (Yamaguchi et al. 1980). This showed that vasopressin was released by ANG II through ANG II receptors. Release of vasopressin by exogenous ANG II may not be due to direct action on the neural lobe, since ANG II receptors are not present in the neural lobe (Israel at al. 1984).

As mentioned above, pharmacological doses of ANG II were required to stimulate the release of vasopressin, when given peripherally or centrally. However, ANG II generated in the brain may act locally at high concentrations. Therefore, the requirement of large doses of peripheral or central ANG II does not exclude the possibility of physiological control of endogenous ANG II over the vasopressin release (Yamaguchi et al. 1980). In particular, increased peripheral ANG II may be of importance for restoring water balance under such conditions as dehydration or hemorrhage.

Potentiation by angiotensin II of the vasopressin response to an increasing plasma osmolality has been reported in rats (Shimizu et al. 1973; Yamaguchi et al. 1985). Vasopressin secretion seems to depend on an interaction between ANG II and plasma osmolality. The interaction of sodium and ANG II may be on receptors or may be by opening the blood–brain barrier (Phillips 1987a). Although ANG II stimulates vasopressin release, vasopressin inhibits renin secretion. There appears to exist a vasopressin–renin negative feedback (Brooks and Malvin 1993; Fig. 8.19).

The receptive sites for ANG II in releasing vasopressin and the transmission of the ANG II stimulus therefrom may be similar to the case of ANG II-induced drinking: ANG II administered peripherally at high doses may reach (1) the OVLT and SFO, which have no blood–brain barrier, as blood-borne ANG II; and (2) the median preoptic nucleus (MnPO) as CSF-borne ANG II (Fig. 8.12). From these receptive sites, the ANG II stimulus may be signalled to the paraventricular nucleus (PVN) and the supraoptic nucleus (SON). Electrophysiologic studies have revealed that the SFO has efferent projections into the PVN (Miselis 1981; Silverman et al. 1981; Tanaka et al. 1985a, b: Gutman et al. 1986; Li and Ferguson 1993a) and to the SON (Miselis 1981; Swanson and Sawchenko 1983; Sgro et al. 1984). The PVN neurons project reciprocally into the SFO (Yamashita et al. 1984). Further it is known that electrical stimulation of SFO releases vasopressin and oxytocin in the rat (Ferguson and Kasting 1986, 1987). Thus, it is considered that ANG II mediates an excitatory input to the PVN and the SON from the SFO.

ANG II administered intraventricularly may reach the MnPO as CSF-borne ANG II or the PVN by diffusion (Fig. 8.12). Endogenous ANG II generated by the brain RAS may reach MnPO as interstitial ANG II or stimulate the PVN directly (Shoji et al. 1989). The PVN contains neurons possessing both ANG II and III receptors (Catt et al. 1986; Harding and Felix 1987; Saavedra 1992).

The OVLT, SFO and MnPO have reciprocal projections between them (Fig. 8.12). Lesions of the SFO, OVLT or MnPO or transection of their neural connections abolished vasopressin release (Bealer et al. 1979; Mangiapane et al. 1983; Iovino and Steardo 1984, 1985; Gardiner et al. 1985). Lesions of the PVN stopped the vasopressin releasing effects of centrally administered ANG II (Gutman et al. 1988). It is considered that information of ANG II administered peripherally or centrally is perceived by the SFO, OVLT or MnPO, through partly different routes (Fig.8.12), and then this information is transferred to the PVN and the SON. However, ANG II produced in the RAS of the brain may stimulate ANG II receptors in the PVN directly to release vasopressin.

Yamaguchi et al. (1988) showed that intraventricularly injected dopamine and ANG II increased plasma vasopressin. However, this response of vasopressin to dopamine and ANG II was inhibited by pretreatment with haloperidol (a dopamine blocker). They suggested that dopaminergic mechanisms are involved in ANG II-induced vasopressin release.

It is well known that oxytocin is released in relation to phenomena such as parturition, lactation and impregnation. Oxytocin release was also stimulated by intracerebroventricular administration of ANG II (Lang et al. 1981). Electric stimulation of the SFO induced both oxytocin and vasopressin release (Ferguson and Kasting 1986, 1987). Salt loading and dehydration also stimulated release of oxytocin and vasopressin (Balment et al. 1980; Edwards and LaRochelle 1984). Under such conditions, oxytocin may coordinate with vasopressin in controlling water balance in the body. The mechanism of oxytocin release by ANG II administered centrally or peripherally may be the same as that for vasopressin, since both vasopressin and oxytocin were released by salt loading and dehydration at the same time, and their secretion was impaired after lesions of the MnPO (Gardiner et al. 1985). It is noteworthy that intraventricular injection of ANG II stimulated oxytocinergic neurosecretory cells as assessed by extracellular recordings of action potentials and that combined application of ANG II and hypertonic NaCl elicited a marked potentiation of the response of the neurosecretory cells (Akaishi et al. 1980).

Coexistence of immunoreactive renin or ANG II and oxytocin or vasopressin has been reported in the rat hypothalamic neurosecretory cells (Kilcoyne et al. 1980; Calza et al. 1982; Fuxe et al. 1982). In the goldfish, intragranular colocalization of arginine vasopressin- and ANG II-like immunoreactivity has been found in the hypothalamo-neurohypophysial system (Yamada et al. 1990). ANG II may have paracrine function and may modulate vasopressin secretion or oxytocin.

In nonmammalian animals, data on the relationship between neurohypophysial hormones and ANG II are very limited. Pharmacological doses of ANG II stimulated the release of arginine vasotocin (AVT), when administered peripherally in the Japanese quail (Kobayashi and Takei 1982), chicken (Goto et al. 1986) and the kelp gull (Gray and Erasmus 1989). Centrally administered ANG II stimulated AVT release in the duck (Gerstberger et al. 1984).

8.6.2 ACTH and CRH

ANG II stimulated ACTH secretion when given intravenously in dogs (Ramsay et al. 1978) and intravenously or intraventricularly in rats (Rivier and Vale 1983; Plotsky et al. 1988; see Saavedra 1992 for review). In man the results are controversial: intravenous infusion of ANG II stimulated ACTH release in some cases (Haller et al. 1986; degli Uberti et al. 1990), but not in others (Calogero et al. 1991). Peripheral administration needed relatively high doses of ANG II to increase plasma ACTH.

Rivier and Vale (1983) observed that stimulation of ACTH release by ANG II was totally abolished by immunoneutralization of endogenous CRH. Moreover, ANG II injected intravenously into rats treated with morphine, chlorpromazine and nembutal, pharmacological blockers of the release of endogenous CRH (Arimura et al. 1967), did not increase ACTH release. These findings suggest that stimulation of ACTH release by ANG II is mediated by CRH. Central and peripheral ANG II has been found to stimulate CRH release (Sumitomo et al. 1991; see Saavedra 1992 for review; Fig. 8.19). Further Rivier and Vale (1983) noted that the ANG II receptor antagonist $[Sar^1, Thr(Me)^8]$- ANG II blocked the stimulation of ACTH release by ANG II in nonanesthetized rats, suggesting that ANG II receptors are involved in ACTH release.

ANG II administered peripherally may stimulate CRH secretion by way of circumventricular organs (Murakami and Ganong 1987). Among circumventricular organs, the SFO seems to be an important receptive site for allowing ANG II to induce CRH secretion into the portal circulation. The SFO contains ANG II receptors (Mendelsohn et al. 1984) and binds ANG II given systemically (van Houten et al. 1980). Since the SFO projects into the PVN (Miselis 1981; Lind et al. 1982; Li and Ferguson 1993a), information of an increase in plasma ANG II concentrations may reach parvocellular CRH-containing neurons in the PVN by way of the SFO, resulting in an increase of ACTH in the blood (Plotsky et al. 1988). Electrical stimulation of the SFO is associated with the release of CRH and ACTH (Plotsky et al. 1988). Paraventricular neurons and SFO neurons in rat brain slices showed excitatory responses through AT_1 receptors to ANG II but less response to ANG III (Li and Ferguson 1993b). In addition to the SFO, the OVLT and the MnPO, which contain ANG II receptors (Tsutsumi and Saavedra 1991), should be considered as receptive sites for ANG II in CRH release, as in the case of ANG II-induced drinking (Fig. 8.12).

It is possible, however, that plasma ACTH increase by ANG II injected cerebroventricularly (Brooks and Malvin 1979; Spinedi and Negro-Vilar 1983; Sumitomo et al. 1991) was caused by CRH released through direct action of ANG II on the neurons in the PVN (Spinedi and Rodriguez 1986; Anguilera et al. 1995). The PVN contains ANG II receptors (Tsutsumi and Saavedra 1991). The CRH-containing cells in the PVN project fibers into the median eminence to release CRH into the portal veins (see Catt et al. 1986 for review).

Enhanced ACTH secretion by ANG II may have been mediated partly by vasopressin released by peripheral or brain ANG II (Sect. 8.6.1). Vasopressin may have stimulated CRH secretion and potentiated the activity of CRH in vitro and in vivo in experimental animals and humans (see Buckingham et al. 1992, p. 134). ANG III stimulated release of ACTH in vitro (Spinedi and Negro-Vilar 1983).

Increased release of ACTH was observed with ANG II in rat pituitary cell cultures (Sobel and Vagnucci 1982). The anterior pituitary is known to contain ANG II receptors (Tsutsumi and Saavedra 1991). Corticotrophs, mammotrophs and presumptive thyrotrophs among dispersed cells from the anterior pituitary of the male rat were specifically labeled with [^{125}I] ANG II. This suggests that ANG II may regulate the secretion of corticotrophs under physiological conditions (Paglin et al. 1984). In addition, ANG II was found to potentiate the corticotropin-releasing activity of CRH in perfused rat anterior pituitary cells (Vale et al. 1993; Schoenenberg et al. 1987). ANG II could be released from the median eminence, where the ANG II-immunoreactive fibers (Lind et al. 1985b) and ANG II receptors (Tsutsumi and Saavedra 1991) are abundant, into the portal circulation where it could act directly on the anterior pituitary, resulting in release of ACTH, or where it could exhibit a synergism with CRH on ACTH release in the pituitary. The physiological significance of ACTH release by peripheral ANG II requires further investigation.

As mentioned above, both peripheral and central angiotensin II stimulate ACTH secretion. However, there is little evidence that ANG II mediates ACTH secretion in response to stress. In sheep the ACTH response to hemorrhage was reduced by intraventricularly injected captopril, suggesting that ANG II is a mediator in the ACTH response (Cameron et al. 1986). Aguilera et al. (1983) reported that intraventricular captopril reduced the expected ACTH response to ether as well as hemorrhage stress. In contrast, in dogs saralasin did not reduce the ACTH response to surgical stress (Ganong and Murakami 1987) or shaking stress (Hirasawa et al. 1990). Buckner et al. (1986) and Ganong (1989) reported that saralasin had no effect on ACTH response to ether stress. Thus, the effects of circulating ANG II on the secretion of ACTH in stress are controversial. However, repeated immobilization stress produced a significant increase in the density of ANG II binding sites in the PVN and SFO in rats (Castren and Saavedra 1988). There findings suggest that brain-generated ANG II may have a role in regulation of the ACTH response to stress.

In the Japanese quail, dehydration increased ANG II, aldosterone, corticosterone and arginine vasotocin in the blood; injection of ANG II also induced increases of the last three hormones. After rehydration, they returned to normal plasma levels. Hemorrhage produced a marked increase in plasma ANG II. Therefore, there is a possibility that increased plasma ANG II following dehydration or hemorrhage may stimulate CRH and ACTH release (Kobayashi and Takei 1982).

Weld and Fryer (1987) were the first to describe the stimulation of ACTH release in teleosts by ANG I and II using pituitary cell columns. Dispersed, superfused goldfish anterior pituitary cell columns were stimulated with pulses of salmon ANG I (sANG I), human ANG I (hANG I) and human ANG II (hANG II). hANG II stimulated the greatest release of ACTH. hANG I and sANG I were about one-tenth as potent as hANG II in stimulating ACTH release. Captopril, which inhibits the conversion of ANG I to ANG II, did not block the stimulation of ACTH release by sANG I. The authors suggest that the ANG receptor of the goldfish corticotroph is less specific than that of the mammalian corticotroph and is able to recognize both ANG I and ANG II. Further they found that three competitive ANG II analogs, [Sar^1, Ala^8]ANG II, [Sar^1, Thr^8]ANG II and [Sar^1, Ile^8]ANG II did not block hANG II-stimulated ACTH release in the goldfish, but did block it in mammals. From these findings, Weld and Fryer concluded that the ACTH-releasing activity of ANG appeared early in the evolution of the vertebrate pituitary.

8.6.3 Prolactin (PRL)

Intraventricular administration of ANG II suppressed PRL secretion from the rat pituitary (see Steele and Ganong 1986b; Denef 1986; Saavedra 1992; Ganong 1993 for reviews). Saralasin pretreatment abolished the suppression by ANG II, and administration of saralasin alone elevated plasma PRL concentration (Steele et al. 1981). Similar results were observed in the rhesus monkey (Dufy-Barbe et al. 1982). These findings suggest that ANG II suppression of PRL release takes place through ANG II receptors in the brain.

It is well known that dopamine inhibits PRL secretion. Blockade of dopamine receptors by domperidone abolished the inhibitory effects of central ANG II on PRL secretion, suggesting the involvement of dopaminergic mechanisms in ANG II suppression of PRL release (Steele et al. 1982; Steele and Ganong 1986a). Supporting this idea, ANG II-containing neurons were found to project directly to the dopaminergic neurons in the arcuate nucleus (see Ganong 1993 for review). Further, intraventricularly injected renin increased dopamine turnover in the tuberoinfundibular neurons and this was due to an increased formation of ANG II (Anderson et al. 1982). Intraventricularly administered ANG II increased dopamine levels in the arcuate and median eminence regions of rats (Steele et al. 1982). Increased dopamine may be released into portal vessels to reach lactotrophs. Recently, it was found that ANG II AT$_1$ receptors are highly localized in the dorsomedial arcuate nucleus (Seltzer et al. 1993). This nucleus contains dopaminergic neurons, whose fibers terminate in the median eminence (Everitt et al. 1992). Therefore, it is most likely that centrally administered ANG II activates AT$_1$ receptors of the dopaminergic neurons in the dorsomedial arcuate nucleus to release dopamine into portal veins, leading to a decrease in PRL secretion.

Intravenous administration of ANG II increased plasma PRL concentration (Steele et al. 1982), which was the opposite action to that of central ANG II. The increase is considered to be caused by a direct action of circulating ANG II on the lactotrophs which contain ANG II receptors (Aguilera et al. 1982). Intravenous injection of domperidone increased PRL secretion, but pretreatment with [Sar^1, Thr^8]ANG II, an ANG II antagonist, attenuated the release of PRL after domperidone. Thus, ANG II is involved in the modulation of the PRL secretion occurring after the removal of dopamine inhibition (Inoue and Ne-gro-Vilar 1989). It is not known which levels the ANG II worked at, in this case, the level of the lactotrophs, the hypothalamus, or both.

ANG II was capable of stimulating PRL release when added to anterior pituitary cell cultures, and saralasin blocked this release (Aguilera et al. 1982; Schramme and Denef 1983). LHRH stimulated pituitary cell aggregates to secrete PRL (Aguilera et al. 1982). This may be due to ANG II released from gonadotrophs, since gonadotrophs contain ANG II which coexists with LH in secretory granules (Deschepper et al. 1986; Naruse et al. 1986). Saralasin inhibited PRL release by LHRH (Jones et al. 1988; Kubota and Aso 1991) in pituitary cell cultures. The PRL release in response to LHRH is considered to be ANG II-mediated paracrine regulation (Denef and Andries 1983; Jones et al. 1988; see Schwartz and Cherny 1992 for review). However, Robberecht et al. (1992) reported that ANG II is not the paracrine factor mediating the effect of LHRH on PRL release in pituitary cell aggregates. The discrepancy between these results may be due to the differences in methods of cell culture, doses of LHRH, age of rats or other experimental conditions.

In vivo LHRH did not stimulate PRL secretion in male rats and in ovariectomized rats, but it increased PRL release in ovariectomized rats treated with estrogen and progesterone (Steele and Myers 1990). Estrogen may increase the sensitivity of lactotrophs to ANG II, which is secreted from LHRH-stimulated gonadotrophs. When ANG II is administered centrally in ovariectomized rats, ANG II did not inhibit PRL secretion, but supressed it in ovariectomized rats treated with estrogen. The endogenous brain ANG II system appears to be activated and involved in inhibiting PRL release (Myers and Steele 1989).

8.6.4 GH and TSH

Steele et al. (1981, 1982) reported that intraventricular injection of ANG II depressed both PRL and GH plasma levels in ovariectomized rats with estrogen treatment. Pretreatment of these rats with saralasin completely blocked the suppression of both PRL and GH by ANG II. Intravenous injection of ANG II produced a transient reduction of GH levels.

ANG II stimulated GH release from hemipituitaries in vitro at 5.0g/flask dose of ANG II, but not at higher or lower doses (Steele et al. 1981). In pituitary cell aggregates, ANG II showed either stimulatory or inhibitory effects on GH release, depending on the presence or absence of glucocorticoid in the culture medium and the age of the animals (Robberecht and Denef 1988). Stimulation of GH release by ANG II or GRH was inhibited by LHRH, but Mr2266, a κ-opioid antagonist, nullified the inhibitory effect of LHRH. GH release by GRH was also enhanced by Mr2266 in pituitary cell aggregates (Robberecht and Denef 1987). It seems that opioid is inhibitory to the secretion of GH. In addition to LH and FSH, gonadotrophs synthesize dynorphin, an opioid peptide (Khachaturian et al. 1986) and ANG II (Deschepper et al. 1986; Naruse et al. 1986). It is possible, thus, that dynorphin and ANG II produced in gonadotrophs may modulate the secretion of GH from somatotrophs through paracrine actions in vivo.

In ovariectomized rats treated with estrogen, intraventricular and intravenous administration of ANG II did not alter the concentration of TSH in plasma. In incubation of hemipituitaries with relatively high doses of ANG II, TSH concentrations were significantly elevated. Here, the effects on TSH may have been due to a direct action of ANG II on the pituitary thyrotrophs (Steele et al. 1981, 1982). Mitsuma and Nogimori (1984) reported that peripheral ANG II stimulated the release of TRH and TSH at the hypothalamus and pituitary levels. Further investigations are necessary to elucidate the physiological role of ANG II in secretion of TRH and TSH.

8.6.5 LH and LHRH

In ovariectomized rats treated with estrogen and progesterone, intraventricular and intravenous injection of ANG II increased plasma LH concentrations (Steele et al. 1981, 1982). Conversely, in ovariectomized rats without estrogen replacement, neither intraventricular nor intravenous injection of ANG II had a clear effect on plasma LH levels (Steele et al. 1982). Later, Steele et al. (1985) found in untreated ovariectomized rats that the mean whole blood concentration of LH was not affected by intraventricular infusion of artificial cerebrospinal fluid or by 15 ng ANG II/h, but was suppressed dose-dependently by infusion of ANG II at doses of 150 or 600 ng/h. Saralasin blocked the inhibitory effect of ANG II. In ovariectomized rats pretreated with only estradiol, infusion of ANG II (150 or 600 ng/h) did not modify mean blood LH levels. However, in ovariectomized rats pretreated with both estradiol and progesterone, intraventricular infusion of ANG II at 150 or 600 ng/h increased mean LH concentrations. Thus, ANG II shows both inhibitory and stimulatory effects on LH secretion, the direction of the effect being determined by gonadal steroids (Steele et al. 1985).

Intraventricularly administered saralasin and enalaprilat, an ANG II-converting enzyme inhibitor, prevented the normal LH surge and ovulation in cycling female rats (see Saavedra 1992; Ganong 1993 for reviews). Increases in plasma LH concentration were observed in intact females if they were given intraventricular ANG II during proestrus (Steele et al. 1983). In fertile, healthy women, intravenous infusion of ANG II induced an increase in plasma LH, but not FSH, in the middle luteal phase, when the circulating concentrations of both estradiol and progesterone were high, but not in the middle follicular phase (degli Uberti et al. 1991). The rate of brain ANG II release was significantly greater on proestrus compared to diestrus day 1 in rats, assessed by the content of ANG II in the cerebrospinal fluid and in the interstitial fluid of the preoptic-anterior hypothalamus. This suggests that brain ANG II may play a role in the regulation of LH release on the day of proestrus (Ghazi et al. 1994). Thus, ANG II appears to be involved in the secretion of LH and the secretion appears to be modulated by gonadal steroids.

The effect of ANG II on LH secretion is thought to be mediated through regulation of LHRH release. Intracerebroventricular infusion of ANG II increased the amount of LHRH in the extracellular fluid of the anterior pituitary gland as measured by in vivo microdialysis. LH release was also observed. The increase was found in ovariectomized rats treated with estrogen and progesterone, but was not detected in ovariectomized rats without the hormone treatment. These data demonstrate that in hormone-treated rats intracerebroventricular infusion of ANG II stimulates the release of LHRH from the median eminence, which, in turn, acts at the anterior pituitary gland to mediate the effects of ANG II on LH secretion (Steele et al. 1992).

The brain site of action for the stimulatory effect of intraventricular ANG II (1.0 µl of 10^{-10} M) on LH in ovariectomized rats treated with estradiol and progesterone is located within the anterior hypothalamus–preoptic area (AHPOA), and the medial preoptic area is a likely candidate to mediate this effect (Steele 1987). Since other areas such as the OVLT and SFO appear to be involved in the control of LH secretion (Saavedra 1992, p. 343; Ganong 1989, pp. 100–101), the role of these organs in modulating LHRH or LH release by ANG II should be investigated. Steele (1987) reported that equimolar injections of ANG I or ANG III (10^{-10} M) into the AHPOA did not significantly affect blood LH concentrations. It is worthwhile to examine further the efficacy of ANG I and ANG III in LH release.

Norepinephrine administered intraventricularly stimulated LH secretion in intact rats (Barraclough and Wise 1982). However, stimulation was changed to inhibition when rats were ovariectomized, but stimulation was restored if the rats were treated with estrogen and progesterone. The effects of ANG II and norepinephrine on LH release are similar in a variety of different experimental circumstances (see Ganong 1993 for review). Administration of drugs that block α_2-adrenergic receptors prevented LH secretion produced by ANG II and norepinephrine (Steele and Ganong 1986a). Depletion of brain norepinephrine with diethyldithiocarbamate, which inhibits norepinephrine formation, blocked the response to ANG II (Ganong 1993). Thus, it seems that brain ANG II acts on LHRH secretion by facilitating the release of norepinephrine in the preoptic area (see Saavedra 1992; Ganong 1993 for reviews). It is known that ANG II fascilitates norepinephrine release from postganglionic noradrenergic neurons in the autonomic nervous system (Langer 1981). Further, norepinephrine stimulates LHRH neurons (Ramirez et al. 1984).

Taken together, it is most probable that intraventricular injection of ANG II facilitates the secretion of norepinephrine and in turn the norepinephrine stimulates LHRH release into the pituitary, resulting in LH secretion therefrom. However, there is a possibility that peripheral ANG II acts directly on the pituitary, since in vitro incubation of hemipituitaries with ANG II at low doses stimulates LH release (Steele et al. 1982). The effects of ANG II and norepinephrine on LH release are modulated by gonadal steroids.

8.6.6. Corticoids

The distribution of corticoids in vertebrates was summarized by Gorbman et al. (1983). The adrenals of mammals secrete cortisol primarily, except in some species of rodents and rabbits where corticosterone is secreted. The major mineralocorticoid is aldosterone. It is now generally accepted that, in mammals, aldosterone biosynthesis and secretion are controlled predominantly by the circulating RAS and to some degree by ACTH. In this regard, the activity of the circulating RAS is modified by the potassium and sodium content and the volume of the renal tubular fluid. Adrenergic nervous control is also involved. Details of aldosterone biosynthesis were reviewed by Carey (1982) and Quinn and Williams (1992). ANG II stimulates aldosterone biosynthesis in rat glomerulosa cells. Sodium restriction and potassium loading increase the sensitivity and the magnitude of the aldosterone response to ANG II. ANG II caused dose-related increases in aldosterone and corticosteroids, but did not alter ACTH in conscious dogs (Keller-Wood et al. 1986). Compared with ANG II, ANG III had the low steroidogenic activity of zona glomerulosa cells in vitro. This is probably related to the rapid degradation of ANG III during incubation. Endothelin-1 potentiated ANG II stimulation of aldosterone production in cultured bovine adrenal glomerulosa cells (Cozza et al. 1992). Parathyroid hormone stimulated aldosterone secretion from bovine glomerulosa cells, both alone and in combination with angiotensin II (Isales et al. 1991). Human platelet-derived growth factor (PDGF) caused a dose-dependent inhibition of ANG II-induced aldosterone synthesis of rat glomerulosa cells. PDGF may play an important role in the regulation of aldosterone synthesis by acting as a potent negative modulator of ANG II action in the adrenal (Natarajan and Nadler 1992).

ANG II stimulated cortisol secretion from bovine cultured zona fasciculata/reticularis cells, being mediated by activation of phospholipase C (Bird et al. 1989). Both ANG II-stimulated cortisol secretion and phosphoinositidase C activity were inhibited by saralasin and Dup753 (AT_1 antagonist), but not by PD123177 (AT_2 antagonist; Clyne et al. 1993). ANG II stimulated [^{3}H] thymidine incorporation into DNA in cultured bovine adrenal cortical cells (Gill et al. 1977) and zona fasciculata/reticularis cells (Clyne et al. 1993). This ANG II effect was also inhibited by saralasin and Dup753, but not by PD123177 (Quali et al. 1992; Clyne et al. 1993). Thus, the steroidogenic and mitogenic effects of ANG II in bovine zona fasciculata/reticularis cells appear to be mediated by the AT_1 receptor. ANG II stimulated corticosterone and cAMP production in bovine zona fasciculata cells, and this effect was inhibited by Dup753, but not by PD123319. Therefore, the ANG II stimulation of cAMP and corticosterone synthesis is regulated by the AT_1 receptor subtype (Rainey et al. 1991).

In birds, the major corticoids are corticosterone and aldosterone. In the duck (*Anas platyrhynchos*), ACTH stimulated the release of corticosterone and aldosterone from slices of the subcapsular (SCZ) and inner (IZ) zones of the adrenal gland in incubation and superfusion systems. Release ratios of corticosterone and aldosterone to ACTH were different between the SCZ and IZ. Superfusion with 10^{-12} to 10^{-5} M ANG II stimulated the release of aldosterone from the SCZ cells but had no detectable effect on the IZ and failed to stimulate corticosterone release from either SCZ or IZ cells (Klingbeil 1985). ANG II selectively induced aldosterone production, but not that of corticosterone and deoxycorticosterone, by superfused duck adrenal steroidogenic tissue (Holmes et al. 1991) and by dispersed steroidogenic cells from embryonic duck adrenal tissue, but not by cells from chick embryo (Collie et al. 1992). ANG II induced aldosterone production, but not corticosterone production in adrenal steroidogenic cells of the domestic turkey, *Meleagris gallopavo* (Kocsis et al. 1994b). Properties of turkey ANG II receptors are pharmacologically distinct from mammalian adrenocortical type-1 receptors (Kocsis et al. 1994a; see Chap. 7 for details).

In reptiles, corticosterone and aldosterone are considered to be the major corticoids. In the freshwater turtle, *Pseudemys scripta*, infusion of native ANG II increased blood pressure and plasma corticosterone concentrations. However, mammalian ACTH failed to alter either blood pressure or corticosterone levels. Neither ANG II nor ACTH had any effect on plasma cortisol or aldosterone concentrations, which fell below the minimal detection levels for the assays employed. It is suggested that corticosterone is the primary mineral corticoid in *P. scripta* and perhaps in reptiles in general (Sanford and Stephens 1988). However, aldosterone, cortisol, deoxycorticosterone, 18-hydroxycorticosterone and others have been found in reptilian adrenal tissue incubated in vitro. The effects of ACTH and ANG II on corticoid production are not consistent among reptilian species. The presence of an adenohypophysis-adrenal axis in reptiles is still uncertain (Sanford and Stephens 1988).

In amphibians, homologous renin extracts stimulated aldosterone and corticosterone in *Rana catesbeiana* (Nishimura 1980a). Specific binding sites for ANG II were present in the adrenal of the semiterrestrial frog, *Rana temporaria* (Kloas and Hanke 1992a), but in the aquatic clawed toad, *Xenopus laevis*, binding sites for ANG II were not detected (Kloas and Hanke 1992b). In the urodele, *Ambystoma mexicanum*, the adrenal tissue lacked ANG II receptors and ANG II did not affect adrenal steroid and catecholamine secretion (Kloas and Hanke 1993). It seems that the difference in environment, aquatic or terrestrial, may be important in the role of ANG II in corticosteroid production.

In bony fishes, cortisol is the major corticoid and aldosterone is typically absent or present in very low concentrations. In the eel, *Anguilla anguilla*, intravascular injections of ACTH, mammalian ANG II and eel renin preparations increased plasma cortisol (Borriraja et al. 1973). In the trout, *Oncorhynchus mykiss*, ACTH, ANG II and the urophysial preparations, urotensins I and II (UI and UII), stimulated cortisol secretion in interrenal preparations of SW- and FW-adapted trout (Arnold-Reed and Balment 1994). After several experiments, these authors concluded that secretion from the interrenal is basically under hypothalamic–pituitary (ACTH) control. However, a number of other blood-borne systemic factors (ANG II, UI, UII) affect this control.

ACTH and ANG II stimulated the secretion of 1α-hydroxycorticosterone, which is the principal corticosteroid produced by the elasmobranch, from an isolated perfused interrenal gland of the dogfish, *Scyliorhynus canicula*, although the stimulatory mechanism was different between ACTH and ANG II (Armour et al. 1993). In the lungfish, *Neoceratodus forsteri*, corticosterone, aldosterone and cortisol were found in plasma unlike in the bony fishes. This finding is interesting, because the lungfish is most closely related to the tetrapod vertebrates. Intramuscular injections of [Asp[1], Ile[5]] ANG II increased aldosterone levels. [Asp[1], Val[5]] ANG II increased plasma corticosterone levels (Joss et al. 1994).

It is not completely clear yet which is more important, the adenohypophysial–adrenal system or the RAS, in the secretion of aldosterone, corticosterone and cortisol especially among lower vertebrates. Further, the regulatory mechanisms of ACTH and ANG II at the cellular or molecular levels in vitro have not been fully studied.

Tissue RAS has been found in the mammalian adrenal cortex in a variety of species. In the rat, mainly renin is present and is synthesized in the zona glomerulosa cells. Renin production is altered by changes in electrolyte intake and ACTH. A positive correlation is present between the concentrations of adrenal renin and aldosterone. Further, in primary culture of zona glomerulosa cells, a converting enzyme inhibitor, lisinopril, reduced aldosterone stimulation by ACTH and potassium. However, these studies are not convincing evidence for a physiological role of the adrenal RAS (Mulrow 1992).

8.7 Other Actions of ANG II

ANG II is involved in many biological phenomena. In previous chapters, drinking behavior and release of several pituitary hormones are discussed in connection with ANG II. These are only part of the actions caused by ANG II. It is difficult to cover all of its biological activities in this book. Therefore, some other reviews and books pertinent to areas not discussed herein will be introduced briefly. The interrelations between renin and atrial natriuretic peptide (ANP) are discussed briefly in Section 5.1.4.3 and there are reviews by Richards and Nicholls (1993) and Schiffrin et al. (1993). The interactions between ANG II and catecholamines were briefly mentioned in Section 5.1.4.2. This topic was reviewed by Phillips (1987b) and Saavedra (1992). These reviews also dealt with interactions between ANG II and serotonin and opioids. The effects of ANG II on renal function were discussed by Zhuo and Mendelsohn (1993), Hall and Brands (1993) and Schalekamp and Derkx (1993). The angiogenic action of ANG II in glomerular growth was reported by Fogo et al. (1990). The involvement of ANG II in biological phenomena or behavior such as ovulation, pregnancy or sexual behavior was described in a review by Wright and Harding (1992). ANG II's role in cognitive processing and analgesia was well documented by Wright and Harding (1992). Psychoactive properties of ANG II and its fragments are discussed by Braszko et al. (1991). Papers on the relationship between ANG II and choroid plexus and cerebrospinal fluid formation can be found in a review by Chodobski et al. (1995).

The paracrine functions of ANG II of the tissue RAS have been discussed extensively in relation to the release of anterior pituitary hormones (Denef 1986; Ganong 1989; Jones et al. 1990; Schwartz and Cherny 1992). Abstracts describing paracrine functions of the tissue RAS were collected in Regulatory Peptides 53:139–159, 1994, and several reviews exist in the book (Vol.1) edited by Robertson and Nicholls (1993).

Chapter 9

General Conclusions

9.1 Comparative Anatomy of the RAS

9.1.1 Invertebrates

An immunoreactive ANG II-like substance has been found in the spinal cord of the amphioxus and in the nervous system of the earthworm. Further, ANG II amide and an ANG I-like molecule were chemically identified and the presence of the RAS was proposed in the nervous system of the leech by Laurent et al. (1995; Sect. 1.1). It seems, therefore, that ANG I- and ANG II-like substances occurred originally in the nervous system of invertebrates, suggesting that the primary actions of these peptides were associated with nervous functions and/ or neurosecretion. Since ANG II is generated by the tissue RAS in the brain of vertebrates, it appears that this peptide has been conserved in the nervous tissue during evolution. As organisms became more complex in structure and function, ANG II has come to be produced not only by tissue RAS in the nervous system but also by other tissues and organs, as a circulating or local hormone in vertebrates. Other components of the RAS should be studied in invertebrates.

9.1.2 Components of JGA

The distribution of the components of the JGA in vertebrates is summarized in Table 2.1. Mammals and birds possess all the components of the JGA: juxta-glomerular (JG) cells, macula densa (MD) and extraglomerular mesangium (EGM). The kidney glomeruli of mammals and birds contain peripolar cells, although it is not clear whether these cells are associated with JGA. In reptiles, JG cells and peripolar cells are present, but the EGM is absent. Although reptiles were thought to lack the MD, Koval'Chuk (1987) found primitive forms of the MD in the lizard, *Lacerta agilis*. Further detailed studies with electron microscopy using more species are needed. Amphibians have JG cells and peripolar cells, but not the EGM. In some amphibian species, the presence of a structure similar to the mammalian MD has been demonstrated. Sarcopter-ygians and holocephalans possess JG cells, but they lack the MD and EGM. Teleosts have JG cells, but there is no clear evidence for the existence of the MD or EGM. In the carp and goldfish, cells resembling mammalian peripolar cells have been observed.

For many years, elasmobranchs were thought to lack JG cells, MD and EGM. However, recent studies demonstrated that elasmobranchs have JG cells containing granules stainable with toluidine blue but not with Bowie's stain. The conflict in the data is probably due to differences in fixatives, stains and species. An MD structure was also identified with electron microscopy. Furthermore, immunoreactive ANG I and II were detected in the plasma of the nurse shark, *Gynclymostoma cirratum*, and ANG I was recently isolated in the dogfish, *Triakis scyllia*. ANG I pressor activity was detected and this activity was blocked by SQ20881 in the spiny dogfish, *Squalus acanthias*. Converting enzyme-like activity, which was inhibited by captopril, was found in the dogfish, *Scyliorhinus canicula*. Thus, elasmobranchs are now known to possess all the components of the RAS. Peripolar cells are also evident, and the EGM seems to be present. It is interesting that elasmobranchs, which have evolved separately from mammals since the Devonian period, have followed the same evolutionary direction as mammals in the development of the JGA. In the cyclostomes, no JG cells, MD, or EGM have been found. However, lamprey renal extracts incubated with canine renin substrate generated pressor substances when assayed in the rat. Further investigations using different fixatives and stains and utilizing more recent techniques are needed for the cyclostomes.

9.1.3 Tissue RAS

Renin secreted from JG cells cleaves angiotensinogen into ANG I in the blood, which is converted into ANG II by converting enzyme in the pulmonary capillary bed. In contrast to this circulating RAS, in mammals RAS components have been found in many tissues and organs, such as the brain, adrenal, pituitary, kidney, vascular smooth muscle and others (Table 4.1). ANG II thus produced may function locally as a paracrine or autocrine mediator. Further physiological and biochemical actions of ANG II produced by tissue RAS should be explored. Tissue RAS has not yet been studied in lower vertebrates. Its presence has been proposed in invertebrates (Sect. 1.1) and should be examined.

9.2 Renin Release

The functioning of the RAS is primarily regulated by plasma renin activity. Regulation of renin release is the most important determinant of RAS activity. Renin release is regulated by three major factors: (1) perfusion pressure in the renal afferent arteriole; (2) Na^+ or Cl^- load to the macula densa; and (3) activity of the renal sympathetic nerve (Fig. 5.1). Other humoral factors are also involved, of which ANP, prostaglandins and adenosines are the most likely candidates. Ca ions are known to be a stimulatory second messenger for hormone secretion in general. However, Ca ions are an inhibitory intracellular messenger for renin secretion, and cAMP is a stimulatory second messenger (Fig. 5.2). A feedback mechanism exists between circulating ANG II and renin secretion from the JG cells (Fig.8.19).

In nonmammals, it is not yet possible to measure plasma renin concentration by direct redioimmunoassay. Thus, renin activity is usually determined indirectly by measuring the amount of ANG I produced after incubation with mammalian angiotensinogen. However, plasma renin activity is low because of the heterologous combination of renin and angiotensinogen. Thus, kidney renin activity is often measured instead of plasma renin activity. However, it is not yet clear whether changes in renal renin activity are caused by increased synthesis or decreased release of renin. For these reasons, little is known about the mechanisms regulating renin release in nonmammalian species. Data obtained to date indicate that the mechanisms sensing changes in blood pressure or blood volume are phylogenetically old regulatory systems for renin release, whereas Na^+ sensors in the kidney seem to have appeared later in vertebrate phylogeny (Fig. 5.3). Homologous radioimmunoassays for nonmammalian renin need to be developed to clarify renin release regulation in nonmammalian species.

9.3 Comparative Biochemistry of the RAS

9.3.1 Components of the RAS

In mammals, the primary structures of renin, converting enzyme and angiotensinogen, as well as ANG I and II, have been determined for their cDNA and genomic DNA. By means of Northern blot analysis or ribonuclease protection assay sequence tissues are being identified which synthesize one or more of the components of the RAS (Table 4.1). Some of the tissues produce all the components, and can locally generate ANG II independently of blood-borne ANG II.

In nonmammalian species, ANG I has been isolated from incubates of plasma with homologous kidney extracts; the primary structure of ANG I has been determined for selected species of birds, reptiles, amphibians, teleosts and elasmobranchs (Table 4.2). However, the primary structures of other components of the RAS are not known to date. Renin-producing JG cells have been identified in the kidney of most of these species, but the presence of renin in other tissues has not yet been examined. The presence of converting enzyme has been assessed by the ability of captopril or SQ20881 to inhibit the conversion of ANG I to ANG II. Converting enzyme has been localized in the lung of birds and the gill of teleost fish; extracts of these tissues have high catalytic activities at Hip-His-Leu, although nonmammalian ANG I does not always have histidine at position 9 (Table 4.2). The gill and lung are ideally located for conversion of ANG I to II in the blood, since they receive the entire output from the heart.

9.3.2 Receptors and Cellular Mechanisms

ANG II receptors have been identified by autoradiography and radioligand binding assay in various tissues of mammals where ANG II actions occur. The affinity of these receptors is comparable to the ED_{50} of ANG II's biological action in vitro; their affinity and capacity are altered by procedures which alter the potency and efficacy of ANG II action on the tissue (Table 7.1). At least two types of ANG II receptos have been identified in mammalian tissues; AT_1 receptors which coupled to GTP-binding protein and utilize the IP_3/Ca system or inhibition of adenylate cyclase as a signal transduction system, and AT_2 receptors whose signal transduction system is still unknown (Table 7.2; Fig. 7.3).

Nonmammalian tissues may have a different type of ANG II receptors. ANG II elicits a contractile response in mammalian vascular smooth muscle via the IP_3/Ca system, but it relaxes smooth muscle in the chicken. Therefore, ANG II receptors in chicken vascular smooth muscle may utilize a unique signal transduction system. However, it was later shown that ANG II receptors cloned in the African clawed frogs were of AT_1 type.

9.4 Biological Actions

9.4.1 Body Fluid Regulation

As mentioned above, angiotensins seem to be distributed widely in invertebrates, especially in the nervous system. The functions of angiotensins in invertebrates are not known, but there are two reports which suggest possibilities. It has been reported that ANG II amide is diuretic in the leech and elsewhere that injection of mammalian ANG II stimulates water absorption through the foot in the terrestrial slug. It appears that ANG seems to be associated with body fluid regulation through nervous function or neurosecretion in invertebrates.

In vertebrates, ANG II exhibits a wide spectrum of biological actions including body fluid balance, cardiovascular homeostasis and growth promotion in relation to angiogenesis and cardiotrophy. Summarizing all the reported actions, it appears that the primary role of ANG II is the regulation of body fluid homeostasis. For example, renin is released in response to decreases in blood volume (Sect. 5.1.1), which then leads to an immediate increase in plasma ANG II. ANG II induces thirst and sodium appetite and increases intestinal absorption of water and Na$^+$. ANG II also reduces loss of water and Na$^+$ through the kidney by decreasing the glomerular filtration rate and increasing tubular reabsorption. Synergistic effects of ANG II with other hormones have also been demonstrated, as in the arousal of sodium appetite with aldosterone. ANG II increases vasopressin and aldosterone release (Fig. 8.19), which further inhibits renal loss of water and Na$^+$. Thus, ANG II seems to act, in concert with vasopressin and aldosterone, to retain water and Na$^+$ in the body of terrestrial animals. In semiterrestrial amphibians, ANG II induces behavior which results in water absorption across the ventral skin. Fishes imbibe surrounding water simply by swallowing, and ANG II appears to induce reflex swallowing in the eel.

Available data indicate that the RAS is the first humoral messenger responsible for body fluid homeostasis. Thus the interaction of the RAS with atrial natriuretic peptide (ANP), a possible first messenger promoting body fluid loss, is of interest (Sect. 5.1.4.3). ANP has been shown to counteract all ANG II actions; it inhibits secretions of renin, vasopressin and aldosterone, reduces drinking and sodium appetite, decreases intestinal water and Na$^+$ absorption, and promotes renal water and Na$^+$ excretion. Investigations of the RAS have been almost exclusively limited to mammals. However, since amphibians and fishes have unique mechanisms for body fluid homeostasis, studies of the biological actions of the RAS and its interaction with other hormonal systems in these semiaquatic and aquatic animals may provide new insights into the function of the RAS in general.

9.4.2 Ecological Aspects of ANG II-Induced Drinking

ANG II stimulates drinking in brackish-water fishes but not in fishes inhabiting only sea water or only fresh water (Sect. 8.2). Fishes living in osmotically constant environments do not need to regulate drinking rate. In contrast, brackish-water fishes start to drink when they encounter sea water (hyperosmotic media), but stop drinking when the ambient water becomes fresh (hyposmotic media). An ANG II-induced drinking mechanism may function to stimulate this sea water drinking. Amphibians do not respond to ANG II by drinking; rather, they have developed a water-absorption behavioral mechanism responding to ANG II. Further, the ventral skin is stimulated by ANG II to absorb water. In most reptiles, birds and mammals, drinking behavior is induced by ANG II; however, those which originated in arid areas where drinking in nature is rare, are unresponsive to ANG II. Carnivorous birds also do not respond to ANG II. They derive water mostly from their prey and usually do not require additional water by drinking in nature. Among the mammals studied, the herbivorous species which ingest needed water from grass do not drink in response to ANG II. All of these findings suggest that in species which drink only a little water in nature, an ANG-thirst mechanism, as part of their water-acquiring behavior, has become attenuated or lost during evolution. It is of interest to examine the distribution of ANG II receptors in the brain in animals showing unresponsiveness to ANG II in drinking, as ANG II receptors are absent in the brain of the possum lacking an ANG II drinking response (Sernia et al. 1990).

Receptive sites for ANG II are the SFO, OVLT, MnPO and perhaps OCPLT in the brain of mammals (Sect. 8.4) and are in similar regions in birds and probably in reptiles. In amphibians, drinking mechanisms induced by ANG II have disappeared. However, receptors for ANG II have developed in the skin and ANG II stimulates water absorption through the skin. The brain seems to have receptive sites for ANG II which are involved in cutaneous water-absorption behavior. In fishes responding to ANG II by drinking, the receptive site for ANG II seems to be in the medulla oblongata, possibly in the swallowing center. The existence of receptive sites in other portions of the brain has not been examined. Thus, a brain mechanism for ANG II-induced water uptake appears to have been preserved during evolution, although the way it functions varies, especially in lower vertebrates.

9.4.3 ANG II and Reaction to Emergency

Involvement of ANG II in natural drinking has been described in Section 8.3. In this section, we consider the contribution of ANG II to emergencies in water balance. Brackish-water fishes may not drink when they are in fresh water, but they drink sea water when they encounter an emergency caused by a sudden increase in osmotic pressure at high tide. This sea water drinking must be induced by increase of plasma ANG II in response to a sudden environmental salinity change. Similarly, eels transferred from fresh water to sea water show an increase in plasma ANG II concentration and start drinking. This is true for terrestrial species, for example, the Japanese quail immediately increases plasma ANG II levels in response to emergencies such as water deprivation or hemorrhage, leading to drinking. At the same time, plasma arginine vasotocin, corticosterone and aldosterone increase similarly when mammals react to emergencies. Since ANG II injection induces increases in these hormones, ANG II seems to function as a trigger for secretion of other hormones involved in emergency reactions. Thus the RAS appears to be a true first messenger for body fluid homeostasis. The role of ANG II in emergency reactions seems to have been preserved during the evolutionary process from fishes to mammals. Involvement of ANG II in emergency reactions should also be studied in amphibians and invertebrates.

References

Abbrecht PJ, Vander AJ (1970) Effects of chronic potassium deficiency on plasma renin activity. J Clin Invest 49: 1510–1516

Abe Y, Okahara T, Kishimoto T, Yamamoto K, Ueda J (1973) Relationship between intrarenal distribution of blood flow and renin secretion. Am J Physiol 225: 319–323

Abraham SF, Baker RM, Blaine EH, Denton DA, McKinley MJ (1975) Water drinking induced in sheep by angiotensin – a physiological or pharmacological effect? J Comp Physiol Psychol 88: 503–518

Abraham SF, Denton DA, Weisinger RS (1976) Effect of an angiotensin antagonist, Sar^1-Ala^8-angiotensin II on physiological thirst. Pharm Physiol Behav 4: 243–247

Aguilera G, Catt K (1981) Regulation of vascular angiotensin II receptors in the rat during altered sodium intake. Circ Res 49: 751–758

Aguilera G, Hauger RL, Catt KJ (1978) Control of aldosterone secretion during sodium restriction: adrenal receptor regulation and increased adrenal sensitivity to angiotensin II. Proc Natl Acad Sci USA 75: 975–979

Aguilera G, Schirar A, Baukal A, Catt KJ (1980) Angiotensin receptors. Properties and regulation in adrenal glomerulosa cells. Circ Res 46 (Suppl I): 118–127

Aguilera G, Hyde LC, Catt KJ (1982) Angiotensin II receptors and prolactin release in pituitary lactotrophs. Endocrinology 111: 1045–1050

Aguilera G, Chiueh CC, Mohan MK, Catt KJ (1983) Role of angiotensin II in the regulation of ACTH secretion. Endocrinology 112: 90A

Aguilera G, Young WS, Kiss A, Bathia A (1995) Direct regulation of hypothalamic corticotropin-releasing-hormone neurons by angiotensin II. Neuroendocrinology 61: 437–444

Akagi H, Hayashi T, Nakayama T, Nakajima T, Watanabe TX, Sokabe H (1992) Comparative studies on angiotensin. VI. Structure of angiotensin I produced by renal renin of the dog, guinea pig, and rabbit, and re-examination of the peptides of the pig, horse and ox using homologous renin sources. Chem Pharm Bull 30: 2498–2502

Akahane K, Umeyama H, Nakagawa S, Moriguchi I, Hirose S, Iizuka K, Murakami K (1985) Three-dimensional structure of human renin. Hypertension 7: 3–12

Akaishi T, Negoro H, Kobayasi S (1980) Responses of paraventricular and supraoptic units to angiotensin II, Sar^1-Ile^8-angiotensin II and hypertonic NaCl administrated into the cerebral ventricle. Brain Res 188: 499–511

Alcorn D, Cheshire GR, Coghlan JP, Ryan GB (1984) Peripolar cell hypertrophy in the renal juxtaglomerular region of the newborn sheep. Cell Tissue Res 236: 197–202

Aldred GP, Chai SY, Song K, Zhuo J, MacGregor DP, Mendelsohn FAO (1993) Distribution of angiotensin II receptor subtypes in the rabbit brain. Regulatory Peptides 44: 119–130

Alexander RW, Brock TA, Gimbrone Jr MA, Rittenhouse SE (1985) Angiotensin increases inositol triphosphate and calcium in vascular smooth muscle. Hypertension 7: 447–451

Allison DJ, Tanigawa H, Assaykeen TA (1972) The effects of cyclic nucleotides on plasma renin activity and renal function in dogs. In: Assaykeen TA (ed) Control of renin secretion. Plenum Press, New York, pp 33–47

Anand-Srivastava MB (1983) Angiotensin II receptors negatively coupled to adenylate cyclase in rat aorta. Biochem Biophys Res Commun 117: 420–428

Anderson B, Eriksson L (1971) Conjoint action of sodium and angiotensin on brain mechanisms controlling water and salt balances. Acta Physiol Scand 81: 18–29

Anderson B, Westbye O (1970) Synergistic action of sodium and angiotensin on brain mechanisms controlling fluid balance. Life Sci 9: 601–608

Anderson B, Leksell G, Lishajko F (1975) Perturbations in fluid balance induced by medially placed forebrain lesions. Brain Res 99: 261–275

Anderson K, Fuxe K, Agnati LF, Ganten D, Zini I, Mascagni EP, Infantillina F (1982) Intraventricular injections of renin increase amine turnover in the tuberoinfundibular dopamine neurons and reduce secretion of prolactin in the male rat. Acta Physiol Scand 116: 317–320

Anderson RC, Herbert PN, Murlow PJ (1968) A comparison of properties of renin obtained from the kidney and uterus of the rabbit. Am J Physiol 215: 774–778

Andre P, Schott C, Nehlig H, Stoclet J-C (1990) Aortic smooth muscle cells are able to convert angiotensin I to angiotensin II. Biochem Biophys Res Commun 173: 1137–1142

Aoi W, Wade B, Rosner DR, Weinberger MH (1974) Renin release by rat kidney slices in vitro: effects of cations and catecholamines. Am J Physiol 227: 630–634

Aoyagi T, Tobe H, Kojima F, Hamada M, Takeuchi T, Umezawa H (1978) Amastatin, an inhibitor of aminopeptidase A, produced by actinomycetes. J Antibiotics 31: 636–638

Arend LJ, Hanamati A, Thompson CI, Spielman WS (1984) Adenosine-induced decrease in renin-release: dissociation from hemodynamic effects. Am J Physiol 247: F447–F452

Arillo A, Uva B, Vallarino M (1981) Renin activity in rainbow trout (*Salmo gairdneri* Rich.) and effects of environmental ammonia. Comp Biochem Physiol 68A: 307–311

Arimura A, Saito T, Shally AV (1967) Assays for corticotropin-releasing factor (CRF) using rats treated with morphine, chlorpromazine and nembutal. Endocrinology 81: 235–245

Armour KJ, O'Toole LB, Hazon N (1993) Mechanisms of ACTH- and angiotensin II-stimulated 1α-hydroxycorticosterone secretion in the dogfish *Scyliorhynus canicula*. J Mol Endocrinol 10: 235–244

Arnold-Reed DE, Balment RJ (1994) Peptide hormones influence in vitro interrenal secretion of cortisol in the trout, *Oncorhynchus mykiss*. Gen Comp Endocrinol 96: 85–91

Assad MM, Antonaccio MJ (1982) Vascular wall renin in spontaneously hypertensive rats: potential relevance to hypertension maintenance and antihypertensive effect by captopril. Hypertension 4: 487–493

Assaf SY, Mogenson GJ (1976) Evidence that the preoptic region is a receptive site for the dipsogenic effects of angiotensin II. Pharm Biochem Behav 5: 697–699

Assaykeen TA, Ganong WF (1971) The sympathetic nervous system and renin secretion. In: Martini L, Ganong WF (eds) Frontiers in neuroendocrinology. Oxford Univ Press, New York, pp 67–102

Assaykeen TA, Clayton PL, Goldfien A, Ganong WF (1970) Effect of alpha- and beta-adrenergic blocking agents on the renin response to hypoglycemia and epinephrine in dogs. Endocrinology 87: 1318–1322

Assaykeen TA, Tanigawa H, Allison DJ (1974) Effect of adrenoceptor-blocking agents on the renin response to isoproterenol in dogs. Eur J Pharmacol 26: 285–297

Asscher AW, Anson SG (1963) A vascular permeability factor of renal origin. Nature 198: 1097–1099

Atlas D, Melamed E, Lahau M (1977) Beta-adrenergic receptors in rat kidney. Direct localization by a fluorescent alpha-blocker. Lab Invest 36: 465–468

Avrith DB, Fitzsimons JT (1980) Increased sodium appetite in the rat induced by intracranial administration of components of the renin-angiotensin system. J Physiol (Lond) 301: 349–364

Avrith DB, Wiselka MJ, Fitzsimons JT (1980) Increased sodium appetite in adrenalectomized or hypophysectomized rats after intracranial injections of renin or angiotensin II. J Endocrinol 87: 109–112

Bailey JR, Randall DJ (1981) Renal perfusion pressure and renin secretion in the rainbow trout, *Salmo gairdneri*. Can J Zool 59: 1220–1226

Bailie MD, Derkx FMK, Schalekamp MADH (1979) Release of active and inactive renin by the porcine kidney. Circ Res 44: 32–37

Baker KM, Aceto JA (1989) Characterization of avian angiotensin II cardiac receptors. J Mol Cell Cardiol 21: 375–382

Baker KM, Aceto JF (1990) Angiotensin II stimulation of protein synthesis and cell growth in chick heart cells. Am J Physiol 259: H610–H618

Baker KM, Campanile CP, Trachte GJ, Peach MJ (1984) Identification and characterization of the rabbit angiotensin II myocardial receptor. Circ Res 54: 286–293

Baker KM, Booz GW, Dostal DE (1992) Cardiac actions of angiotensin II: role of an intracardiac renin-angiotensin system. Annu Rev Physiol 54: 227–241

Baldwin BA, Thornton SN (1986) Operant drinking in pigs following intracerebroventricular injections of hypertonic solutions and angiotensin II. Physiol Behav 36: 325–328

180

Balment RJ, Carrick S (1985) Endogenous renin-angiotensin system and drinking behaviour in flounder. Am J Physiol 248: R157–R160

Balment RJ, Loveridge JP (1989) Endocrines and osmoregulatory mechanisms in the Nile crocodile, *Crocodylus niloticus*. Gen Comp Endocrinol 73: 361–367

Balment RJ, Henderson IW, Oliver JA (1975) The effects of vasopressin on pituitary oxytocin content and plasma renin activity in rats with hypothalamic diabetes insipidus (Brattleboro strain). Gen Comp Endocrinol 26: 468–477

Balment RJ, Brimble MJ, Forsling ML (1980) Release of oxytocin induced by salt loading and its influence on renal excretion in the male rat. J Physiol (Lond) 308: 439–449

Barajas L (1964) The innervation of the juxtaglomerular apparatus. An electron microscopic study of the innervation of the glomerular arterioles. Lab Invest 13: 916–929

Barajas L (1970) The ultrastructure of the juxtaglomerular apparatus as disclosed by three-dimensional reconstructions from serial sections. The anatomical relationship between the tubular and vascular components. J Ultrastruct Res 33: 116–147

Barajas L (1978) Innervation of the renal cortex. Fed Proc 37: 1192–1201

Barajas L (1979) Anatomy of the juxtaglomerular apparatus. Am J Physiol 237: F333–F343

Barajas L, Latta H (1963) Three-dimensional study of the juxtaglomerular apparatus in the rat: light and electron microscopic observation. Lab Invest 12: 257–269

Barajas L, Latta H (1967) Structure of the juxtaglomerular apparatus. Circ Res 21/22: II15–II28

Barajas L, Müller J (1973) The innervation of the juxtaglomerular apparatus and surrounding tubules: a quantitative analysis by serial section electron microscopy. J Ultrastruct Res 43: 107–132

Barajas L, Müller J (1980) Structure of the juxtaglomerular apparatus. In: Johnson JA, Anderson RR (eds) The renin-angiotensin system. Advances in experimental medicine and biology. Plenum Press, New York, pp 85–134

Barraclough CA, Wise PM (1982) The role of catecholamines in the regulation of pituitary luteinizing hormone and follicle-stimulating hormone release. Endocr Rev 3: 91–119

Bartholomew GA, Cade TJ (1963) The water economy of land birds. Auk 80: 504–539

Bartoli E, Earley LE (1973) Measurements of nephron filtration rate in the rat with and without occlusion of the proximal tubule. Kidney Int 3: 372–380

Baumbach L, Leyssac PP (1977) Studies on the mechanisms of renin release from isolated superfused rat glomeruli: effects of calcium, calcium ionophore and lanthanum. J Physiol (Lond) 273: 745–764

Baxter CR, Horvath JS, Duggin GG, Tiller DJ (1980) Effect of age on specific angiotensin II binding sites in rat brain. Endocrinology 106: 995–999

Bealer SL, Phillips MI, Johnson AK, Schmid PG (1979) Anteroventral third ventricle lesions reduce antidiuretic responses to angiotensin II. Am J Physiol 236: E610–E615

Bean JW (1942) Specificity in the renin-hypertensinogen reaction. Am J Physiol 136: 731–742

Beasley D, Shier DN, Malvin RL, Smith G (1986) Angiotensin-stimulated drinking in marine fish. Am J Physiol 250: R1034–R1038

Beaufils M, Sraer J, Lepreux C, Ardaillou R (1976) Angiotensin II binding to renal glomeruli from sodium-loaded and sodium depleted rats. Am J Physiol 230: 1187–1193

Beck N, Kim KS, Davis B (1975) Catecholamine-dependent cyclic adenosine monophosphate and renin in the dog kidney. Circ Res 36: 401–405

Beierwaltes WH, Schryver S, Olson PS, Romero JC (1980) Interaction of prostaglandin and renin-angiotensin systems in isolated rat glomeruli. Am J Physiol 236: F602–F608

Beierwaltes WH, Prada J, Carretero OA (1985) Effect of glandular kallikrein on renin release in isolated rat glomeruli. Hypertension 7: 27–31

Beldent V, Michaud A, Wei L, Chauvet M-T, Corvol P (1993) Proteolytic release of human angiotensin-converting enzyme. J Biol Chem 268: 26428–26434

Bell FR, Doris PA, Wood TJ (1985) The coincidental effects of dehydration and rehydration on plasma and cerebrospinal fluid angiotensin II levels in unrestrained steers. Brain Res 325: 143–150

Bellocci M, Picardi R, Martino CD (1971) Juxtaglomerular apparatus in mesonephros of newt (*Triturus cristatus*). A morphologic study. Z Zellforsch 114: 203–219

Bencsath P, Szaloy L, Debreczeni LA, Vajada L, Takacs L, Fischer A (1972) Denervation diuresis and renin secretion in the anesthetized dog. Eur J Clin Invest 2: 422–425

Bennett JP Jr, Snyder SH (1976) Angiotensin II binding to mammalian brain membranes. J Biol Chem 251: 7423–7430

Bennett JP Jr, Snyder SH (1980) Regulation of receptor binding interactions of ^{125}I-angiotensin II and [125]I-[Sarcosine[1], leucine[8]] angiotensin II, an angiotensin antagonist, by sodium ion. Eur J Pharmacol 67: 1–10

Bentley PJ (1971) Endocrines and osmoregulation. Zoophysiology and ecology 1. Springer, Berlin Heidelberg New York, pp 138–141

Bentley PJ, Yorio T (1979) Do frogs drink? J Exp Biol 79: 41–46

Beresford MJ, Fitzsimons JT (1992) Intracerebroventricular angiotensin II-induced thirst and sodium appetite in rat are blocked by AT_1 receptor antagonist, losartan (DuP753), but not by the AT_2 antagonist, CGP42112B. Exp Physiol 77: 761–764

Berger C (1966) Mikroskopische und histochemische Untersuchungen an der Niere von Columba livia oberratio domestica. Z Mikrosk Anat Forsch 74: 436–456

Bergsma DJ, Ellis C, Nuthulaganti PR, Nambi P, Scaife K, Kumar C, Aiyar N (1993) Isolation and expression of a novel angiotensin II receptor from Xenopus laevis heart. Mol Pharmacol 44: 277–284

Berl T, Henrich WL, Erikson AL, Schrier RW (1979) Prostaglandins in the beta-adrenergic and baroreceptor mediated secretion of renin. Am J Physiol 236: F472–F477

Bernstein KE, Alexander RW (1992) Counterpoint; molecular analysis of the angiotensin II receptor. Endocr Rev 13: 381–386

Bernstein KE, Martin BM, Berk BC, Bernstein EA (1989) Mouse angiotensin-converting enzyme is a protein composed of two homologous domains. J Biol Chem 264: 11945–11951

Berridge MJ, Irvine RF (1984) Inositol triphosphate, a novel second messenger in cellular signal transduction. Nature 312: 315–321

Bianchi C, Gutkowska J, Charbonneau C, Ballak M, Anad-Srivastava MB, de Lean A, Genest J, Cantin M (1986) Internalization and lysosomal association of [^{125}I] angiotensin II in norepinephrine-containing cells of the rat adrenal medulla. Endocrinology 119: 1873–1875

Biava C, West M (1966) Fine structure of normal juxtaglomerular cells. II. Specific and nonspecific cytoplasmic granules. Am J Pathol 49: 955–979

Billah MM, Michell RH (1979) Phosphatidylinositol metabolism in rat hepatocytes stimulated by glycogenolytic hormones. Biochem J 182: 661–668

Bing J, Faarup P (1966) A qualitative and quantitative study of renin in the different layers of the rabbit uterus. Acta Pathol Microbiol Scand 67: 169–179

Bing J, Kazimierczak J (1962) Localization of renin in the kidney. IV. Renin content of different parts of the juxtaglomerular apparatus. Acta Path Microbiol Scand 54: 80–84

Bing J, Kazimierczak J (1963) Location of renin. In: Williams PC (ed) Hormones and the kidney. Academic Press, New York, pp 255–261

Bingel A, Claus R (1910) Weitere Untersuchungen über die blutdrucksteigernde Substanz der Niere. Dtsch Arch Klin Med 100: 412–420

Bingel A, Strauss E (1909) Über die blutdrucksteigernde Substanz der Niere. Dtsch Arch Klin Med 96: 476–492

Bird IM, Meikle I, Williams BC, Walker SW (1989) Angiotensin II-stimulated cortisol secretion is mediated by a hormone-sensitive phospholipase C in bovine adrenal fasciculata/reticularis cells. Mol Cell Endocrinol 64: 45–53

Blackshear JL, Spielman WS, Knox FG, Romero JC (1979) Dissociation of renin release in renal vasodilation by prostaglandin synthesis inhibitors. Am J Physiol 237: F20–F24

Blaine EH, Davis JO, Witty RT (1970) Renin release after hemorrhage and after suprarenal aortic constriction in dogs without sodium delivery to the macula densa. Circ Res 27: 1081–1089

Blaine EH, Davis JO, Prewitt Rt (1971) Evidence for a renal vascular receptor in control of renin secretion. Am J Physiol 220: 1593–1597

Blair ML (1983) Stimulation of renin secretion by alpha-adrenoceptor agonists. Am J Physiol 244: E37–E44

Blair-West JR, Harding R, McKenzie JS (1968) Effect of sodium concentration on the vaso-constrictor action of angiotensin in the rabbit ear. Eur J Pharmacol 4: 77–82

Blair-West JR, Coghlan JP, Denton DA, Funder JW, Scoggins BA, Wright RD (1971a) Inhibition of renin secretion by systemic and intrarenal angiotensin infusion. Am J Physiol 220: 1309–1315

Blair-West JR, Coghlan JP, Denton DA, Funder JW, Scoggins BA, Wright RD (1971b) The effect of the heptapeptide (2–8) and the hexapeptide (3–8) fragments of angiotensin II on aldosterone secretion. J Clin Endocrinol 32: 575–578

Blair-West JR, Coghlan JP, Denton DA, Gibson AP, Oddie CJ, Sawyer WH, Scoggins BA (1977) Plasma renin activity and blood corticosteriods in the Australian lungfisch, *Neoceratodus forsteri*. J Endocrinol 74: 137–142

Blair-West JR, Gibson A, McKinley MJ, Nelson JF (1983) Water drinking and the effect of angiotensin and renin in a dasyurid marsupial *Antechinus stuartii*. Gen Comp Endocrinol 52: 388–394

Blair-West JR, Denton DA, McKinley MJ, Weisinger RS (1988) Angiotensin-related sodium appetite and thirst in cattle. Am J Physiol 255: R205–R211

Blantz RC, Konnan KS, Tucker BJ (1976) Angiotensin II effects upon the glomerular micro-circulation and ultrafiltration coefficient of the rat. J Clin Invest 57: 419–434

Blendstrup K, Leyssac PP, Poulsen K, Skinner SL (1975) Characteristics of renin release from isolated superfused glomeruli in vitro. J Physiol (Lond) 246: 653–672

Block CH, Santos RAS, Brosnihan KB, Ferrario CM (1989) Immunocytochemical localization of angiotensin-(1–7) in the rat forebrain. Peptides 9: 1395–1401

Block ML, Vallier GH, Glickman SE (1974) Elicitation of water ingestion in the Mongolian gerbil (*Meriones unguiculatus*) by intracranial injections of angiotensin II and L-nor-epinephrine. Pharm Biochem Behav 2: 235–242

Bohle A (1954) Kritischer Beitrag zur Morphologie einer endokrinen Nierenfunktion und deren Bedeutung für den Hochdruck. Arch Kreislaufforsch 20: 193–246

Bohle A (1959) Elektronen mikroskopische Untersuchungen über die Struktur des Gefasspols der Niere. Verh Dtsch Ges Pathol 43: 219–225

Bohle A, Walvig F (1964) Beitrag zur vergleichenden Morphologie der epithelioiden Zellen der Nierenarteriolen unter besonderer Berücksichtigung der epithelioiden Zellen in den Nieren von Seewasserfischen. Klin Wochenschr 42: 415–421

Bolger PM, Eisner GM, Ramwell PW, Slotkoff LB (1976) Effect of prostaglandin synthesis on renal function and renin in the dog. Nature 259: 244–245

Bondar N, Cadnapaphornchai P, McDonald FD, Taher S (1984) Mechanism of effect of dibutyl cyclic adenocine 3'5'-monophosphate on canine renal renin release. J Physiol (Lond) 355: 33–41

Bonjour JP, Malvin RL (1970) Stimulation of ADH release by the renin-angiotensin system. Am J Physiol 218: 1555–1559

Booth DA (1968) Mechanism of action of norepinephrine in eliciting an eating response on injection into the rat hypothalamus. J Pharm Exp Ther 160: 336–348

Booz GW, Conrad KM, Hess AL, Singer HA, Baker KM (1992) Angiotensin II-binding sites on hepatocyte nuclei. Endocrinology 130: 3641–3649

Borghese E (1966) Studies on the nephron of an elasmobranch fish, *Scyliorhinus stellaris*. Z Zellforsch 72: 88–99

Borriraja V, Henderson IW, Chester Jones I (1973) Renal fractions affecting the concentration of plasma cortisol in the eel (*Anguilla anguilla* L.). J Endocrinol 57: p xiii

Bott E, Denton DA, Weller S (1967) The effect of angiotensin II infusion, renal hypertension and nephrectomy on salt appetite of sodium deficient sheep. Aust J Exp Biol Med Sci 45: 595–612

Bottani SP, de Gasparo M, Steckelings UM, Levens NR (1993) Angiotensin II receptor sub-types: characterization, signaling mechanisms, and possible physiological implications. Front Neuroendocrinol 14: 123–172

Boucher R, Menard J, Genest J (1967) A micromethod for measurement of renin in the plasma and kidney of rats. Can J Physiol Pharmacol 45: 881–891

183

Bouhnik J, Clauser E, Strosberg D, Frenoy J-P, Menard J, Corvol P (1981) Rat angiotensinogen and des (angiotensin I) angiotensinogen: purification, characterization and partial sequencing. Biochemistry 20: 7010-7015

Bowie DJ (1935-1936) A method for staining pepsinogen granules in gastric glands. Anat Res 64: 357-367

Boyd GW (1977) An inactive higher-molecular-weight renin in normal subjects and hypertensive patients. Lancet 1: 215-218

Brandt CR, Pumfery AM, Miales B, Bindley CD, Lyons GE, Sramek SJ, Wallow IHL (1994) Renin mRNA is synthesized locally in rat ocular tissues. Curr Eye Res 13: 755-763

Brasszko JJ, Własienko J, Koziołkiewicz W, Janecka A, Wiśneiwski K (1991) The 3-7 fragment of angiotensin II is probably responsible for its psychoactive properties. Brain Res 542: 49-54

Braun-Menéndez E, Page IH (1958) Suggested revision of nomenclature - angiotensin. Science 127: 242

Braun-Menéndez E, Fasciolo JC, Leloir LF, Muñoz JM (1940a) The substance causing renal hypertension. J Physiol (Lond) 98: 283-298

Braun-Menéndez E, Fasciolo JC, Leloir LF, Muñoz JM (1940b) La substance hypertensine extraite du sang des reins ischémiés. C R Soc Biol 133: 731-733

Bravo EL, Khosla MC, Bumpus FM (1975) Action of [1-des-(aspartic acid), 8-isoleucine] angiotensin II upon pressor and steroidogenic activity of angiotensin II. J Clin Endocrinol Metab 40: 530-533

Brenner BM, Ballermann BJ, Gunning ME, Zeidel ML (1990) Diverse biological actions of atrial natriuretic peptide. Physiol Rev 70: 665-699

Breuhaus BA, Chimoskey JE (1990) Hemodynamic and behavioral effects of angiotensin II in conscious sheep. Am J Physiol 258: R1230-R1237

Brinton GS, Jubiz W, Lagerquist LD (1975) Hypertension in primary hyperparathyroidism: the role of the renin-angiotensin system. J Clin Endocrinol Metab 41: 1025-1029

Britton SL, Thomas G, Daniel C, Ronau TF (1983) Kinase II-dependent formation of angiotensin II and III in the hepatic circulation. Am J Physiol 245: H849-854

Brock TA, Lewis LJ, Smith JB (1982) Angiotensin increases Na^+ entry and Na^+/K^+ pump activity in cultures of smooth muscle from rat aorta. Proc Natl Acad Sci USA 79: 1438-1442

Brock TA, Alexander WA, Ekstein LS, Atkinson WJ, Gimbrone MA Jr (1985) Angiotensin increases cytosolic free calcium in cultured vascular smooth muscle cells. Hypertension (Suppl I): 105-109

Brooks DP, Share L, Crofton JT, Rockhold RW, Matsui K (1984) Effect of vertebral artery infusions of oxytocin on plasma vasopressin concentration, plasma renin activity, blood pressure and heart rate and their responses to hemorrhage. Neuroendocrinology 38: 382-386

Brooks VL, Malvin RL (1979) An intracerebral physiological role for angiotensin: effects of central blockade. Fed Proc 38: 2272-2275

Brooks VL, Malvin RL (1993) Interrelations between renin and vasopressin. In: Robertson JIS, Nicholls MG (eds) The renin-angiotensin system, vol 1. Gower Medical Publ, London, New York, pp 35.1-35.14

Brophy PD, Levitt RA (1974) Dose responses analysis of angiotensin- and renin-induced drinking in the cat. Pharm Biochem Behav 2: 509-514

Broughton Pipkin F, Kirkpatrick SML, Lumbers ER, Mott CJ (1974) Renin and angiotensin-like levels in foetal, new-born and adult sheep. J Physiol (Lond) 241: 575-589

Brown CA, Zusman RM, Haber E (1980) Identification of an angiotensin receptor in rabbit renomedullary interstitial cells in tissue culture. Correlation with prostaglandin biosynthesis. Circ Res 46: 802-807

Brown GP, Douglas JG (1982) Angiotensin II binding sites on isolated rat renal brush border membranes. Endocrinology 111: 1830-1836

Brown GP, Douglas JG (1983) Angiotensin II-binding sites in rat and primate isolated renal tubular basolateral membranes. Endocrinology 112: 2007-2014

Brown GP, Douglas JG, Krontiris-Litowitz J (1980) Properties of angiotensin II receptors of isolated rat glomeruli: factors influencing binding affinity and comparative binding of angiotensin II analogues. Endocrinology 106: 1923-1929

184

Brown JA, Oliver JA, Henderson IW, Jackson BA (1980) Angiotensin and single nephron glomerular function in the trout *Salmo gairdneri*. Am J Physiol 239: R509–R514

Brown JA, Taylor SM. Gray CJ (1990) Glomerular binding of angiotensin II in the rainbow trout, *Salmo gairdneri*. Cell Tissue Res 259: 479–482

Brown JJ, Davies DL, Lever AF, Robertson JIS (1964) Variations in plasma renin during the mensutrual cycle. Br Med J 2: 1114–1115

Brown JJ, Davies DL, Lever AF, Robertson JIS (1965) Plasma renin concentration in human hypertension I: relationship between renin, sodium, and potassium. Br Med J 2: 144–148

Brown-Séquard CE, d'Arsonval JA (1892) Des injections souscutanée au intraveineuses d'extraits liquides de nombre d'organes, comme méthode thérapeutique. C R Acad Sci 114: 1399–1404

Brunner HR, Baer L, Sealey JE, Ledingham JGG, Laragh JH (1970) The influence of potassium administration and of potassium deprivation on plasma renin in normal and hypertensive subjects. J Clin Invest 49: 2128–2138

Brunner HR, Chang P, Wallach R, Sealey JE, Laragh JH (1972) Angiotensin II vascular receptors: their avidity in relationship to sodium balance, the autonomic nervous system, and hypertension. J Clin Invest 51: 58–67

Bryant RW, Epstein AN, Fitzsimons JT, Fluharty SJ (1980) Arousal of a specific and persistent sodium appetite in the rat with continuous intracerebro-ventricular infusion of angiotensin II. J Physiol (Lond) 301: 365–382

Bucher O, Reale E (1962) Zur elektronen mikroskopischen Untersuchungen der Niere: III Mitteilung die epithelioiden Zellen der Arteriola afferens. Z Zellforsch 56: 344–358

Buckingham JC, Smith T, Loxley HD (1992) The control of ACTH secretion. In: James VHT (ed) The adrenal gland, 2nd edn. Raven Press, New York, pp 133–137

Buckley JP (1972) Actions of angiotensin on the central nervous system. Fed Proc 31: 1332–1337

Buckley JP, Ferrario CM (1977) Central actions of angiotensin and related hormones. Pergamon Press, New York

Buckner FS, Chen FN, Wade CE, Ganong WF (1986) Centrally administered inhibitors of generation and action of angiotensin II do not attenuate the increase in ACTH secretion produced by ether stress in rats. Neuroendocrinology 42: 97–101

Buggy J (1977) Drinking elicited by angiotensin or hyperosmotic stimulation of the rat anteroventral third ventricle: single or separate neural substrates? In: Buckley J, Ferrario CM (eds) Central actions of angiotensin and related hormones. Pergamon Press, New York, pp 315–323

Buggy J, Johnson AK (1977) Anteroventral third ventricle periventricular ablation: temporary adipsia and persisting thirst deficits. Neurosci Lett 5: 177–182

Buggy J, Johnson AK (1978) Angiotensin-induced thirst: effects of third ventricle obstruction and periventricular ablation. Brain Res 149: 117–128

Buijs RM, Geffard M, Pool CW, Hoorneman EMD (1984) The dopaminergic innervation of the supraoptic and paraventricular nucleus. A light and electron microscopical study. Brain Res 323: 65–72

Bumpus FM, Schwartz H, Page IH (1957) Synthesis and pharmacology of the octapeptide angiotensin. Science 125: 886–887

Bumpus FM, Catt KJ, Chiu AT, deGasparo M, Goodfriend T, Husain A, Peach MJ, Taylor DG Jr, Timmermans PBMWM (1991) Nomenclature of angiotensin receptors. Hypertension 17: 720–721

Bunag RD, Page IH, McCubbin JW (1966) Neural stimulation of release of renin. Circ Res 19: 851–858

Bunag RD, Page IH, McCubbin JW (1967) Inhibition of renin release by vasopressin and angiotensin. Cardiovasc Res 1: 67–73

Bunnemann B, Fuxe K, Bjelke B, Ganten D (1991) The brain renin-angiotensin system and its possible involvement in volume transmission. In: Fuxe K, Agnati LF (eds) Volume transmission in the brain: novel mechanisms for neural transmission. Raven, New York, pp 131–158

Bunnemann B, Fuxe K, Ganten D (1993) Extrarenal renin systems: the brain. In: Robertson JIS, Nicholls MG (eds) The renin-angiotensin system, vol 1. Gower Medical publ, London, pp 41.1–41.17

Buonassisi V, Venter JC (1976) Hormone and neurotransmitter receptors in an established vascular endothelial cell line. Proc Natl Acad Sci USA 73: 1612–1616

Burnett JC Jr, Granger JP, Opgenorth TJ (1984) Effects of synthetic atrial natriuretic factor on renal function and renin release. Am J Physiol 247: F863–866

Bührle CP, Nobiling R, Mannek E, Schneider D, Hakkenthal E, Taugner R (1984) The afferent glomerular arteriole: immunocytochemical and electrophysiological investigations. J Cardiovas Pharmacol 6: S383–S393

Butler DG, Oudit GY, Cadinouche MZA (1995) Angiotensin I- and II- and norepinephrine-mediated pressor responses in an ancient holostean fish, the bowfin (Amia calcva). Gen Comp Endocrinol 98: 289–302

Cade TJ, Dybas A Jr (1962) Water economy of the budgerigah. Auk 79: 345–364

Caldwell RPB, Seegal BC, Hsu KC, Das M, Soffer RL (1976) Angiotensin-converting enzyme: vascular endothelial localization. Science 191: 1050–1051

Calogero AE, Fornito MC, Aliffi A, Vicari E, Moncada ML, Mantero F, Polosa P, D'Agata R (1991) Role of peripherally infused angiotensin II on the human hypothalamic-pituitary-adrenal axis. Clin Endocrinol 34: 183–186

Calza L, Fuxe K, Agnati LF, Zini I, Ganten D, Lang RE, Poulsen K, Hökfelt T (1982) Presence of renin-like immunoreactivity in oxytocin immunoreactive nerve cells of the paraventricular and supraoptic nuclei in rat hypothalamus. Acta Physiol Scand 116: 313–316

Cameron VA, Espiner EA, Nicholls MG, MacFarlane MR, Sadler WA (1986) Intracerebroventricular captopril reduces plasma ACTH and vasopressin responses to hemorrhagic stress. Life Sci 38: 553–559

Camilleri JP, Phat VN, Bariety J, Corvol P, Menard J (1980) Use of a specific antiserum for renin detection in human kidney. J Histochem Cytochem 28: 1343–1346

Campagnole-Santos MJ, Diz DI, Santos RAS, Khosla MC, Brosnihan KB, Ferrario CM (1989) Cardiovascular effects of angiotensin-(1–7) injected into the dorsal medulla of rats. Am J Physiol 257: H314–H319

Campanile CP, Crane JK, Peach MJ, Garrison JC (1982) The hepatic angiotensin II receptor. I. Characterization of the membrane-binding site and correlation with physiological response in hepatocytes. J Biol Chem 257: 4951–4958

Campbell DJ (1987a) Tissue renin-angiotensin system: sites of angiotensin formation. J Cardiovasc Pharmacol 10: S1–S8

Campbell DJ (1987b) Circulating and tissue angiotensin system. J Clin Invest 79: 1–6

Campbell DJ, Habener JF (1986) Angiotensinogen gene is expressed and differentially regulated in multiple tissues of the rat. J Clin Invest 78: 31–39

Campbell DJ, Bouhnik J, Menard J, Corvol P (1984) Identity of angiotensinogen precursors of rat brain and liver. Nature 308: 206–208

Campbell DJ, Bouhnik J, Coezy E, Menard J, Corvol P (1985) Characterization of precursor and secreted forms of human angiotensinogen. J Clin Invest 75: 1880–1893

Campbell WB, Zimmer J (1980) Insulin-induced renin release: blockade by indomethacin in the rat. Clin Sci 58: 415–418

Campbell WB, Graham RM, Jackson EK (1979) Role of renal prostaglandins in sympathetically mediated renin release in the rat. J Clin Invest 64: 448–456

Cantalamessa F, de Caro G, Massi M, Micossi LG (1982) Drinking stimulation by a new angiotensin, Crinia-angiotensin II, in rats and pigeons. Pharm Biochem Behav 17: 741–747

Cantin M, Gutkowska J, Lacasse J, Ballak M, Ledoux S, Inagami T, Beuzeron J, Genest J (1984) Ultrastructural immunocytochemical localization of renin and angiotensin II in the juxtaglomerular cells of the ischemic kidney in experimental renal hypertensin. Am J Pathol 115: 212–224

Capelli JP, Wesson LG, Aponte GE (1968) The effect of sodium on renal renin and on glucose-6-phosphate dehydrogenase in the kidneys, salivary glands and adrenal glands. Nephron 5: 106–123

Capelli JP, Wesson LG Jr, Aponte GE (1970) A phylogenetic study of the renin-angiotensin system. Am J Physiol 218: 1171-1178

Capponi AM, Catt KJ (1979) Angiotensin II receptors in adrenal cortex and uterus. J Biol Chem 254: 5120-5127

Capponi AM, Catt KJ (1980) Solubilization and characterization of adrenal and uterine angiotensin II receptors after photoaffinity labeling. J Biol Chem 255: 12081-12086

Capponi AM, Vallotton MB (1976) Renin release by rat kidney slices incubated in vitro. Role of sodium and alpha- and beta-adrenergic receptors, and effect of vincristine. Circ Res 39: 200-203

Capponi AM, Gourjon M, Vallotton MB (1977) Effects of beta-blocking agents and angiotensin II on isoproterenol-stimulated renin release from rat kidney slices. Circ Res 40 (Suppl I): 89-93

Capponi AM, Lew PD, Vallotton MB (1985) Cytosolic free calcium levels in monolayers of cultured rat aortic smooth muscle cells. J Biol Chem 269: 7836-7842

Capréol SV, Sutherland LE (1968) Comparative morphology of juxtaglomerular cells. I. Juxtaglomerular cells in fish. Can J Zool 46: 249-256

Carey RM (1982) Neuroendocrine regulation of the renin-angiotensin-aldosterone system. In Müller EE, MacLeod RM (eds) Neuroendocrine perspectives 1. Elsevier Biomed Press Amsterdam pp 253-303

Cargill RI, Coutie WJ, Lipworth BJ (1994) The effects of angiotensin II on circulating levels of natriuretic peptides. Br J Clin Pharmacol 38: 139-142

Carlson W, Karplus M, Haber E (1985) Construction of a model for the three-dimentional structure of human renal renin. Hypertension 7: 13-26

Carrick S, Balment RJ (1983) The renin-angiotensin system and drinking in the euryhaline flounder, *Platichthys flesus*. Gen Comp Endocrinol 51: 423-433

Carrol RG, Opdyke DF (1982) Evolution of angiotensin II-induced catecholamine release. Am J Physiol 243: R65-R69

Cash JR (1924) A preliminary study of the blood-pressure following reduction of renal substance with a note on simultaneous changes in blood-chemistry and blood-volume. Bull Johns Hopkins Hosp 35: 168-180

Castren E, Saavedra JM (1988) Repeated stress increases the density of angiotensin II binding sites in rat paraventricular nucleus and subfornical organ. Endocrinology 122: 370-372

Castro R, Phillips MI (1985) Neuropeptide action in nucleus tractus solitarius: angiotensin specificity and hypertensive rats. Am J Physiol 249: R341-R347

Catanzaro DF, Mesterovic N, Morris BJ (1985) Studies of the regulation of mouse renin genes by measurement of renin messenger ribonucleic acid. Endocrinology 117: 872-878

Catt KJ, Abbott A (1991) Molecular cloning of angiotensin II receptors may presage further receptor subtypes. Trends Pharmacol Sci 12: 279-281

Catt KJ, Wynn PC, Millan MA, Mendelsohn FAO, Aguilera G (1986) Brain receptors for angiotensin II and corticotropin-releasing factor. Front Neuroendocrinology 9: 225-253

Cauraud PO (1987) Anti-angiotensin II anti-idiotypic antibodies bind to angiotensin II receptor. J Immunol 138: 1164-1168

Celio M, Inagami T (1981a) Renin in human kidney. Immunohistochemical localization. Histochemistry 72: 1-10

Celio M, Inagami T (1981b) Angiotensin II immunoreactivity coexists with renin in the juxtaglomerular granular cells of the kidney. Proc Natl Acad Sci USA 78: 3897-3900

Chamley-Campbell JH, Campbell GR, Ross R (1979) Smooth muscle cell in culture. Physiol Rev 59: 1-61

Chan MY, Holmes WN (1971) Studies on a "renin-angiotensin" in the normal and hypophysectomized pigeon (*Columba livia*) Gen Comp Endocrinol 16: 304-311

Chandra S, Hubbard JC, Skelton FR, Bernadis LI, Kamura S (1965) Genesis of juxtaglomerular cell granules. A physiologic light and electron microscopic study concerning experimental renal hypertension. Lab Invest 14: 1835-1842

Chang JJ, Kisaraji M, Okamoto H, Inagami T (1981) Isolation and activation of inactive renin from human kidney and plasma. Hypertension 3: 509-515

187

Changaris DG, Demars LM, Keil LC, Severs WB (1977) Immunopharmacology of angiotensin I in brain. In: Buckley JP, Ferrario CM, Lokhandwala MF (eds) Central actions of angiotensin and related hormones. Pergamon Press. New York, pp 233–243

Chansel D, Ardaillow N, Nivez MP, Ardaillou R (1982) Angiotensin II receptors in human isolated renal glomeruli. J Clin Endocrinol Metab 55: 961–966

Chao HS, Poisner AM, Poisner R, Handwerger S (1994) Endothelin-1 modulates renin and prolactin release from human decidua by different mechanisms. Am J Physiol 267: E842–E846

Chappell MC, Brosnihan KB, Diz DI, Ferrario CM (1989) Identification of angiotensin-(1–7) in rat brain. Evidence for differential processing of angiotensin peptides. J Biol chem 264: 16518–16523

Charest R, Prpič V, Exton JH, Blackmore PF (1985) Stimulation of inositol triphosphate formation in hepatocytes by vasopressin, adrenaline and angiotensin II and its relationship to changes in cytosolic free Ca^{2+}. Biochem J 227: 79–90

Chen DS, Poisner AM (1976) Direct stimulation of renin release by calcium. Proc Soc Exp Biol Med 152: 565–567

Chen M, Schnermann J, Smart AM, Brosius FC, Killen PD, Briggs JP (1993) Cyclic AMP selectively increases renin mRNA stability in cultured juxtaglomerular granular cells. J Biol Chem 268: 24138–24144

Chiu AT, Herblin WF, McCall DE, Ardecky RJ, Carini DJ, Duncia JV, Pease LJ, Wong PC, Wexier RR, Johnson AL, Timmermans PBMWM (1989) Identification of angiotensin II receptor subtypes. Biochem Biophys Res Commun 165: 196–203

Chodobski A, Szmdynger-Chodobska J, Epstein MH, Johanson CE (1995) The role of angiotensin II in the regulation of blood flow to choroid plexuses and cerebrospinal fluid formation in the rat. J Cereb Blood Flow Metab 15: 143–151

Chokshi DS, Yeh BK, Samet P (1972) Effects of dopamine and isoproterenol on renin secretion in the dog. Proc Soc Exp Biol Med 140: 54–57

Christensen JA, Morild I, Mikeler E, Bohle A (1982) Juxtaglomerular apparatus in the domestic fowl (Gallus domesticus). Kidney Int 22: S24

Christensen JA, Taugner R, Meyer DS, Bohle A (1987) The granular epithelioid cells in the kidney of the lemon sole (Pleuronectes microcephalus Donovani), Cell Tissue Res 249: 137–143

Churchill PC (1985) Second messengers in renin secretion. Am J Physiol 249: F175–F184

Churchill PC, Churchill MC (1980a) Separate and combined effects of ouabain and extracellular potassium on renin secretion from rat renal cortical slices. J Physiol (Lond) 300: 105–114

Churchill PC, Churchill MC (1980b) Vanadate inhibits renin secretion from rat kidney slices. J Pharmacol Exp Ther 213: 144–149

Churchill PC, Churchill MC (1982) Ca-dependence of the inhibitory effect of K-depolarization on renin secretion from rat kidney slices. Arch Int Pharmacodyn Ther 224: 68–72

Churchill PC, Churchill MC (1985) A_1 and A_2 adenosine receptor activation inhibits and stimulates renin secretion of rat renal cortical slices. J Pharmacol Exp Ther 232: 589–594

Churchill PC, Lyons HJ (1976) Effect of intrarenal arterial infusions of magnesium on renin release in dogs. Proc Soc Exp Biol Med 152: 6–10

Churchill PC, McDonald FD (1974) Effect of oubain on renin secretion in anesthetized dogs. J Physiol (Lond) 242: 635–646

Churchill PC, Malvin RL, Opava SC (1974) Evidence for baroreceptor control of renin release. Nephron 13: 383–389

Churchill PC, Churchill MC, McDonald FD (1979a) Effects of saline and mannitol on renin and distal tubule Na in rats. Circ Res 45: 786–792

Churchill PC, McDonald FD, Churchill MC (1979b) Phenytoin stimulates renin secretion from rat kidney slices. J Pharmacol Exp Ther 211: 615–619

Churchill PC, McDonald FD, Churchill MC (1981) Effect of diltiazem, a calcium antagonist, on renin secretion from rat kidney slices. Life Sci 29: 383–389

Churchill PC, Churchill MC, McDonald FD (1983) Comparison of the effects of rubidium and potassium on renin secretion from rat kidney slices. Endocrinology 112: 777–781

Cipolle MD, Zehr JE (1984) Characterization of the renin-angiotensin system in the turtle *Pseudemys scripta.* Am J Physiol 247: R15–R23

Cipolle MD, Zehr JE (1985) Renin release in turtles: effects of volume depletion and furosemide administration. Am J Physiol 249: R100–R105

Cipolle MD, Zehr JE, Reinhart GA (1986) Effect of autonomic agents on renin release in the turtle, *Pseudemys scripta.* Am J Physiol 251: R1103–R1108

Ciuffo GM, Heemskerk FMJ, Saavedra JM (1993) Purification and characterization of angiotensin II AT$_2$ receptors from neonatal rat kidney. Proc Natl Acad Sci USA 90: 11009–11013

Clauser E, Bouhnik J, Jaramillo HN, Auzan C, Corvol P, Menard J (1985) Angiotensinogen production and consumption in the adrenalecomized rat. Endocrinology 116: 274–280

Clouston WM, Lyons IG, Richard RI (1989) Tissue-specific and hormonal regulation of angiotensinogen minigenes in transgenic mice. EMBO J 8: 3337–3343

Clyne CD, Nicol MR, MacDonald S, Williams BC, Walker SW (1993) Angiotensin II stimulates growth and steroidogenesis in zona fasciculata/reticularis cells from bovine adrenal cortex via the AT$_1$ receptor subtype. Endocrinology 132: 2206–2212

Cobb CS, Brown JA (1992) Angiotensin II binding to tissues of the rainbow trout, *Oncorhynchus mykiss,* studied by autoradiography. J Comp Physiol B 162: 197–202

Cobb CS, Brown JA (1993) Characterization of putative glomerular receptors for angiotensin II in the rainbow trout *Oncorhynchus mykiss* using the antagonists losartan, PD123177, and saralasin. Gen Comp Endocrinol 92: 123–131

Cohen AJ, Laurens P, Fray JCS (1983) Suppression of renin secretion by insulin: dependence on extracellular calcium. Am J Physiol 245: E531–E534

Cohen S, Taylor JM, Murakami K, Michelakis AM, Inagami T (1972) Isolation and characterization of renin-like enzymes from mouse submaxillary glands. Biochemistry 11: 4286–4293

Cole FE, Blakesley HL, Graci KA, Frohlich ED, MacPhee AA (1980) Brain angiotensin II receptor affinity and capacity in SHR and WKY rats: effects of acute dietary changes in NaCl. Brain Res 190: 272–277

Collie MA, Holmes WN, Cronshaw J (1992) A comparison of the responses of dispersed steroidogenic cells derived from embryonic adrenal tissue from the domestic chicken (*Gallus domesticus*), the domestic Pekin duck, the wild Mallard duck (*Anas platyrhnchos*), and the domestic Muscovy duck (*Cairina moschata*). Gen Comp Endocrinol 88: 375–387

Cook WF (1971) Cellular localization of renin. In: Fisher JW (ed) Kidney hormones. Academic Press, New York, pp 117–128

Cooke CR, Brown TC, Zacherle BJ, Walker GW (1970) Effect of altered sodium concentration in the distal nephron segments on renin release. J Clin Invest 49: 1630–1638

Cooling MJ, Day MD (1975) Angiotensin induced drinking in the cat. In: Peters G, Fitzsimons JT, Peters-Haefeli L (eds) Control mechanisms of drinking. Springer-Verlag, Heidelberg, pp 132–137

Cornell MJ, Williams TA, Lamango NS, Coates D, Corvol P, Soubrier F, Hoheisel J, Lehrach H, Isaac RE (1995) Cloning and expression of an evolutionary conserved single-domain angiotensin converting enzyme from *Drosophila melanogaster.* J Biol Chem 270: 13613–13619

Corvol P, Ménard J (1991) Renin purification and cloning. Hypertension 18: 252–256

Corwin E, Malvin GM, Katz S, Malvin RL (1984) Temperature sensitivity of the renin-angiotensin system in *Ambystoma tigrinum.* Am J Physiol 246: R510–R515

Coviello A, Brauckmann ES (1973) Hydrosmotic effect of angiotensin II: Isolated toad skin. Acta Physiol Latinoam 23: 18–23

Coviello A, Orce G, Causarano J (1974) Effect of a competitive antagonist (8-Leu-angiotensin II) of angiotensin II on sodium and water transport in toad skin. Acta Physiol Latinoam 24: 409–413

Coviello A, Brauckmann ES, DeAtenor MSB, Apud JA, Causarano J (1975) Hydrosmotic effect of angiotensin II in the toad skin: role of cyclic AMP. Acta Physiol Latinoam 25: 379–386

Coviello A, Elso G, Fernandez FM (1976) Effect of angiotensin II on short-circuit current in amphibian membranes. Biochem Pharmacol 25: 106–107

Cox HM, Munday KA, Poat JA (1983) The binding of ^{125}I-angiotensin II to rat renal epithelial cell membraines. Br J Pharmacol 79: 63–70

Cox HM, Munday KA, Poat JA (1984) Location of [^{125}I]-angiotensin II receptors on rat kidney cortex epithelial cells. Br J Pharmacol 82: 891–895

Cox HM, Munday KA, Poat JA (1986) Identification of selective, high affinity [^{125}I]-angiotensin and [^{125}I]-bradykinin binding sites in rat intestinal epithelia. Br J Pharmacol 87: 201–209

Cozza EN, Chiou S, Gomez-Sanchez CE (1992) Endothelin-1 potentiation of angiotensin II stimulation of aldosterone production. Am J Physiol 262: R85–R89

Crane JK, Campanile CP, Garrison JC (1982) The hepatic angiotensin II receptor II. Effect of guanine nucleotides and interaction with cyclic AMP production. J Biol Chem 257: 4959–4965

Creba JA, Downes CP, Hawkins PT, Brewster G, Michell RH, Kirk CJ (1983) Rapid breakdown of phosphatidylinositol 4-phosphate and phosphatidylinosytol 4,5-bisphosphate in rat hepatocytes stimulated by vasopressin and other Ca^{2+}-mobilizing hormones. Biochem J 212: 733–747

Crockett DR, Gerst JW, Blankenship S (1973) Absence of juxtaglomerular cells in the kidneys of elasmobranch fishes. Comp Biochem Physiol 44A: 673–675

Crozat A, Penhoat A, Saez JM (1986) Processing of angiotensin II (AII) and (Sarl, Ala^8) AII by cultured bovine adrenocortical cells. Endocrinology 118: 2312–2318

Cunningham JT, Johnson AK (1991) The effects of central norepinephrine infusions on drinking behavior induced by angiotensin after 6-hydroxydopamine injections into the anteroventral region of the third ventricle (AV3V). Brain Res 558: 112–116

Cuthbert AW, Shutter MC (1965) The effects of drugs on the relation between the action of potential discharge and tension in a mammalian vein. Br J Pharmacol 25: 592–601

Dalton AI (1951) Structural details of some of the epithelial cell types in the kidney of the mouse as revealed by the electron microscope. J Natl Cancer Inst 11: 1163–1185

Danzler WH (1989) Comparative physiology of the vertebrate kidney. Springer, Berlin Heidelberg New York

Davalos M, Frega NO, Saker B, Leaf A (1978) Effect of exogenous and endogenous angiotensin II in the isolated perfused rat kidney. Am J Physiol 235: F605–F610

Davies NT, Munday KA, Parson BJ (1970) The effect of angiotensin on rat intestinal fluid transfer. J Endocrinol 48: 39–46

Davis JO, Freeman RH (1976) Mechanisms regulating renin release. Physiol Rev 56: 1–56

Davis JO, Johnson JA, Witty RT, Shade RE, Braverman B (1972) New observation on renin release by the non-filtering kidney. In: Genest J, Koiw E (eds) Hypertension, 7,2. Springer, Berlin Heidelberg New York, pp 56–63

Day RP, Luetscher JA (1975) Occurrence of big renin in human plasma, amniotic fluid and kidney extracts. J Clin Endocrinol Metab 40: 1078–1084

Day RP, Reid IA (1976) Renin activity in dog brain. Enzymological similarity to cathepsin D. Endocrinology 99: 93–100

De Arruda Camargo LA, Saad WA, Renzi A, De Luca LA Jr, Gonzalves JR, Menani JV (1991) Hypothalamic lesions increase saline ingestion induced by injection of angiotensin II into AV3V in rats. Am J Physiol 261: R647–R651

de Caro G, Mariotti M, Massi M, Micossi LG (1980) Dipsogenic effect of angiotensin II, bombesin and tachykinins in the duck. Pharm Biochem Behav 13: 229–233

de Caro G, Massi M, Micossi LG, Perfumi M (1982) Angiotensin II antagonists vs drinking induced by bombesin or eledoisin in pigeons, Columba livia. Peptides 3: 631–636

de Caro G, Epstein AN, Massi M (1986) The physiology of thirst and sodium appetite. NATO ASI Ser, Ser A: Life Sci Vol 105. Plenum Press, New York

de Gasparo M, Levens NR (1994) Pharmacology of angiotensin II receptors in the kidney. Kidney Int 46: 1486–1491

De Jong W, Loverberg W, Sjoerdsma A (1972) Renin-like activity in submaxillary gland in several strains of rats including the spontaneously hypertensive rat. Biochem Pharmacol 21: 2123–2129

De Muylder CG (1952) The "neurility" of the kidney. A monogragh on nerve supply to the kidney. Blackwell, Oxford

De Vito E, Gordon SB, Cabrera RR, Fasciolo JC (1970) Renin release by rat kidney slices. Am J Physiol 219: 1036–1041

Decavel C, Geffard M, Calas A (1987) Comparative study of dopamine- and noradrenaline-immunoreactive terminals in the paraventricular and supraoptic nuclei of the rat. Neurosci Lett 77: 149–154

Defendini R, Zimmerman EA, Weare JA, Alhenc-Gelas F, Erdös EG (1983) Angiotensin converting enzyme in epithelial and neuroepithelial cells. Neuroendocrinology 37: 32–40

degli Uberti EC, Trasforini G, Margutti A, Rossi R, Ambrosio MR, Pansini R (1990) Stimulation of growth hormone and corticotropin release by angiotensin II in man. Metabolism 39: 1063–1067

degli Uberti EC, Trasforini G, Margutti A, Ambrosio MR, Rossi R, Pansini R (1991) Stimulatory effect of angiotensin II upon luteinizing hormone release in normal women. Neuroendocrinology 53: 204–208

Demassieux S, Boucher R, Crisé C, Genest J (1976) Purification and characterization of tonin. Can J Biochem 54: 788–795

Dempsey PJ, McCallum ZT, Kent KM, Cooper T (1971) Direct myocardial effects of angiotensin II. Am J Physiol 220: 477–481

Denbow DM (1985) Food and water intake response of turkeys to intracerebroventricular injections of angiotensin II. Poultry Sci 64: 1996–2000

Denef C (1986) Paracrine interactions in the anterior pituitary. Clin Endocrinol Metab 15: 1–32

Denef C, Andries M (1983) Evidence for paracrine interaction between gonadotrophs and lactotrophs in pituitary cell aggregates. Endocrinology 112: 813–821

Dengler H (1956) Über ein reninartigen Wirkstoff in Arterienextrakten. Naunyn-Schmiedeberg's Arch Exp Pathol Pharmakol 227: 481–487

Denton D (1982) The hunger for salt. Springer, Berlin Heidelberg New York, pp 115–136, 382–416

Denton DA (1991) Mineral appetite: an overview. In: Ramsey DJ, Booth DA (eds) Thirst-physiological and psychological aspects. Springer, Berlin Heidelberg New York pp 131–146

Denton DA, Nelson JF, Tarjan E (1985) Water and salt intake of wild rabbits (Oryctolagus cuniculus) (L.) following dipsogenic stimuli. J Physiol (Lond) 362: 285–301

Deschepper CF, Seidler CD, Steele MK, Ganong WF (1985) Further studies on the localization of angiotensin-II-like immunoreactivity in the anterior pituitary gland of the male rat, comparing various antisera to pituitary hormones and their specificity. Neuroendocrinology 40: 471–475

Deschepper CF, Crumrine DA, Ganong WF (1986) Evidence that the gonadotrophs are the likely site of production of angiotensin II in the anterior pituitary of the rat. Endocrinology 119: 36–43

Deth R, van Breemen C (1974) Relative contributions of Ca^{2+} influx and cellular Ca^{2+} release during drug induced activation of the rabbit aorta. Pflügers Arch Eur J Physiol 348: 13–22

Deth R, Van Breemen C (1977) Agonist induced release of intracellular Ca^{2+} in the rabbit aorta. J Membr Biol 30: 363–380

Devynck M-A, Meyer P (1976) Angiotensin receptors in vascular tissue. Am J Med 61: 758–767

Devynck M-A, Rouzarie-Dubois B, Chevillotte E, Meyer P (1976) Variations in number of uterine receptors following changes in plasma angiotensin levels. Eur J Pharmacol 40: 27–37

Devynck M-A, Pernollet M-G, Matthews PG, Khosla MC, Bumpus FM, Meyer P (1977) Specific receptors for des-Asp¹-angiotensin II ("angiotensin III") in rat adrenals. Proc Natl Acad Sci USA 74: 4029–4032

Di Nicolantonio R, Mendelsohn FAO (1986) Plasma renin and angiotensin in dehydrated and rehydrated rats. Am J Physiol 250: R898–R901

Ding YA, Kenyon CJ, Semple PF (1984) Receptors for angiotensin II on platelets from men. Clin Sci 66: 725–731

Doi Y, Mulrow PJ (1984) Effects of sodium, potassium ACTH and nephrectomy on adrenal renin. Jpn Circ J 48: 1280–1281

Doi Y, Franco-Saenz R, Murlow PJ (1984) Evidence for an extrarenal source of inactive renin in rats. Hypertension 6: 627–632

Doležel S (1966) Monoaminergic innervation of the arteries and veins of the kidney observed using fluorscence reaction. Folia Morphol (Warsaw) 14: 168–174

Doležel S, Edvinsson L, Owman CH, Owman T (1976) Fluroescence histochemistry and autoradiography of adrenergic nerves in the renal juxtaglomerular complex of mammals and man, with special regard to the efferent arteriole. Cell Tissue Res 169: 211–220

Douglas JG, Brown GP (1982) Effects of prolonged low dose infusion of angiotensin II and aldosterone on rat smooth muscle and adrenal angiotensin II receptors. Endocrinology 111: 988–992

Douglas JG, Catt KJ (1976) Regulation of angiotensin II receptors in the rat adrenal cortex by dietary electrolytes. J Clin Invest 58: 834–843

Douglas JG, Bartley P, Kondo T, Catt K (1978) Formation of des-Asp[1]-angiotensin II is not an obligatory step in the steroidogenic action of angiotensin II in the canine adrenal. Endocrinology 102: 1921–1924

Douglas JG, Michailov M, Khosla MC, Bumpus FM (1980) Comparative receptor-binding properties of heptapeptide and octapeptide antagonists of angiotensin II in rat adrenal glomerulosa and uterine smooth muscle. Endocrinology 106: 120–124

Douglas JG, Brown G, White C (1982) Influence of cations on kinetics of angiotensin II binding to adrenal, renal, and smooth muscle receptors. Hypertension 4 (Suppl III): 79–84

Douglas JG, Khosla MC, Bumpus FM (1985) Efficacy of octa-and heptapeptide antagonists of angiotensin II as inhibitors of angiotensin II binding in the rat adrenal glomerulosa. Endocrinology 116: 1598–1602

Dourish C, Duggan JA, Banks RJA (1992) Drinking induced by subcutaneous injection of angiotensin II in the rat is blocked by the selective AT_1 receptor antagonist Dup753 but not by the selective AT_2 receptor antanonist WL19. Eur J Pharmacol 211: 113–116

Dufy-Barbe L, Rodriguez F, Arsaut J, Verrier D, Vincent J-D (1982) Angiotensin II stimulates prolactin release in the rhesus monkey. Neuroendocrinology 35: 242–247

Dunham EW, Zimmerman BG (1970) Release of prostaglandin-like material from dog kidney during nerve stimulation. Am J Physiol 219: 1279–1285

Dzau VJ (1984) Vascular wall renin-angiotensin pathway in control of the circulation. A hypothesis. Am J Med 77: 31–36

Dzau VJ (1987) Vascular angiotensin pathways: a new therapeutic target. J Cardiovasc Pharmacol 10 (Suppl): S9–S16

Dzau VJ (1988) Circulating versus local renin-angiotensin system in cardiovascular homeostasis. Circulation 77 (Suppl) : 4–13

Dzau VJ, Brenner A, Emmet N, Haber E (1980) Identification of renin and renin-like activity in rat brain by a renin-specific antibody. Clin Sci 59: 45s–47s

Dzau VJ, Ellison KE, Brody T, Ingelfinger J, Pratt RE (1987) A comparative study of the distributions of renin and angiotensin messenger ribonucleic acid in rat and mouse tissues. Endocrinology 120: 2334–2338

Edelman R, Hartroft PM (1961) Localization of renin in juxtaglomerular cells of rabbit and dog through use of fluorescent antibody technique. Circ Res 9: 1069–1077

Edwards BR, LaRochelle FT (1984) Antidiuretic effect of endogenous oxytocin in dehydrated Brattleboro homozygous rats. Am J Physiol 247: F453–F465

Edwards JG (1940) The vascular pole of the glomerulus in the kidney of vertebrates. Anat Rec 76: 381–389

Ehlers MRW, Riordan JF (1990) Angiotensin-converting enzyme. Biochemistry and molecular biology. In: Laragh JH, Brenner BM (eds) Hypertension: pathophysiology, diagnosis, and management. Raven Press, New York, pp 1217–1231

Ehlers MRW, Fox EA, Strydom DG, Riordan JF (1989) Molecular cloning of human testicular angiotensin-converting enzyme: the testis isozyme is identical to the C-terminal half of endothelial angiotensin-converting enzyme. Proc Natl Acad Sci USA 86: 7741–7745

Elfont RM, Fitzsimons JT (1985) The effect of captopril on sodium appetite in adrenalectomized and deoxycorticosterone-treated rats. J Physiol 365: 1–12

Elfont RM, Epstein AN, Findlay ALR (1980) The role of the subfornical organ in angiotensin-induced drinking in the North American opossum. J Physiol 301: 49 p

Elger M, Wahlqvist I, Hentschel H (1984) Ultrastructure and adrenergic innervation of preglomerular arterioles in the euryhaline teleost, *Salmo gairdneri*. Cell Tissue Res 237: 451–458

Elton TS, Stephan CC, Taylor GR, Kimball MG, Martin MM, Durand JN, April S (1992) Isolation of two distinct type 1 angiotensin II receptor genes. Biochem Biophys Res Commun 184: 1067–1073

Endo M (1975) Mechanism of action of caffeine on the sarcoplasmic reticulum of skeletal muscle. Proc Jpn Acad 51: 479–494

Epstein AN (1982) Mineralocorticoids and cerebral angiotensin may act together to produce sodium appetite. Peptides 3: 483–484

Epstein AN, Hsiao S (1975) Angiotensin as a dipsogen. In: Peters G, Fitzsimons JT, Peters-Haefeli L (eds) Control mechanisms of drinking. Springer, Berlin Heidelberg New York, pp 108–116

Epstein AN, Massi M (1987) Salt appetite in the pigeon in response to pharmacological treatments. J Physiol (Lond) 393: 555–568

Epstein AN, Fitzsimons JT, Simons BJ (1969) Drinking caused by the intracranial injection of angiotensin into the rat. J Physiol (Lond) 200: 98p–100p

Epstein AN, Fitzsimons JT, Rolls BJ (1970) Drinking induced by injection of angiotensin into the brain of the rat. J Physiol 210: 457–474

Epstein AN, Zhang DM, Schultz J, Rosenberg M, Kupsha P, Stellar E (1984) The failure of ventricular sodium to control of sodium appetite in the rat. Physiol Behav 32: 683–685

Erdös EG (1975) Angiotensin I converting enzyme. Circ Res 36: 247–255

Erdös EG (1976) Conversion of angiotensin I to angiotensin II. Am J Med 60: 749–759

Eriksson L, Fyhrquist F (1976) Plasma renin activity following central infusion of angiotensin II and altered CSF sodium concentration in the conscious goat. Act Physiol Scand 98: 209–216

Ernsberger P, Zhou J, Damon TH, Douglas JG (1992) Angiotensin II receptor subtypes in cultured rat ranal mesangial cells. Am J Physiol 263: F411–F416

Erspamer V, Melchiorri P, Nakajima T, Yasuhara T, Endean R (1979) Amino acid composition and sequence of Crinia-angiotensin, and angiotensin II-like endecapeptide from the skin of the Australian frog *Crinia georgiana*. Experientia 35: 1132–1133

Esther CR, Thomas KE, Bernstein KE (1994) Chicken lacks the testis specific isozyme of angiotensin converting enzyme found in mammals. Biochem Biophys Res Commun 205: 1916–1921

Evered MD, Fitzsimons JT (1976) Drinking induced by angiotensin in the pigeon (*Columba livia*). J Physiol (Lond) 263: 193P–194P

Evered MD, Fitzsimons JT (1981) Drinking and changes in blood pressure in response to angiotensin II in the pigeon *Columba livia*. J Physiol (Lond) 310: 337–352

Evered MD, Robinson MM (1984) Increased or decreased thirst caused by inhibition of angiotensin-converting enzyme in the rat. J Physiol (Lond) 348: 573–588

Evered MD, Robinson MM, Richardson MA (1980) Captopril, given intracerebroventricularly, subcutaneously or by gavage inhibits angiotensin-converting enzyme activity in the rat brain. Eur J Pharm 68: 443–449

Everitt BJ, Meister B, Hökfelt T (1992) The organization of monoaminergic neurons in the hypothalamus in relation to neuroendocrine integration. In: Demeroff CH (ed) Neuroendocrinology. CRC Press, Boca Raton, pp 87–128

Faarup P (1967) Renin location in the different parts of the juxtaglomerular apparatus in the cat kidney. I. The afferent arteriole and the macula densa. Acta Path Microbiol Scand 71: 509–521

Faarup P (1968) Renin location in the different parts of the juxtaglomerular apparatus in the cat kidney. II. Fractions of the afferent arteriole, the cell group of Goormaghtigh, the efferent arteriole, and the glomerulus. Acta Pathol Microbiol Scand 72: 109–117

Falck B, Hillarp A, Thieme G, Thorp A (1962) Fluorescence of catecholamines and related compounds condensed with formaldehyde. J Histochem Cytochem 10: 348–354

Faraggiana T, Gresik E, Tanaka T, Inagami T, Lupo A (1982) Immunohistochemical localization of renin in the human kidney. J Histochem Cytochem 30: 459–465

Farese RV, Larson RE, Davis JS (1984) Rapid effects of angiotensin-II on polyphosphoinositide metabolism in the rat adrenal glomerulosa. Endocrinology 114: 302–304

Fasciolo JC, Houssay BA, Taquini AC (1938) The blood-pressure raising secretion of the ischaemic kidney. J Physiol (Lond) 94: 281–293

Favrod-Coune CA, Capponi AM, Gailard RC, Muller AF (1982) Characterization of angiotensin II (AII) binding and corticotropin-releasing activity in rat anterior pituitary cells. Experientia 38: 713

Feldstein JB, Sumners C, Raizada MK (1986) Sodium increases angiotensin II receptors in neuronal cultures from brains of normotensive and hypertensive rats. Brain Res 370: 265–272

Felix D, Akert K (1974) The effect of angiotensin II on neurones of the cat subfornical organ. Brain Res 76: 350–353

Felix D, Schlegel W (1978) Angiotensin receptive neurones in the subfornical organ. Structure-activity relations. Brain Res 149: 107–116

Felix D, Gambino C, Yong Y, Schelling P (1986) Angiotensin sensitive sites in the central nervous system. In: de Caro G, Epstein AN, Massi M (eds) The physiology of thirst and sodium appetite. Plenum Press, New York, pp 136–140

Felix D, Khosla MC, Barnes KL, Imboden H, Mantani B, Ferrario CM (1991) Neurophysiological responses to angiotensin-(1–7). Hypertension 17: 1111–1114

Ferguson AV, Kasting NW (1986) Electrical stimulation in subfornical organ increases plasma vasopressin in the conscious rat. Am J Physiol 251: R425–R428

Ferguson AV, Kasting NW (1987) Activation of subfornical organ efferents stimulates oxytocin secretion in the rat. Regul Pept 18: 93–100

Fernandez-Cruz A, Noth RH Jr, Helder RG, Mulrow PJ (1975) Glucagon stimulation of plasma renin activity in humans. J Clin Endocrinol Metab 41: 183–184

Fernley RT, John M, Niall HD, Coghlan JP (1986) Purification and characterization of ovine angiotensinogen. Eur J Biochem 154: 597–601

Ferris TF, Gorlden P, Murlow PJ (1967) Rabbit uterus as a source of renin. Am J Physiol 212: 698–706

Feuerstein G, Klausz M, Cohen S, Gutman Y, Khosla MC, Bumpus FM (1978) Angiotensin II mediation of adrenal catecholamine secretion induced by intrarenal isoprenaline infusion. Eur J Pharmacol 52: 375–377

Findley ARL, Epstein AN (1980) Increased sodium intake is somehow induced in rats by intravenous angiotensin II. Horm Behav 14: 86–92

Findlay ALR, Elfont RM, Epstein AN (1980) The site of the dipsogenic action of angiotensin II in the North American opossum *Didelphis virginiana*. Brain Res 198: 85–94

Fischer-Ferraro C, Nahmod VE, Goldstein DJ, Finkielman S (1971) Angiotensin and renin in rat and dog brain. J Exp Med 133: 353–361

Fisher CD (1979) Geographic variation and evolution in the Australian Ringneck parrots (*Barnardius*). Thesis, Diss Abst, Univ of Michigan Press, Ann Arbor p 6737-B

Fisher ER (1961) Correlation of juxtaglomerular granulation, pressor activity, and enzymes of macula densa in experimental hypertension. Lab Invest 10: 707–718

Fishman MC, Zimmerman EA, Slater EE (1981) Renin and angiotensins: the complete system within the neuroblastoma x glioma cell. Science 214: 921–923

Fitzsimons JT (1964) Drinking caused by constriction of the inferior vena cava in the rat. Nature 204: 479–480

Fitzsimons JT (1966) Hypovolaemic drinking and renin. J Physiol (Lond) 186: 130P–131P

Fitzsimons JT (1967) The kidney as a thirst receptor. J Physiol (Lond) 191: 128P–129P

Fitzsimons JT (1969) The role of a renal thirst factor in drinking induced by extracellular stimuli. J Physiol (Lond) 201: 349–368

Fitzsimons JT (1970) The renin-angiotensin system in the control of drinking. In: Martin L, Motta M, Fraschini F (eds) The hypothalamus. Academic Press, New York, pp 195–212

Fitzsimons JT (1971) The physiology of thirst: a review of the extraneural aspects of the mechanisms of drinking. In: Stellar E, Sprague JM (eds) Progress in physiological psychology, vol 4. Academic Press, New York, pp 119–201

Fitzsimons JT (1975) The renin-angiotensin system and drinking behavior. In: Gipsen WH, van Wimersma Greidanus JJB, Bohus B, de Wied D (eds) Hormones, homeostasis and the brain. Progress in brain research vol 42. Elsevier, Amsterdam, pp 215–233

Fitzsimons JT (1979) The physiology of thirst and sodium appetite. Cambridge Univ Press, Cambridge, pp 266, 276–282, 287, 308–328, 471–478

Fitzsimons JT (1993) Renin in thirst and sodium appetite. In: Robertson JIS, Nicholls MG (eds) The renin-angiotensin system 1. Gower Medical Publ, London, pp 32.1–32.8

Fitzsimons JT, Kaufman S (1977) Cellular and extracellular dehydration and angiotensin as stimuli to drinking in the common iguana *Iguana iguana*. J Physiol (Lond) 265: 443–463

Fitzsimons JT, Kucharczyk J (1978) Drinking and haemodynamic changes induced in the dog by intracranial injection of components of the renin-angiotensin system. J Physiol (Lond) 276: 419–434

Fitzsimons JT, Simons BJ (1968) The effect of angiotensin on drinking in the rat. J Physiol (Lond) 196: 39P–41P

Fitzsimons JT, Simons BJ (1969) The effect on drinking in the rat of intravenous infusion of angiotensin given alone or in combination with other stimuli of thirst. J Physiol (Lond) 203: 45–57

Fitzsimons JT, Stricker EM (1971) Sodium appetite and the renin-angiotensin system. Nature 231: 58–60

Fitzsimons JT, Wirth JB (1978) The renin-angiotensin system and sodium appetite. J Physiol (Lond) 274: 63–80

Fitzsimons JT, Epstein AN, Johnson AK (1978a) Peptide antagonists of the renin-angiotensin system in the characterization of receptors for angiotensin-induced drinking. Brain Res 153: 319–331

Fitzsimons JT, Kucharczyk J, Richards G (1978b) Systemic angiotensin-induced drinking in the dog: a physiological phenomenon. J Physiol (Lond) 276: 435–448

Fitzsimons JT, Massi M, Thornton SN (1982) The effects of changes in osmolality and sodium concentration on angiotensin-induced drinking and excretion in the pigeon. J Physiol (Lond) 330: 1–15

Fluharty SJ, Epstein AN (1983) Sodium appetite elicited by intracerebroventricular infusion of angiotensin II in the rat: II. Synergic interaction with systemic mineralocorticoids. Behav Neurosci 97: 746–758

Fogo A, Yoshida Y, Yared A, Ichikawa I (1990) Importance of angiogenic action of angiotensin II in the glomerular growth of maturing kidneys. Kidney Int 38: 1068–1074

Force T, Kyriakis JM, Avruch J, Bonventre JV (1991) Endothelin, vasopressin, and angiotensin II enhance tyrosine phosphorylation by protein kinase C-dependent and -independent pathways in glomerular mesangial cells. J Biol Chem 266: 6650–6656

Forget G, Heisler S (1976) Preparation and characterization of adrenocortical plasma membrane angiotensin II receptors. Can J Physiol Pharmacol 54: 698–707

Forshaw JM (1969) Australian parrots. Lansdowne Press, Melbourne, 173 pp

Fowler NO, Holmes JC (1964) Coronary and myocardial actions of angiotensin. Circ Res 14: 191–201

Francisco LL, Osborn JL, DiBona GF (1982) Prostaglandins in renin release during sodium deprivation. Am J Physiol 243: F537–F542

Fray JCS (1977) Stimulation of renin release in perfused kidney by low calcium and high magnesium. Am J Physiol 232: F377–F382

Fray JCS (1978) Stretch receptor control of renin release in perfused rat kidnsy: effect of high perfusate potassium. J Physiol (Lond) 282: 207–217

Fray JCS (1980) Mechanisms by which renin secretion from perfused rat kidneys is stimulated by isoprenaline and inhibited by high perfusion pressure. J Physiol (Lond) 308: 1–13

Fray JCS, Park CS, Valentine AND (1987) Calcium and the control of renin secretion. Endocr Rev 70: 53–93

Frederiksen O, Leyssac PP, Skinner SL (1975) Sensitive osmometer function of juxtaglomerular cells in vitro. J Physiol (Lond) 252: 669–679

Freeman RH, Rostorfer HH (1972) Hepatic changes in renin substrate biosynthesis and alkaline phosphatase activity in the rat. Am J Physiol 223: 364–370

Freeman RH, Davis JO, Goteshall RW, Johnson JA, Spielman WS (1974) The signal perceived by the macula densa during changes in renin release. Circ Res 35: 307–315

Freeman RH, Davis JO, Lohmeier TE (1975) Des-1-Asp-angiotensin II. Possible intrarenal role in homeostasis in the dog. Circ Res 37: 30–34

Freeman RH, Davis JO, Dietz D, Villarreal D, Seymour AA, Echtenkamp SF (1982) Renal prostaglandins and the control of renin release. Hypertension 4 (Suppl II): 106–112

Freeman RH, Davis JO, Villarreal D (1984) Role of renal prostaglandins in the control of renin release. Circ Res 54: 1–9

Fregly MJ (1980) Effect of the angiotensin converting enzyme inhibitor, captopril, on NaCl appetite of rats. J Pharmacol Exp Ther 215: 407–412

Fregly MJ, Rowland NE (1985) Role of renin-angiotensin-aldosterone system in NaCl appetite of rats. Am J Physiol 248: R1–R11

Fregly MJ, Rowland NE (1991) Effect of a nonpeptide angiotensin II receptor antagonist, Dup753, on angiotensin-related water intake in rats. Brain Res Bull 27: 97–100

Fregly MJ, Waters IW (1965) Effect of propiothiouracil on the preference threshold of rats for NaCl solutions. Proc Soc Exp Biol Med 120: 637–640

Fregly MJ, Shechtman O, Vanbergen P, Reeber C, Papanek PE (1991) Changes in blood pressure and dipsogenic responsiveness to angiotensin-II during chronic exposure of rats to cold. Pharm Behav Biochem 38: 837–842

Friedman M, Kaplan A (1942) Studies concerning the site of renin formation in the kidney. I. The absence of renin in the aglomerular kidney of the midshipman fish. J Exp Med 75: 127–134

Friedman M, Kaplan A, Williams E (1942) Studies concerning site of renin formation in the kidney. II. Absence of renin in glomerular kidney of marine fish. Proc Soc Exp Biol Med 50: 199–202

Fruchter J, Yang M, Pang PKT (1980) The renin-angiotensin system in bullfrogs and mud-puppies. Fed Proc 39: 946

Fukamizu A, Nishi K, Cho T, Saitoh M, Nakayama K, Ohkubo H, Nakanishi S, Murakami K (1988) Structure of rat renin gene. J Mol Biol 201: 443–450

Fukamizu A, Takahashi S, Seo MS, Tada M, Tanimoto K, Uehara S, Murakami K (1990) Structure and expression of the human angiotensinogen gene: identification of a unique and highly active promoter. J Biol Chem 265: 7576–7582

Fukamizu A, Hatae T, Kon Y, Sugiura M, Hasegawa T, Yokoyama M, Nomura T, Katsuki M, Murakami K (1991) Human renin in transgenic mouse kidney is localized to juxtaglo-merular cells. Biochem J 278: 601–603

Fuxe K, Agnati LF, Ganten D, Lang RE, Calza L, Poulsen K, Infantellina F (1982) Morpho-metric evaluation of the coexistence of renin-like and oxytocin-like immunoreactivity in nerve cells of the paraventricular hypothalamic nucleus of the rat. Neurosci Lett 33: 19–24

Gadbut AP, Cash SA, Noble JA, Radice TR, Weyhenmeyer JA (1991) The effect of Ca^{2+} channel antagonists (cadmium, omega-conotoxin, GIVA, and nitrendipine) on the release of an-giotensin II from fetal rat brain in vitro. Neurosci Lett 123: 91–94

Gaillard-Sanchez I, Mattei MG, Clauser E, Corvol P (1990) Assignment by in situ hybridization of the angiotensinogen gene to chromosome band 1q4, the same region as the human renin gene. Human Genet 84: 341–343

Galen F-X, Devaux C, Guyenre T, Menard J, Corvol P (1979) Multiple forms of human renin. J Biol Chem 254: 4848–4855

Galen F-X, Devaux C, Houot AM, Menard J, Corvol P, Corvol MT, Gubler MC, Mounier F, Camilleri JP (1984) Rennin biosynthesis by human tumoral juxtaglomerualr cells. Evidence for a renin precursor. J Clin Invest 73: 1144–1155

Gall JAM, Alcorn D, Butkus A, Coghlan JP, Ryan GB (1986) Distribution of glomerular peripolar cells in different mammalian species. Cell Tissue Res 244: 203–208

Galla JH, Kochen TA, Luke RG (1977) Failure of sodium iodide loading to inhibit renin in the rat. Proc Soc Exp Biol Med 154: 30–32

Gallardy R, Podhasky P, Olson KR (1984) Angiotensin-converting enzyme activity in tissues of the rainbow trout. J Exp Zool 230: 155–158

Galli-Phillips SM (1991) Evidence for the presence of an active angiotensin system in the nurse shark, *Gynclymostoma cirrratum*. Book of Abstr, 5th Int Symp on Fish Physiol, p 25

Ganong JA, Rice MK, Flamenbaum W (1974) Effect of angiotensin converting enzyme inhibition on renal autoregulation. Proc Soc Exp Biol Med 146: 414–418

Ganong WF (1981) The brain and the renin angiotensin system. In: Buckley JD, Ferrario CM (eds) Central nervous system mechanisms of hypertension. Raven Press, New York, pp 283–292

Ganong WF (1984) The brain renin-angiotensin system. Annu Rev Physiol 46: 17–31

Ganong WF (1989) Angiotensin II in the brain and pituitary: contrasting roles in the regulation of adenohypophyseal secretion. Horm Res 31: 24–31

Ganong WF (1993) Blood, pituitary, and brain renin-angiotensin systems and regulation of secretion of anterior pituitary gland. Front Neuroendocrinol 14: 233–249

Ganong WF, Murakami K (1987) The role of angiotensin II in the regulation of ACTH secretion. Ann NY Acad Sci 512: 176–186

Ganong WF, Rudolph CD, Zimmerman H (1979) Neuroendocrine components in the regulation of blood pressure and renin secretion. Hypertension 1: 207–218

Ganten D, Merquez-Julio A, Granger P, Hayduk K, Karsunsky KP, Boucher R, Genest J (1971) Renin in dog brain. Am J Physiol 221: 1733–1737

Ganten D, Ganten U, Kubo S, Granger P, Nowaczynski W, Boucher R, Genest J (1974) Influence of sodium, potassium and pituitary hormones on iso-renin in rat adrenal glands. Am J Physiol 227: 224–229

Ganten D, Schelling P, Vecsei P, Ganten U (1976) Iso-renin of extrarenal origin. "The tissue angiotensigenase systems". Am J Med 60: 760–772

Ganten D, Prinz M, Phillips MI, Scholkens BA (1982) The renin angiotensin system in the brain. Springer, Berlin Heidelberg New York

Ganten D, Hermann K, Bayer C, Unger T, Lang RE (1983) Angiotensin synthesis in the brain and increased turnover in hypertensive rats. Science 221: 869–871

Ganten D, Lang RE, Lehmann E, Unger T (1984) Brain angiotensin: on the way to becoming a well-studied neuropeptide system. Biochem Pharmacol 33: 3523–3528

Ganten D, Unger T, Lang RE (1988) The brain renin-angiotensin system: basic and functional consideration. In: Harding JW, Wright JW, Speth RC, Barnes CD (eds) Angiotensin and blood pressure regulation. Academic Press, San Diego

Garcia NH, Garvin JL (1994) Angiotensin-(1–7) has a biphasic effect on fluid absorption in the proximal straight tubule. J Am Soc Nephrol 5: 1133–1138

Gardiner DS, Lindop GBM (1985) The granular peripolar cell of the human glomerulus: a new component of the juxtaglomerular apparatus? Histopathology 9: 675–685

Gardiner DS, More IAR, Lindop GBM (1986) The granular peripolar cells of the human glomerulus: an ultrastructural study. J Anat 146: 31–43

Gardiner TW, Stricker (1985) Impaired drinking responses of rats with lesions of nucleus medianus: circadian dependence. Am J Physiol 248: R224–R230

Gardiner TW, Verbalis JG, Stricker EM (1985) Impaired secretion of vasopressin and oxytocin in rats after lesion of nucleus medianus. Am J Physiol 249: R681–688

Garland HO, Henderson IV (1975) Influence of environmental salinity on renal and adrenocortical function in the toad, Bufo marinus. Gen Comp Endocrinol 27: 136–143

Garrison JC, Borland MK, Florio VA, Twible DA (1979) The role of calcium ion as a mediator of the effects of angiotensin II, catecholamines, and vasopressin on the phosphorylation and activity of enzymes in isolated hepatocytes. J Biol Chem 254: 7147–7156

Gaynes RP, Szidon JP, Oparil S (1978) In vivo and in vitro conversion of des-1-Asp angiotensin I to angiotensin III. Biochem Pharmacol 27: 2871–2877

Geary KM, Hunt MK, Peach MJ, Gomez RA, Carey RM (1992) Effects of angiotensin converting enzyme inhibition, sodium depletion, calcium, isoproterenol, and angiotensin II on renin secretion by individual renocortical cells. Endocrinology 131: 1588–1594

Geber JG, Nies AS, Olsen RD (1981) Control of canine renin release: macula densa requires prostaglandin synthesis. J Physiol (Lond) 319: 419–429

Gehlert DR, Speth RC, Wamsley JK (1984) Autoradiographic localization of angiotensin II receptors in the rat brain and kidney. Eur J Pharmacol 98: 145–146

Geisterfer AAt, Peach MJ, Owens GK (1988) Angiotensin II induces hypertrophy, not hyperplasia, of cultured rat aortic smooth muscle cells. Circ Res 62: 749–756

197

Gerstberger R, Gray DA, Simon E (1984) Circulatory and osmoregulatory effects of angiotensin II perfusion of the third ventricle in a bird with salt glands. J Physiol (Lond) 349: 167–182

Gerstberger R, Hearly DP, Hammel HT (1987) Possible role of periventricular angiotensin II receptors in the regulation of salt gland function in saltwater-acclimated ducks. Wiss Z Karl-Marx-Univ Leipzig 36: 192–194

Gerstberger R, Müller AR, Simon-Oppermann C (1992) Functional hypothalamic angiotensin II and catecholamine receptor systems inside and outside the blood-brain barrier. Prog Brain Res 91: 423–433

Ghazi KN, Grove KL, Wright JW, Phillips MI, Speth RC (1994) Variations in angiotensin-II release from the rat brain during the estrous cycle. Endocrinology 135: 1945–1950

Ghosh A, Ghosh S (1972) Effect of dehydration on the neurosecretory system of an arid zone avian species. Gegenbauers Morphol Jahrb 118: 414–422

Ghouse AM, Parsa B, Boylan JW, Brennan JC (1969) The anatomy, micro-anatomy, and ultrastructure of the kidney of the dogfish, *Squalus acanthias*. Bull Mt Desert Isl Biol Lab 8: 22–29

Gibbons GH, Dzau VJ, Farhi ER, Barger AC (1984) Interaction of signals influencing renin release. Annu Rev Physiol 46: 291–308

Gibson IW, More IAR, Lindop GBM (1989) A scanning electron microscopic study of the peripolar cell of the rat renal glomerulus. Cell Tissue Res 257: 201–206

Gill JM, Stephens GA (1983) Sympathetic innervation of reptilian kidney. Fed Proc 42: 473 (Abst)

Gill GN, Ill CR , Simonian MH (1977) Angiotensin stimulation of bovine adrenocortical cell growth. Proc Natl Acad Sci USA 74: 5569–5573

Gimbrone Jr MA, Majeau GR, Atkinson WJ, Sadler W, Cruise SA (1979) Angiotensin converting enzyme EC-3.4.15.1 activity in isolated brain microvessels. Life Sci 25: 1075–1084

Ginesi LM, Munday KA, Noble AR (1983) Secretion control for active and inactive renin: effects of calcium and potassium on rabbit kidney cortex slices. J Physiol (Lond) 344453–463

Glossmann H, Baukal AJ, Catt KJ (1974) Properties of angiotensin II receptors in the bovine and rat adrenal cortex. J Biol Chem 249: 825–834

Goldblatt H (1947) The renal origin of hypertensin. Physiol Rev 27: 120–165

Goldblatt H (1948) The renal origin of hypertensin. Charles C Thomas, Springfield, Illinois

Goldblatt H, Lynch J, Hanzal RF, Summerville WW (1934) Studies on experimental hypertension I. The production of persistent elevation of systolic blood pressure by means of renal ischemia. J Exp Med 59: 347–379

Goldstein DJ, Diaz A, Finkielamn S, Nahmod VE, Fisher-Ferraro C (1972) Angiotensinase activity in rat and dog brain. J Neurochem 19: 2451–2452

Goodfriend TL, Fyhrquist F, Gutmann F, Knych E, Hollemans H, Allmann D, Kent K, Cooper T (1972) Clinical and conceptual use of angiotensin receptors. In: Genest J, Koiw E (eds) Hypertension '72. Springer, New York pp 549–563

Goodwin FJ, Knowlton, AI, Laragh JH (1970) Absence of renin suppression by deoxycorticosterone in rats. Am J Physiol 216: 1476–1480

Goormaghtigh N (1932) Les segments neuro-myo-arteriels juxtaglomerulaires du rein. Arch Biol Paris 43: 575–591

Goormaghtigh N (1939) Existence of an endocrine gland in the media of the renal arterioles. Proc Soc Exp Biol Med 42: 688–689

Goormaghtigh N (1945) La fonction endocrine des artérioles rénales: son rôle dans la pathogénie de l'hypertension artérielle. Rev Belg Sci Med 16: 65–155

Gorbman A, Dickhoff WW, Vigna SR, Clark NB, Ralph CL (1983) Comparative endocrinology. John Wiley & Sons, New York pp 423–425

Gordon P, Ferris TF, Murlow PJ (1967) Rabbit uterus as a possible site of renin synthesis. Am J Physiol 212: 703–706

Gordon RD, Kuchel O, Liddle GW, Island DP (1967) Role of the sympathetic nervous system in regulating renin and aldosterone production in man. J Clin Invest 46: 599–605

Gorgas K (1978a) Innervation of the juxtaglomerular apparatus. In: Coupland RE, Forssman WG (eds) Peripheral neuroendocrine interaction. Springer, Berlin Heidelberg New York, pp 144–152

Gorgas K (1987b) Struktur und Innervation des juxtaglomerularen Apparates der Ratte. Adv Anat Embryol Cell Biol 54: 1–84

Goto K, Koike TI, Nelson HL, McKay DW (1986) Peripheral angiotensin II stimulates release of vasotocin in conscious chickens. Am J Physiol 251: R333–R339

Gould AB, Goodman SA, Onesti G (1974) A new biological function of the renin system: the control of erythropoietin. Circulation 50: 111–32

Grafflin AL (1929) The pseudoglomeruli of the kidney of *Lophius piscatorius*. Am J Anat 44: 441–454

Granger P, Rojo-Ortega JM, Gruner A, Dahlheim H, Thurau K, Boucher R, Genset J (1972) On the intrarenal role of the renin angiotensin system. In: Assaykeen TA (ed) Control of renin secretion. Plenum Press, New York, pp 131–144

Gray CJ, Brown JA (1985) Renal and cardiovascular effects of angiotensin II in the rainbow trout *Salmo gairdneri*. Gen Comp Endocrinol 95: 375–381

Gray DA, Erasmus T (1989) Control of plasma arginine vasotocin in kelp gulls (*Larus domesticanus*): roles of osmolality, volume, and plasma angiotensin II. Gen Comp Endocrinol 74: 110–119

Gray DA, Simon E (1985a) Control of plasma angiotensin II in a bird with salt glands (*Anas platyrhynchos*). Gen Comp Endocrinol 60: 1–13

Gray DA, Simon E (1985b) Extracellular volume and sodium handling in saltwater-adapted Pekin ducks. Pflügers Arch Eur J Physiol 403: R19

Gray DA, Simon E (1987) Dehydration and arginine vasotocin and angiotensin II in CSF and plasma of Pekin ducks. Am J Physiol 253: R285–R291

Gray DA, Naude RJ, Erasmus T (1988) Plasma arginine vasotocin and angiotensin II in the water deprived ostrich (*Struthio camelus*). Comp Biochem Physiol 89A: 251–256

Gray DA, Gerstberger R, Simon E (1989) Role of angiotension II in aldosterone regulation in the Pekin duck. J Endocrinol 123: 445–452

Gray DA, Schüts H, Gerstburger R (1991) Interaction of atrial natriuretic factor and osmoregulatory hormones in the Pekin duck. Gen Comp Endocrionl 81: 246–255

Gronan RJ, York DH (1979) Effects of chronic intraventricular administration of angiotensin II on drinking behavior and blood pressure. Pharm Biochem Behav 10: 121–126

Grossman SP (1990) Thirst and sodium appetete. Physiological basis. Academic Press, San Diego, pp 177–201

Guglielmone R, Daneo-Sisto L (1978) Innervation of the kidney in the lizard. Electron microscopical investigations in *Lacerta muralis* (Laur.) and *Lacerta sicula* (Raf.) (Reptilia Lacertidae). Monitore Zool ital (N Ser) 12: 229–242

Guillemette G, Balla T, Baukal AJ, Spät A, Catt KJ (1987) Intracellular receptors for inositol 1, 4, 5-triphosphate in angiotensin target tissues. J Biol Chem 262: 1010–1015

Gunther S (1984) Characterization of angiotensin II receptor subtypes in rat liver. J Biol Chem 259: 7622–7629

Gunther S, Gimbrone MA Jr, Alexander RW (1980a) Identification and characterization of the high affinity vascular angiotensin II receptor in rat mesenteric artery. Circ Res 47: 278–286

Gunther S, Gimbrone MA Jr, Alexander WR (1980b) Regulation by angiotensin II of its receptors in resistance blood vessels. Nature 287: 230–232

Gunther S, Alexander RW, Atkinson WJ, Gimbrone WA Jr (1982) Functional angiotensin II receptors in cultured vascular smooth muscle cells. J Cell Biol 92: 289–298

Gurchinoff S, Khairallah PA, Devynck MA, Meyer P (1976) Angiotensin II binding to zona glomerulosa cells from rabbit adrenal glands. Biochem Pharmacol 25: 1031–1034

Gutkind JS, Kurihara M, Saavedra JM (1988) Increased angiotensin II receptors in brain nuclei of DOCA-salt hypertensive rats. Am J Physiol 255: H464–H650

Gutman MB, Cirriello J, Mogenson GJ (1986) Electrophysiological identification of forebrain connection of subfornical organ. Brain Res 382: 119–128

Gutman MB, Jones DL, Ciriello J (1988) Effects of paraventricular lesions on drinking and pressor responses to Ang II. Am J Physiol 255: R882–R887

Gutman Y, Benzakein F (1971) Effect of an increase and a lack of antidiuretic hormone on plasma renin activity in the rat. Life Sci 10: 1081–1085

Gutman Y, Benzakein F, Chaimovitz M (1967) Kidney factors affecting water consumption in the rat. Isr J Med Sci 3: 910–911

Gutman Y, Tamir N, Benzakein F (1973) Effect of lithium on plasma renin activity. Eur J Pharmacol 24: 347–351

Haas E, Lamfrom H, Goldblatt H (1954) A simple method for the extraction and partial purification of renin. Arch Biochim Biophys 48: 256–260

Haber E, Carlson W (1983) The biochemistry of the renin-angiotensin system. In: Genest J, Kuchel O, Hamet P, Cantin M (eds) Hypertensin, 2nd edn, Sec 3, renin-angiotensin system, Chap 12. MaGraw-Hill, New York, pp 171–184

Haber E, Zusman R, Burton J, Dzau VJ, Barger AC (1983) Is renin a factor in the etiology of essential hypertension? Hypertension 5 (Suppl V): 8–15

Hackenthal E, Koch C, Bergmann T, Gross F (1972) Partial purification and characterization of a renin-like enzyme from rat submandibular gland. Biochem Pharmacol 21: 2779–2792

Hackenthal E, Hackenthal R, Hilgenfeldt U (1978) Purification and partial characterization of rat brain acid protease (isorenin). Biochim Biophys Acta 522: 561–573

Hackenthal E, Paul M, Ganten D, Taugner R (1990) Morphology, physiology, and molecular biology of renin secretion. Physiol Rev 70: 1067–1116

Hagiwara H, Sugiura N, Wakita K, Hirose S (1989) Purification and characterization of angiotensin-binding protein from porcine liver cytosolic fraction. Eur J Biochem 185: 405–410

Haimoto H, Takahashi Y, Koshikawa T, Nagura H, Kato K (1985) Immunohistochemical localization of γ-enolase in normal human tissues other than nervous and neuroendocrine tissues. Lab Invest 52: 257–263

Haimoto H, Takahashi M, Koshikawa T, Asai J, Kato K (1986) Enolase isozymes in renal tubules and renal cell carcinoma Am J Pathol 124: 488–495

Hall JE, Brands MW (1993) Intrarenal and circulating angiotensin II and renal function. Gower Medical Publ, London, pp 26.1–26.43

Haller H, Hensen J, Bahr V, Oelkers W (1986) Effects of angiotensin II infusion on the early morning surge of ACTH and on o-CRH provoked ACTH secretion in normal men. Acta Endocrinol 112: 150–156

Hanke W, Maser C (1985) Regulation of interrenal function in amphibians. In: Lofts B, Holmes WN (eds) Current trends in comparative endocrinology. Hong Kong Univ Press, Hong Kong pp 447–449

Hanner RH, Ryan GB (1980) Ultrastructure of the renal juxtaglomerular complex and peripolar cells in the axolotl (Ambystoma mexicanum) and the toad (Bufo marinus). J Anat 130: 445–455

Harada E, Rubin RP (1978) Stimulation of renin secretion and calcium efflux from the isolated perfused cat kidney by noradrenaline after prolonged calicum deprivation. J Physiol (Lond) 274: 367–379

Harding JW, Felix D (1987) Angiotensin-sensitive neurons in the rat paraventricular nucleus: relative potencies of angiotensin II and angiotensin III. Brain Res 410: 130–134

Harding JW, Stone LP, Wright JW (1981) The distribution of angiotensin II binding sites in rodent brain. Brain Res 205: 265–274

Harding JW, Wright JW, Speth RC, Barnes CD (1988) Angiotensin and blood pressure regulation. Academic press, San Diego

Harding JW, Jensen LL, Quirk WS, Dewey AL, Wright JW (1989) Brain angiotensin: critical role in the ongoing regulation of body fluid homeostasis and cardiovascular function. Peptides 10: 261–264

Harding JW, Jensen LL, Hanesworth JM, Roberts KA, Page TA, Wright JW (1992) Release of angiotensins in paraventricular nucleus of rat in response to physiological and chemical stimuli. Am J Physiol 262: F17–F23

Harland D, Gardiner SM, Bennett T (1988) Cardiovascular and dipsogenic effects of angiotensin II administered I. C. V. in Long-Evans and Brattleboro rats. Brain Res 455: 58–64

Harper RA, Stephens GA (1985) Blockade of the pressor response to angiotensin I and II in the bullfrog, Rana catesbeiana. Gen Comp Endocrionol 60: 227–235

Harris PJ, Navar LG (1985) Tubular transport responses to angiotensin. Am J Physiol 248: F621–F630

Harrison TR, Blalock A, Mason MF (1936–37) Effects on blood pressure of injection of kidney extracts of dogs with renal hypertension. Proc Soc Exp Biol Med 35: 38–40

Hartroft PM (1966) Electron microscopy of nerve endings associated with juxtaglomerular cells and macula densa. Lab Invest 15: 1127–1128

Hartroft PM, Newmark LN (1961) Electron microscopy of renal juxtaglomerular cells. Anat Rec 139: 185–199

Hartroft PM, Sutherland LE, Hartroft WS (1964) Juxtaglomerular cells as the source of renin: further studies with the fluorescent antibody technique and the effect of passive transfer of antirenin. Can Med Assoc J 90: 163–166

Hartwich A, Hessel G (1932) Experimentelle Untersuchungen zur Kreislaufwirkung körpereigener Stoffe I. Teil, Die Wirkung frischer und autolysierter Organpreßsäfte auf den Blutdruck. Zentralbl Inn Med 53: 612–626

Hasegawa K, Nishimura H, Khosla MC (1993) Angiotensin-induced endothelium-dependent relaxation of fowl aorta. Am J Physiol 264: R903–R911

Hasegawa Y, Nakajima T, Sokabe H (1983a) Chemical structure of angiotensin formed with kidney renin in the Japanese eel, *Anguilla japonica*. Biomed Res 4: 417–420

Hasegawa Y, Watanabe TX, Sokabe H, Nakajima T (1983b) Chemical structure of angiotensin in the bullfrog, *Rana catesbeiana*. Gen Comp Endocrinol 50: 75–80

Hasegawa Y, Cipolle M, Watanabe TX, Nakajima T, Sokabe H, Zehr JE (1984a) Chemical structure of angiotensin in the turtle, *Pseudemys scripta*. Gen Comp Endocrinol 53: 159–162

Hasagawa Y, Watanabe TX, Nakajima T, Sokabe H (1984b) Chemical structure of angiotensin formed by incubating plasma with the corpuscles Stannius in the Japanese goosefish, *Lophius litulon*. Gen Comp Endocrinol 54: 264–269

Hatt PY (1967) The juxtaglomerular apparatus. In: Dalton AJ, Haguenau F (eds) Ultrastructure of the kidney. Academic Press, New York, pp 101–141

Hauger RL, Aguilera G, Baukal AJ, Catt KJ (1982) Characterization of angiotensin II receptors in the anterior pituitary gland. Mol Cell Endocrinol 255: 203–212

Hauger-Klevene JH (1970) ACTH, cyclic AMP, dexamethazone and actinomycin D effect on renin release. Acta Physiol Latinoam 20: 373–381

Hauger-Klevene JH (1976) The effect of growth hormone on renin production and release by rat kidney slices. Acta Physiol Latinoam 26: 82–84

Hauger-Klevene JH, Brown H, Fleischer N (1969) ACTH stimulation and glucocorticoid inhibition of renin release in the rat. Proc Soc Exp Biol Med 131: 539–542

Hayashi T, Nakayama T, Nakajima T, Sokabe H (1978) Comparative studies on angiotensins. V. Structure of angiotensin formed by the kidney of Japanese goosefish and its identification by dansyl method. Chem Pharmacol Bull 26: 215–219

Hayduk K, Boucher R, Genest J (1970) Renin activity content in various tissues of dogs under different physiopathological states. Proc Soc Exp Biol Med 134: 252–255

Haynes FW, Forsham PH, Hume DM (1953) Effects of ACTH, cortisone, desoxycorticosterone, and epinephrine on the plasma hypertensinogen and renin concentration of dogs. Am J Physiol 172: 265–275

Hazon N, Henderson IW (1985) Factors affecting secretory dinamics of 1α-hydroxycorticosterone in the dogfish, *Scyrhiolinus canicula*. Gen Comp Endocrinol 59: 50–55

Hazon N, Balment RJ, Perrott M, O'Toole LB (1989) The renin-angiotensin system and vascular and dipsogenic regulation in elasmobranchs. Gen Comp Endocrinol 74: 230–236

Heagerty AM (1991) Angiotensin II: vasoconstrictor or growth factor? J Cardiovasc Pharmacol 18 (Suppl 2): S14–S19

Healy DP, Printz MP (1984) Distribution of immunoreactive angiotensin II, angiotensin I, angiotensinogen and renin in the central nervous system of intact and nephrectomized rats. Hypertension 6 (Suppl I): 130–136

Healy DP, Maciejewski AR, Printz MP (1985) Autoradiographic localization of $[^{125}I]$-angiotensin II binding sites in the rat adrenal gland. Endocrinology 116: 1221–1223

Healy DP, Ye M-Q, Yuan L-X, Schachter BS (1992) Stimulation of angiotensinogen mRNA levels in rat pituitary by estradiol. Am J Physiol 263: E355–E361

Hedeland H, Dymling JF, Hokfelt B (1972) The effect of insulin induced hypoglycemia on plasma renin activity and urinary catecholamines before and following clonidine (Catapresan) in man. Acta Endocrinol 71: 321–330

Helmer OM (1962) Presence of renin in plasma of patients with arterial hypertension. Circulation 25: 169–173

Helmer OM, Griffith RS (1951) The effect of the administration of estrogens on the renin-substrate (hypertensinogen) content of rat plasma. Endocrinology 51: 421–426

Helmer OM, Page IH (1939) Purification and some properties of renin. J Biol Chem 127: 757–763

Helwig JJ, Schleiffer R, Judes C, Gairard A (1984) Distribution of parathyroid hormone-sensitive adenylate cyclase in isolated rabbit renal cortex microvessels and glomeruli. Life Sci 35: 2649–2657

Helwig JJ, Musso MJ, Judes C, Nikcols GA (1991) Parathyroid hormone and calcium: interactions in the control of renin secretion in the isolated, nonfiltering rat kidney. Endocrinology 129: 1233–1242

Hems DA (1977) Short-term hormonal control of hepatic carbohydrate and lipid catabolism. FEBS Lett 80: 237–245

Henderson IW, Balment RJ (1975) Renal response to the hypophysectomized rat to infusion of bovine growth hormone. J Endocrinol 67: 60P

Henderson IW, Deacon CF (1993) Phylogeny and comparative physiology of the renin-angiotensin system. In: Robertson JIS, Nicholls MG (eds) The renin-angiotensin system, Vol 1. Gower Medical Publ, London, pp 2.1–2.28

Henderson IW, Jotisankasa V, Mosely W, Oguri M (1976) Endocrine and enviornmental influences upon plasma cortisol concentrations and plasma renin activity of the eel. *Anguilla anguilla* L. J Endocrinol 70: 81–95

Henderson IW, Oliver JA, McKeever A, Hazon N (1981) Phylogenetic aspects of the renin-angiotensin system. In: Pethes G, Freny VL (eds) Advances in animal and comparative physiology. Pergamon Press, Oxford, pp 355–363

Henrich WL (1981) Role of prostaglandins in renin secretion. Kidney Int 19: 822–830

Henrich WL, Campbell WB (1984) Relationship between PG and beta-adrenergic pathways to renin release in rat renal cortical slices. Am J Physiol 247: E343–E348

Henrich WL, Needleman P, Campbell WB (1986) Effect of atriopeptin III on renin release in vitro. Life Sci 39: 993–1001

Henrich WL, McAllister EA, Smith PB, Lipton J, Campbell WB (1987) Direct inhibitory effect of atriopeptin III on renin release in primate kidney. Life Sci 41: 259–264

Henrich WL, McAllister EA, Smith PB, Campbell WB (1988) Guanosine 3', 5' monophosphate as a mediator of inhibition of renin release. Am J Physiol 255: F474–F478

Henry DP, Aoi W, Weinberger MH (1977) The effects of dopamine on renin release in vitro. Endocrinolgy 101: 279–283

Herman K, Bayer C, Ganten D (1982) Definite evidence for the presence of (isolucine-5) angiotensin I (Ile-ANG I) and (isolucine-5) angiotensin II (Ile-ANG II) in the brain of rats. Naunyn-Schmiedeberg's Arch Pharmacol 319 (Suppl): R48

Hernesniemi J, Kawana E, Bruppacher H, Sandri C (1972) Afferent connections of the subfornical organ and of the supraoptic crest. Acta Anat 81: 321–336

Hesse B, Nielsen I (1977) Suppression of plasma renin activity by intravenous infusion of antidiuretic hormone in man. Clin Sci Mol Med 52: 357–360

Hill AGS, Cruikshank B, Crossland A (1953) A study of antigenic compounds of kidney tissue. Brit J Exp Pathol 34: 27–34

Hill PA, Coghlan JP, Scoggins BA, Ryan GB (1983) Ultrastructural changes in the sheep renal juxtaglomerular apparatus in response to sodium depletion or loading. Pathology 15: 463–473

Hill PA, Coghlan JP, Scoggins BA, Ryan GB (1984) Functional and morphologic studies of the adrenal cortex and kidney in ovine toxaemia of pregnancy. J Pathol 144: 1–13

Hirano T (1974) Some factors regulating water intake by the eel, *Anguilla japonica*. J Exp Biol 61: 737–747

Hirano T, Hasegawa S (1984) Effects of angiotensins and other vasoactive substances on drinking in the eel *Anguilla japonica*. Zool Sci 1: 106–113

Hirano T, Takei Y, Kobayashi H (1978) Angiotensin and drinking in the eel and the frog. In: Barker Jorgensen C, Skadhauge E (eds) Osmotic and volume regulation. Alfred Benzon Symp XI. Munksgaard, Copenhagen, pp 123–128

Hirasawa R, Hashimoto K, Ota Z (1990) Role of central angiotensinergic mechanism in shaking stress-induced ACTH and catecholamine secretion. Brain Res 533: 1–5

Hirose S, Murakami K (1992) Molecular biology of renin. Blood Vessel Endothel 2: 65–69 (in Japanese)

Hirose S, Yokosawa H, Inagami T (1978) Immunohistochemical identification of renin in rat brain and distinction from acid protease. Nature 274: 392–393

Hirose S, Yokosawa H, Inagami T, Workman RJ (1980) Renin and prorenin in hog brain: ubiquitous distribution and high concentration in the pituitary and pineal. Brain Res 191: 489–499

Hirose S, Ohsawa T, Inagami T, Murakami K (1982) Brain renin from bovine anterior pituitary. Isolation and properties. J Biol Chem 257: 6316–6321

Hisa H, Tomura Y, Satoh S (1989) Atrial natriuretic factor suppresses neural stimulation of renin release in dogs. Am J Physiol 257: E332–E335

Ho BYM, Sham JSK, Chiu KW (1984) The vasopressor action of the renin-angiotensin system in the rat snake, *Ptyas korros*. Gen Comp Endocrinol 56: 313–320

Hofbauer KG, Zschiedrich H, Hackenthal E, Gross F (1974) Function of renin-angiotensin system in the isolated perfused rat kidney. Circ Res 34/35 (Suppl 1): 193–201

Hofbauer KG, Konrads A, Schwarz K, Werner U (1978) Role of cyclic AMP in the regulation of renin release from the isolated perfused rat kidney. Klin Wochenschr 56 (Suppl I): 51–59

Hoff KS, Hillyard SD (1991) Angiotensin II stimulates cutaneous drinking in the toad *Bufo punctatus*. Physiol Zool 64: 1165–1172

Hoff KS, Hillyard SD (1993) Inhibition of cutaneous water absorption in dehydrated toads by saralasin is associated with changes in barometric pressure. Physiol Zool 66: 89–98

Hoffman WE, Phillips MI (1976a) The effect of subfornical organ lesions and ventricular blockade on drinking induced by angiotensin II. Brain Res 108: 59–73

Hoffman WE, Phillips MI (1976b) Regional study of cerebral ventricle sensitive sites to angiotensin II. Brain Res 110: 313–330

Hogarty DC, Speakman EA, Puig V, Phillips MI (1992) The role of angiotensin, AT_1 and AT_2 receptors in the pressor, drinking and vasorpressin responses to central angiotensin. Brain Res 586: 289–294

Holmes WN, A1-Ghawas SC, Cronshaw J, Rhode KE (1991) The structural organization and the steroidogenic responsiveness in vitro of adrenal gland tissue from the neonatal mallard duck (*Anas platyrhynchos*). Cell Tissue Res 263: 557–566

Hooper NM (1991) Angiotensin converting enzyme: implications from molecular biology for its physiological functions. Int J Biochem 23: 641–647

Horio T, Hohno M, Takeda T (1992) Effects of arginine vasopressin, angiotensin II and endothelin-1 on the release of brain natriuretic peptide in vivo and in vitro. Clin Exp Pharmacol Physiol 19: 575–582

Horky K, Broulik P, Pacovsky V (1986) The effect of parathyroid hormone on plasma renin activity in humans and hypertension in patients with primary hyperparathyroidism. J Hypertens 4 (Suppl 6): 585s–587s

Hosie KF, Brown JJ, Harper AM, Lever AF, MacGreger RF, Robertson JIS (1970) The release of renin into the renal circulation of the anesthetized dog. Clin Sci 38: 157–174

Houssay BA, Taquini AC (1938) Action vasso-constrictrice du sang veineux de rein ischemie. C R Soc Biol 128: 1125–1128

Hsiao S, Epstein AN, Camardo JS (1977) The dipsogenic potency of peripheral angiotensin II. Horm behav 8: 129–140

Hubert C, Houot AM, Corvol P, Soubrier F (1991) Structure of the angiotensin I-converting enzyme gene. J Biol Chem 266: 15377–15383

Husain A, Bumpus FM, Smeby RR, Brosnihan KB, Khosla MC, Speth RC, Ferrario CM (1983) Evidence for the existence of a family of biologically active angiotensin I-like peptides in the dog central nervous system. Circ Res 52: 460–464

Hwang BH, Wu J-Y, Severs WB (1986) Effects of chronic dehydration on angiotensin II receptor binding in the subfornical organ, paraventricular hypothalamic nucleus and adrenal meddula of Long-Evans rats. Neurosci Lett 65: 35–40

Igic R, Erdös EG, Yeh HSJ, Sorrells K, Nakajima T (1972) The angiotensin I converting enzyme of the lung. Circ Res 31 (Suppl II): 51–61

Ikemoto F, Takaori K, Iwao H, Yamaoto K (1982) Intrarenal localization of renin binding substance in rats. Life Sci 31: 1011–1016

Illanes A, Perez-Olea J, Quevedo M, et al (1967) Influence of angiotensin on the effects of tyramine upon the contractile force of isolated rabbit atria. J Pharmacol Exp Ther 158: 487–493

Imai T, Miyazaki H, Hirose S, Hori H, Hayashi T, Kageyama R, Ohkubo H, Nakanishi S, Murakami K (1983) Cloning and sequence analysis of cDNA for human renin precursor. Proc Natl Acad Sci USA 80: 7405–7409

Imbs JL, Schmidt M, Schwartz J (1975) Effect of dopamine on renin secretion in the anesthetized dog. Eur J Pharmacol 33: 151–157

Inagaki T, Inagami T (1984) A new form of inactive renin in rat brain. A latent renin. Hypertension 6: I137–142

Inagami T (1991) Purification of renin and prorenin. Hypertension 18: 241–251

Inagami T (1993) Intracellular renin: validation and function. In: Robertson JIS, Nicholls MG (eds) The renin-angiotensin system, vol 1. Gower Medical Publ, London, pp 30.1–30.8

Inagami T, Murakami K (1977) Purification of high molecular weight forms of renin from hog kidney. Circ Res 41 (Suppl II): 11–16

Inagami T, Kawamura M, Naruse K, Okamura T (1986) Localization of components of the renin-angiotensin system within the kidney. Fed Proc 45: 1414–1419

Inagami T, Mizuno K, Higashimori K (1991) Juxtaglomerular cells as a source of intrarenal angiotensin II production. Kidney Int 39 (Suppl 32): S20–S22

Inagami T, Iwai N, Sasaki K, Yamano Y, Bardhan S, Chaki S, Guo D-F, Furuta H (1992a) Cloning, expression and regulation of angiotensin II receptors. J Hypertens 10: 713–716

Inagami T, Mizuno K, Kawamura M, Okamura T, Clemens DL, Higashimori K (1992b) Localization of components of the renin-angiotensin system within the kidney and sustained release of angiotensins from isolated and perfused kidney. Tohoku J Exp Med 166: 17–26

Ingelfinger JR, Zuo WM, Fon EA, Ellison KE, Dzau VJ (1990) In situ hybridization evidence for angiotensinogen messenger RNA in the rat proximal tubule. J Clin Invest 85: 417–423

Inoue T, Negro-Vilar A (1989) Evidence that vasoactive intestinal peptide, angiotensin II and LHRH modulate the release of prolactin induced by dopamine blockade. Soc Neurosci Abstr 15: 724

Iovino M, Steardo L (1984) Vasopressin release to central and peripheral angiotensin II in rats with lesions of the subfornical organ. Brain Res 322: 365–368

Iovino M, Steardo L (1985) Thirst and vasopressin secretion following central administration of angiotensin II in rats with lesions of the septa and subfornical organ. Neuroscience 15: 61–67

Isales CM, Barret PQ, Brines M, Bollag W, Rasmussen H (1991) Parathyroid hormone modulates angiotensin II-induced aldosterone secretion from the adrenal glomerulosa cell. Endocrinology 129: 489–495

Israel A, Correa FMA, Niwa M, Saavedra JM (1984) Quantitative determination of angiotensin II binding sites in rat brain and pituitary gland by autoradiography. Brain Res 322: 341–345

Israel A, Saavedra JM, Plunkett L (1985) Water deprivation upregulates angio-tensin II receptors in rat anterior pituitary. Am J Physiol 248: E264–E267

Ito S, Carretero OA, Abe K, Beierwartes WH, Yoahinaga K (1989) Effects of prostanoids on renin release from rabbit afferent arterioles with and without macula densa. Kidney Int 35: 1138–1144

Ito T, Eggena P, Barrett JD, Katz D, Metter J, Sambhi MP (1980) Studies on angiotensinogen in plasma and cerebrospinal fluid in normal and hypertensive human. Hypertension 2: 432–436

Itskovitz HD, McGiff JC (1974) Hormonal regulation of the renal circulation. Circ Res 34–35 (Suppl I): 65–73

Iwai N, Inagami T (1992) Identification of two subtypes in the rat type I angiotensin II receptor. FEBS Lett 298: 257–260

Iwai N, Yamano Y, Chaki S, Konishi F, Bardhan S, Tibbetts C, Sasaki K, Hasegawa M, Matsuda Y, Inagami T (1991) Rat angiotensin II receptor: cDNA sequence and regulation of the gene expression. Biochem Biophys Res Commun 177: 299–304

Iwao H, Abe Y, Yamamoto K (1974) Effect of intrarenal arterial infusions of calcium on renin release in dogs. Jap J Pharmacol 24: 482–484

Jackson BA, Brown JA, Oliver JA, Henderson IW (1977) Actions of angiotensin on single nepphron rate of trout, *Salmo gaidrneri*, adapted to fresh- and sea-water environments. J Endocrinol 75: 32P

Jackson EK (1991) Adenosine: a physiological brake on renin release. Annu Rev Pharmacol Toxicol 31: 1–35

James SK, Hall RC (1974) The nature of renin released in the dog following hemorrhage and furosemide. Pfluegers Arch Eur J Physiol 347: 323–328

Jarecki M, Thoren PN, Donald DE (1978) Release of renin by the carotid baroreflex in anesthetized dogs. Role of cardiopulmonary vagal afferents and renal arterial pressure. Circ Res 42: 614–619

Jhamandas JH, Lind RW, Renaud LP (1989) Angiotensin II may mediate excitatory neurotransmission from the subfornical organ to the hypothalamic supraoptic nucleus: an anatomical and electrophysiological study in the rat. Brain Res 487: 52–61

Ji H, Sandberg K, Catt KJ (1991) Novel angiotensin II antagonists distinguish amphibian form mammalian angiotensin II receptors expressed in *Xenopus laevis* oocytes. Mol Pharmacol 39: 120–123

Jimenez E, Vinson GP, Montiel M (1994) Angiotensin II (AII)-binding sites in nuclei from rat liver: partial characterization of the mechanism of AII accumulation in nuclei. J Endocrinol 143: 449–453

Jiménez Díaz C, Linazasoro JM, Merchante A (1959) Further study of the part played by the kidneys in the regulation of thirst. Bull Inst Med Res, Univ Madrid 12: 60–67

Johnson AK, Anderson RR (1980) The renin-angiotensin system. Plenum Press, New York

Johnson AK, Epstein AN (1975) The cerebral ventricles as the avenue for the dipsogenic action of intracranial angiotensin. Brain Res 86: 399–418

Johnson AK, Mann JFE, Rascher W, Johnson JK, Ganten D (1981) Plasma angiotensin II concentrations and experimentally induced thirst. Am J Physiol 240: R229–234

Johnson AK, Robinson MM, Mann JFE (1986) The role of the renal renin-angiotensin system in thirst. In: de Caro G, Epstein AN, Massi M (eds) The physiology of thirst and sodium appetite. Plenum Press, New York, pp 161–180

Johnson EM, Marshall GR, Needleman P (1974) Modification of responses to sympathetic nerve stimulation by the renin-angiotensin system in rats. Br J Pharmacol 51: 541–547

Johnson JA, Davis JO, Witty RT (1971) Effects of catecholamines and renal nerve stimulation on renin release in the nonfiltering kidney. Circ Res 29: 646–653

Johnson JA, Davis JO, Braverman B, Gotshall RW, Davis JL, Lohmeier TE, Freeman RH (1974) Inhibition of renin release in Na-depleted dogs by intrarenal infusion of propranolol. Fed Proc 33: 339

Johnson MD (1984) Circulating epinephrine stimulates renin secretion in anesthetized dogs by activation of external adrenoceptors. Am J Physiol 246: F676–F684

Johnson OW (1979) Urinary organs. In: King AS, McLelland J (eds) Form and function in birds, vol 1. Academic Press, New York, pp 183–235

Johnson OW, Mugaas JN (1970) Some histological features of avian kidneys. Am J Anat 127: 423–436

Johnson RH, Park DM (1973) Effect of change of posture on blood pressure and plasma renin concentration in men with spinal transections. Clin Sci 44: 539–546

Jones TH, Brown BL, Dobson PRM (1988) Evidence that angiotensin II is paracrine agent mediating gonadotropin-releasing hormone-stimulated inositol phosphate production and prolactin secretion in the rat. J Endocrinol 116:367–371

Jones TH, Brown BL, Dobson PRM (1990) Paracrine control of anterior pituitary hormone secretion. J Endocrinol 127: 5–13

205

Joss JMP, Arnold-Reed DE, Balment RJ (1994) The steroidogenic response to angiotensin II in the Australian lungfish, *Neoceratodus forsteri*. J Comp Physiol B 164: 378–382

Kageyama R, Ohkubo H, Nakanishi S (1984) Primary structure of human preangiotensinogen deduced from the cloned cDNA sequence. Biochemistry 23: 3603–3609

Kakar SS, Riel KK, Neill JD (1992) Differential expression of angiotensin II receptor subtype mRNAs (AT-1A and AT-1B) in the brain. Biochem Biophys Res Commun 185: 688–692

Kaley G, Donshik PC (1965) Specificity and quantitative aspects of the "renin-angiotensin" system in lower vertebrates. Biol Bull 129: 411 (Abstr)

Kaloyanides GJ, Bastron Rd, Dibona GF (1973) Effect of ureteral clamping and increased renal areterial pressure on renin release. Am J Physiol 225: 95–99

Kambayashi Y, Bardhan S, Takahashi K, Tsuzuki S, Inui H, Hamakubo T, Inagami T (1993) Molecular cloning of a novel angiotensin II receptor isoform involved in phosphotyrosine phosphatase inhibition. J Biol Chem 268: 24543–24546

Kasuya Y, Karakida T, Okawara Y, Yamaguchi K, Kobayashi H (1985) Daily drinking patterns and plasma angiotensin II in the budgerigar (*Melopsittacus undulatus*) and the Japanese quail (*Coturnix coturnix japonica*). J. Yamashina Inst Ornithol 17: 32–43

Kasuya Y, Karakida T, Okawara Y, Kobayashi H (1987a) Comparative studies of food intake and water balance following water deprivation in the budgerigar (*Melopsittacus undulatus*) and the Japanese quail (*Coturnix coturnix japonica*). J. Yamashina Inst Ornithol 19: 89–101

Kasuya Y, Okawara Y, Seki K, Kobayashi H (1987b) Daily drinking patterns and drinking response to angiotensin II in young and adult budgerigars *Melopsittacus undulatus*. J Yamashina Inst Ornithol 19: 1–12

Kato A, Sugiura N, Hagiwara H, Hirose S (1994) Cloning, amino acid sequence and tissue distribution of porcine thimet oligopeptidase. A comparison with soluble angiotensin-binding protein. Eur J Biochem 221: 159–165

Katz FH, Roper EF (1977) Testosterone effect on renin system in rats. Proc Soc Exp Biol Med 155: 330–333

Katz FH, Popovtzer MM, Pinggera WF, Halgrimson CG, Starzl TE (1972) Acute alteration of plasma renin activity by large dose of intravenous prednisone. Proc Soc Exp Biol Med 141: 887–889

Katz SA, Malvin RL, Lee J, Kim SH, Murray RD, Opsahl JA, Abraham PA (1991) Analysis of active renin heterogeneiety. Proc Soc Exp Biol Med 197: 387–392

Kaufman S, Peters G (1980) Regulatory drinking in the pigeon *Columba livia*. Am J Physiol 239: R219–R225

Kawamura M, Nakamura M, Inagami T (1985) Evidence for existence of angiotensins I and II in mature renin granules from rat kidney cortex. Biochem Biophys Res Commun 131: 628–633

Keatinge WR (1966) Electrical and mechanical responses of vascular smooth muscle to vasodilator agents and vasoactive polyepetides. Circ Res 18: 641–649

Keeton TK, Campbell WB (1981) The pharmacologic alteration of renin release. Pharmacol Rev 32: 81–227

Keller-Wood M, Kimura B, Shinsako J, Phillips MI (1986) Interaction between CRF and angiotensin II in control of ACTH and adrenal steroids. Am J Physiol 250: R396–R402

Kelly G, Downie I, Gardiner DS, More IR, Lindop GBM (1990) The peripolar cell: a distinctive cell type in the mammalian glomerulus. Morphological evidence from a study of sheep. J Anat 168: 217–227

Kelsch RC, Light GS, Luciano JR, Oliver WJ (1971) The effect of prednisone on plasma norepinephrine concentration and renin activeity in salt-depleted man. J Lab Clin Med 77: 267–277

Kem DC, Johnson EIM, Capponi AM, Chardonnens D, Lang U, Blondel B, Koshida H, Vallotton MB (1991) Effect of angiotensin II on cytosolic free calcium in neonatal rat cardiomyocytes. Am J Physiol 261: C77–C85

Kempton RT (1943) Studies on the elasmobranch kidney. I. The structure of renal tubule of the spiny dogfish (Squalus acanthias). J Morphol 73: 247–263

Kenyon CJ, McKeever A, Oliver JA, Henderson IW (1985) Control of renal and adrenocortical function by the renin-angiotensin system in two euryhaline teleost fishes. Gen Comp Endocrinol 58: 93–100

Khachaturian H, Sherman TG, Lloyd RV, Civelli O, Douglass J, Herbert E, Akil H, Watson SJ (1986) Pro-dynorphin is endogenous to the anterior pituitary and is co-localized with LH and FSH in the gonadotrophs. Endocrinology 119: 1409–1412

Khairallah PA, Vadaparampil GJ, Page IH (1965) Effects of ions on angiotensin interaction with smooth muscle. Arch Int Pharmacodyn Ther 158: 155–164

Khosla MC, Bumpus FM, Yasuhara T, Nakajima T (1981) Synthesis of Ala-Pro-Gly-[Ile^3, Val^5]-angiotensin II isolated from the skin of the Australian frog *Crinia georgiana*. J Med Chem 24: 885–887

Khosla MC, Nishimura H, Hasegawa Y, Bumpus FM (1985) Identification and synthesis of [1-asparagine, 5-valine, 9-glycine] angiotensin I produced from plasma of American eel, *Anguilla rostrata*. Gen Comp Endocrinol 57: 223–233

Kierkegaad-Hansen A (1974) The effect of lithium on blood pressure and on plasma renin substrate and renin in rats. Acta Pharmacol Toxicol 35: 370–378

Kifor I, Roth T, Dzau VJ (1987) Endothelial renin-angiotensin pathway: evidence for intracellular synthesis and secretion of angiotensin. Circ Res 60: 422–428

Kilcoyne MM, Hoffman DL, Zimmerman EA (1980) Immunocytochemical localization of angiotensin II and vasopressin in rat hypothalamus; evidence for production in the same neuron. Clin Sci 59: 57s–60s

Kim S, Hiruma M, Ikemoto F, Yamaoto K (1988) Importance of glycosilation for hepatic clearance of renal renin. Am J Physiol 255: E642–E651

Kirby RF, Thunhorst RL, Johnson AK (1992) Effects of a non-peptide angiotensin receptor antagonist on drinking and blood pressure responses to centrally administered angiotensins in the rat. Brain Res 576: 348–350

Kiron MAR, Soffer RL (1989) Purification and properties of a soluble angiotensin II-binding protein from rabbit liver. J Biol Chem 264: 4138–4142

Kitamura E, Kikkawa R, Fujiwara Y, Imai T, Shigeta Y (1986) Effect of angiotensin II infusion on glomerular angiotensin II receptor in rats. Biochim Biophys Acta 885: 309–316

Klett C, Ganten D, Hellmann W, Kaling M, Ryffel GU, Weimer-Ehl T, Hackenthal E (1992) Regulation of hepatic angiotensinogen synthesis and secretion by steroid hormones. Endocrinology 130: 3660–3668

Klingbeil CK (1985) Corticosterone and aldosterone dose-dependent responses to ACTH and angiotensin II in the duck (*Anas platyrhynchos*). Gen Comp Endocrinol 59: 382–390

Kloas W, Hanke W (1992a) Angiotensin II binding sites in frog kidney and adrenal. Peptides 13: 349–354

Kloas W, Hanke W (1992b) Localization and quantification of angiotensin II (AII) binding sites in the kidney of *Xenopus laevis* – lack of AII receptors in the adrenal tissue. Gen Comp Enocrinol 86: 173–183

Kloas W, Hanke W (1993) Receptors for atrial natriuretic factor (ANF) in kidney and adrenal tissue of urodeles – lack of angiotensin II (AII) receptors in these tissues. Gen Comp Endocrinol 91: 235–249

Knox FG, Cuche JL, Ott CE, Diaz-Buxo JA, Marchand G (1975) Regulation of glomerular filtration and proximal tubule reabsorption. Circ Res 33 (Suppl I): 107–119

Kobayashi H (1980) Evolution of the metabolic endocrine system: its phylogenetic significance. In: Ishii S, Hirano T, Wada M (eds) Hormones, adaptation and evolution. Jpn Sci Soc Press, Tokyo/Springer, Berlin, pp 15–22

Kobayashi H (1981) Angiotensin-induced drinking in parrots. Gen Comp Endocrinol 43: 399–401

Kobayashi H, Takei Y (1982) Mechanisms for induction of drinking with special reference to angiotensin II. Comp Biochem Physiol 71A: 485–494

Kobayashi H, Uemura H, Wada M, Takei Y (1979) Ecological adaptation of angiotensin-induced thirst mechanism in tetrapods. Gen Comp Endocrinol 38: 93–104

Kobayashi H, Uemura H, Takei Y (1980) Physiological role of the renin-angiotensin system during dehydration. In: Epple A, Stetson MH (eds) Avian endocrinology. Academic Press, New York, pp 319–330

Kobayashi H, Uemura H, Takei Y, Itatsu N, Ozawa M, Ichinohe K (1983) Drinking induced by angiotensin II in fishes. Gen Comp Endocrinol 49: 295–306

Koch-Weser J (1965) Nature of the inotropic action of angiotensin on ventricular myocardium. Circ Res 16: 230–237

Kocsis JF, McIlroy PJ, Chiu AT, Schimmel RJ, Carsia RV (1994a) Properties of angiotensin II receptors of domestic turkey (*Meleagris gallopavo*) adrenal steroidogenic cells. Gen Comp Endocrinol 96: 92–107

Kocsis JF, Boyette MH, McIlroy PJ, Carsia RV (1994b) Regulation of aldosteronogenesis in domestic turkey (*Meleagris gallopavo*) adrenal steroidogenic cells. Gen Comp Endocrinol 96: 108–121

Koella WP, Sutin J (1967) Extra-blood-brain-barrier brain structures. Int Rev Neurobiol 10: 31–55

Kohara K, Brosnihan KB, Chappell MC, Khosla MC, Ferrario CM (1991) Angiotensin-(1–7). A member of circulating angiotensin peptides. Hypertension 17: 131–138

Kojima I, Kojima K, Kreutter D, Rasmussen H (1984) The temporal integration of the aldosterone secretory response to angiotensin occurs via two intracellular pathways. J Biol Chem 259: 14448–14457

Kojima I, Kojima K, Rasmussen H (1985a) Characterisitics of angiotensin II- K^+- and ACTH-induced calcium influx in adrenal glomerulosa cells. J Biol Chem 260: 9171–9176

Kojima I, Kojima K, Rasmussen H (1985b) Role of calcium fluxes in the sustained phase of antiotensin II-mediated aldosterone secretion from adrenal glomerulosa cells. J Biol Chem 260: 9177–9184

Kokubu T, Akutsu H, Fujimoto S, Ueda E, Hiwada K, Yamamura Y (1969) Purification and properties of endopeptidase from rabbit red cells and its process of degradation of angiotensin. Biochim Biophys Acta 191: 668–676

Kon V, Ichikawa I (1985) Hormonal regulation of glomerular filtration. Annu Rev Med 36: 515–531

Kon Y, Hashimoto Y, Kitagawa H, Kudo N (1984) Morphology and quantification of juxtaglomerular cells of the chicken kidney. Jpn J Vet Sci 46: 189–196

Kon Y, Hashimoto Y, Kitagawa H, Kudo N, Murakami K (1986) Immunohistochemical demonstration of juxtaglomerular cells in the kidneys of domestic mammals and fowls. Jpn J Vet Res 34: 111–123

Kon Y, Hashimoto Y, Kitagawa H, Kudo N (1987) Morphological and immunohistochemical studies of juxtaglomerular cells in the carp, *Cyprinus carpio*. Jpn J Vet Sci 49: 323–331

Kondo K, Garcia R, Boucher R, Genest J (1980) Effects of intracerebroventricular administration of tonin on water intake and blood pressure in the rat. Brain Res 200: 437–441

Konrads A, Hofbauer KG, Werner U, Gross F (1978) Effects of vasopressin and its deamino-*d*-arginone analogue on renin release in the isolated perfused rat kidney. Pfluegers Arch Eur J Physiol 377: 81–85

Kopp UC, DiBona GF (1984) Interaction between neural and nonneural mechanisms controlling renin secretion rate. Am J Physiol 246: F620–F626

Kotchen TA, Maull KI, Luke RG, Rees D, Flamenbaum W (1974) Effect of acute and chronic calcium administration on plasma renin. J Clin Invest 54: 1279–1286

Kotchen TA, Galla JH, Luke RG (1976) Failure of $NaHCO_3$ and $KHCO_3$ to inhibit renin in the rat. Am J Physiol 231: 1050–1056

Kotchen TA, Galla JH, Luke RG (1978) Contribution of chloride to the inhibition of plasma renin by sodium chloride in the rat. Kidney Int 13: 201–207

Koval'Chuk LE (1987) Ultrastructure of the juxtaglomerular complex of the kidney and peripolar cells in the sand lizard *Lacerta agilis*. Arkh Anat Gistol Embryol 93: 93–98

Koval'Chuk LE, Likhacheva LM (1990) Ultrastructure of the renal corpuscle in freshwater fishes. Arkh Anat Gistol Embriol 99: 69–74

Krag B, Skadhauge E (1972) Renal water and salt excretion in the budgerigah (*Melopsittacus undulatus*). Comp Biochem Physiol 41A: 667–683

Krakoff LR, Selvadurai R, Sutter E (1975) Effect of methylpredonisolone upon arterial pressure and the renin angiotensin system in the rat. Am J Physiol 228: 613–617

Kraly FS, Corneilson R (1990) Angiotensin II mediates drinking elicited eating in the rat. Am J Physiol 258: R436–R442

Kramer BK, Ritthaler T, Ackermann M, Holmer S, Schricker K, Riegger GA, Kurtz A (1994) Endothelium-mediated regulation of renin secretion. Kidney Int 46: 1577–1579

Krege JH, John SWM, Langenbach LL, Hodgin JB, Hagaman JR, Bachman ES, Jennette JC, O'Brien DA, Smithies O (1995) Male-female differences in fertility and blood pressure in ACE-deficient mice. Nature 375: 146–148

Krishnamurthy VG, Bern HA (1969) Correlative histologic study of the corpuscles of Stannius and the juxtaglomerular cells of teleost fishes. Gen Comp Endocrinol 13: 313–335

Krishnamurthy VG, Bern HA (1973) Juxtaglomerular cell changes in the euryhaline freshwater fish, Tilapia mossambica, during adaptation to sea water Acta Zool 54: 9–14

Krulewitz AH, Bauer WE, Fanburg BL (1984) Hormonal influence on endothelial cell angiotensin-converting enzyme activity. Am J Physiol 247: C163–C168

Kubota T, Aso T (1991) Role of angiotensin on paracrine prolactin release in the pituitary gland and its possible effects on ovarian function. Horm Res 35 (Suppl 1): 13–20

Kucharczyk J, Assaf SY, Mogenson GJ (1976) Differential effects of brain lesions on thirst induced by the administration of angiotensin II to the preoptic region, subfornical organ and anterior third ventricle. Brain Res 108: 327–337

Kuchel O, Debinski W, Racz K, Buu NT, Garcia R, Kusson JR, Larochelle P, Cantin M, Genest J (1987) An emerging relationship between peripheral sympathetic nervous activity and atrial natriuretic factor. Life Sci 40: 1545–1551

Kumar RS, Kusari J, Roy SN, Soffer RL, Sen GC (1989) Structure of testicular angiotensin-converting enzyme. A segmental mosaic isozyme. J Biol Chem 264: 16754–16758

Kurtz A (1989) Cellular control of renin secretion. Rev Physiol Pharmacol 113: 1–40

Kurtz A, Bruna RD (1991) Determinants of renin secretion and renin synthesis in isolated mouse juxtaglomerular cells. Kid Int 39 (Suppl 32): S13–S15

Kurtz A, Bruna RD, Pfeilschifter J, Bauer C (1988) Role of cGMP as second messenger of adenosine in the inhibition of renin release. Kidney Int 33: 798–803

Kuzmina LK, Faleeva TI, Rakitskaya VV (1986) Monoaminergic innervation of the carp kidney. Arch Anat Histol Embryol 90: 38–41

Kwok YC, Moore GJ (1984) Photoaffinity labeling of the rat isolated portal vein: determination of affinity constants and "spare" receptors for angiotensins II and III. J Pharmacol Exp Ther 231: 137–140

Lachance D, Garcia R (1988) Atrial natriuretic factor release by angiotensin II in the conscious rat. Hypertension 11: 502–508

Lacy ER, Reale E (1989) Granulated peripolar epithelial cells in the renal corpuscle of marine elasmobranch fish. Cell Tissue Res 257: 61–67

Lacy ER, Reale E (1990) The presence of a juxtaglomerular apparatus in elasmobranch fish. Anat Embryol 182: 249–262

Lacy MP, McIntosh RP, McIntosh JEA (1989) Angiotensin II stimulates an endogenous response in Xenopus laevis ovarian follicles. Biochem Biophys Res Commun 159: 658–663

Lacy MP, Murray-McIntosh RP, McIntosh JEA (1992) Angiotensin II and acetylcholine differentially activate mobilization of inositol phosphates in Xenopus laevis ovarian follicles. Pfluegers Arch Eur J Physiol 420: 127–135

LaFontaine JJ, Nivez MP, Ardillou R (1979) Hepatic binding sites for angiotensin II in the rat. Clin Sci 56: 33–40

Lagois MD (1968) Granular epitheloid cell involution in the renal arteries of a euryhaline fish, Cymatogaster, adapted to hypertonic salinities Gen Comp Endocrinol 11: 248–250

Lagois MD (1974) Granular epithelioid (juxtaglomerular) cell and renovascular morphology of the coelacanth Latimeria chalumnae Smith (Crossopterygii) compared with that of other fishes. Gen Comp Endocrinol 22: 296–307

LaGrange RG, Sloop CH, Schmid HE (1973) Selective stimulation of renal nerves in the anesthetized dog. Circ Res 33: 704–712

Lamers APM, van Dongen WJ, van Kemenade JAM (1973) The morphology of the juxtaglomerular apparatus in the toad, Bufo bufo. A light microscopic study. Z Zellforsch 138: 545–555

Lamers APM, van Dongen WJ, van Kemenade JAM (1974) An ultrastructural study of the juxtaglomerular apparatus in the toad, Bufo bufo. Cell Tissue Res 153: 449–464

Lamers APM, Stadhouders AM, Verhofstad AAJ, Michelakis AM (1985a) Immunoelectron microscopic localization of renin in the juxtaglomerular cells of the amphibian *Bufo bufo*. Gen Com Endocrinol 60: 380–389

Lamers APM, Verhofstad AAJ, Stadhouders AM, Michelakis AM (1985b) Immunohistochemical demonstration of renin in the juxtaglomerular apparatus of three Bufo species. Cell Tissue Res 239: 677–682

Landis SC (1984) Neurotransmitter plasticity and coexistence during the development of cholinergic sympathetic neurons in culture and in vivo. In: Chan-Palay V, Palay SL (eds) Coexistence of neuroactive substances in neurons. John-Wiley & Sons, New York, pp 205–216

Landas S, Phillips MI, Stamler JF, Raizada MK (1980) Visualization of specific angiotensin II binding sites in the brain by fluorescent microscopy. Science 210: 791–793

Lang RE, Rascher W, Heil J, Unger T, Wiedemann G, Ganten D (1981) Angiotensin stimulates oxytocin release. Life Sci 29: 1425–1428

Langer SZ (1981) Presynaptic regulation of the release of catecholamines. Pharm Rev 32: 337–362

Laribi C, Legendre P, Dupouy B, Vincent JD, Simonnet G (1985) Characterization of two angiotensin II binding sites in cultured mouse spinal cord neurons. Brain Res 347: 94–103

Lassegue B, Alexander RW, Clark M, Griendling KK (1991) Angiotensin II-induced phosphatidylcholine hydrolysis in cultured vascular smooth-muscle cells. Biochem J 276: 19–25

Latta H, Maunsbach AB (1962) Juxtaglomerular apparatus as studied electron microscopically. J Ultrastruct Res 6: 547–561

Lattion A, Soubrier F, Allergrini J, Hubert C, Corvol P, Alhenc-Gelas F (1989) Testicular transcript of the angiotensin I-converting enzyme encodes for anscestral, non-duplicated form of the enzyme. FEBS Lett 252: 99–104

Laurent V, Bulet P, Salzet M (1995) A comparison of the leech *Theromyzon tessulatum* angiotensin I-like molecule with forms of vertebrate angiotensinogens: a hormonal system conserved in the course of evolution. Neurosci Lett 190: 175–178

Lawrence AC, Clarke IJ, Campbell DJ (1992) J Neuroendocrinol 4: 237–244

Le Morvan P, Palaic D (1975) Characterization of the angiotensin receptor in guinea-pig aorta strip. J Pharmacol Exp Ther 195: 167–175

Le Morvan P, Palaic D, Park WK (1974) Blockade of angiotensin action by 8-Leu-angiotensin on the guinea-pig aortic strip. Can J Physio Pharmacol 52: 236–239

Le Noble FAC, Hekking JWM, Van Straaten HWM, Slaaf DW, Struyker Boudier HAJ (1991) Angiotensin II stimulates angiogenesis in the chorio-allantoic membrane of the chick embryo. Eur J Pharmacol 195: 305–306

Leary WPP, Ledingham JG (1969) Removal of angiotensin by isolated perfused organ of the rat. Nature 222: 959–960

LeBrie SJ, Bolecskevy BD (1979) The effects of furosemide on renal function and renin in water snakes. Comp Biochem Physiol 63C: 223–228

Lee ME, Thrasher TN, Ramsay DJ (1981) Is angiotensin essential in drinking induced by water deprivation and caval ligation? Am J Physiol 240: R75–R80

Lee ME, Thrasher TN, Ramsay DJ (1984) Elevated cardiac pressure inhibits renin release after arterial hypotension in conscious dogs. Am J Physiol 247: R953–R959

Legendre P, Simonnet G, Vincent JD (1984) Electrophysiological effects of antiotensin II on cultured mouse spinal cord neurons. Brain Res 297: 287–296

Lehr D, Goldman HW, Casner P (1973) Renin-angiotensin role in thirst: paradoxical enhancement of drinking by angiotensin converting enzyme inhibitor. Science 182: 1031–1034

Levens NR (1985) Control of intestinal absorption by the renin-angiotensin system. Am J Physiol 249: G3–G15

Lewandowsky M (1899) Zur Frage der inneren Secretion von Nebenniere und Niere. Z Klin Med 37: 535–545

Lewicki JA, Fallon JH, Prinz MP (1978) Regional distribution of angiotensinogen in rat brain. Brain Res 158: 359–371

Lewis JL, Serikawa T, Warnock DG (1993) Chromosomal localization of angiotensin II type 1 receptor isoforms in the rat. Biochem Biophys Res Commun 194: 677–682

210

Lewis SJ, Allen AM, Verberne AJM, Figdor R, Jarrott B, Mendelsohn FAO (1986) Angiotensin II receptor binding in the rat nucleus solitarii is reduced after unilateral nodose ganglionectomy or vagotomy. Eur J Pharmacol 125: 305–307

Li Z, Ferguson AV (1993a) subfornical organ efferents to paraventricular nucleus utilize angiotensin as a neurotransmitter. Am J Physiol 265: R302–R309

Li Z, Ferguson AV (1993b) Angiotensin II responsiveness of rat paraventricular and subfornical organ neurones in vitro. Neuroscience 55: 197–207

Lilly LS, Pratt RE, Alexander RW, Gimbrone MA, Dzau VJ (1983) Cultured vascular endothelial cells contain the complete renin angiotensin system. Clin Res 31: 332A

Lin H, Sangmal M, Smith MJ Jr, Young DB (1993) Effect of endothelin-1 on glomerular hydraulic pressure and renin release in dogs. Hypertension 21: 845–851

Lin S-Y, Goodfriend TL (1970) Angiotensin receptors. Am J Physiol 218: 1319–1328

Linas SL (1984) Role of prostaglandin in renin secretion in the isolated kidney. Am J Physiol 246: F811–F818

Linazasoro JM, Jimenez Diaz C, Castro Mendoza H (1954) The kidney and thirst regulation. Bull Inst Med Res, Univ Madrid 7: 53–61

Lind RW (1988) Sites of action of angiotensin in the brain. In: Harding JW, Wright JW, Speth RC, Barnes CD (eds) Angiotensin and blood pressure regulation. Academic Press, New York, pp 135–163

Lind RW, Johnson AK (1982a) Central and peripheral mechanisms mediating angiotensin-induced thirst. In: Ganten D, Printz M, Phillips MI, Scholkens B (eds) The renin angiotensin system in the brain. Springer, Berlin Heidelberg New York, pp 353–364

Lind RW, Johnson AK (1982b) Subfornical organ-median preoptic connections and drinking and pressor responses to angiotensin II. J Neurosci 2: 1043–1051

Lind RW, van Hoesen GW, Johnson AK (1982) An HRP study of the connections of the subfornical organ of the rat. J Comp Neurol 210: 265–277

Lind RW, Swanson LW, Bruhn TO, Ganten D (1985a) The distribution of angiotensin II-immunoreactive cells and fibers in the paraventriculo-hypophysial system of the rat. Brain Res 338: 81–89

Lind RW, Swanson LW, Ganten D (1985b) Organization of angiotensin II immunoreactive cells and fibers in the rat central nervous system. Neuroendocrinology 40: 2–24

Lindop GBM, Lever AF (1986) Anatomy of the renin-angiotensin system in the normal and pathological kidney. Histochemistry 10: 335–362

Liu F-Y, Cogan MG (1987) Angiotensin II: a potent regulator of acidification in the rat early proximal convoluted tubule. J Clin Invest 80: 272–275

Liu F-Y, Cogan MG (1990) Role of angiotensin II in glomerulotubular balance. Am J Physiol 259: F72–F79

Livon C (1898) Secretions internes; glandes hypertensives. C R Soc Biol 1: 98–99

Ljungqvist A, Wågermark J (1970) The adrenergic innervation of intrarenal glomerular and extraglomerular circulatory routes. Nephron 7: 218–229

Llach F, Weidmann P, Reinhart R, Maxwell MH, Coburn JW, Massery SG (1974) Effect of acute and long standing hypocalcemia on blood pressure and plasma renin activity in man. J Clin Endocrinol Metab 38: 841–847

Logan AG, Tenyi I, Quesada T, Peart WS, Breathnach AS, Martin BG (1975) Blockade of renin release by lanthanum. Clin Sci Mol Med 48: 31s–32s

Lokhandwala MF, Buckley JP, Jandhyala BS (1978) Reduction of plasma renin activity by centrally administered angiotensin II in anesthetized cats. Clin Exp Hypertens 1: 167–175

Lopetz-Novoa JM, Garcia JC, Cruz-Soto MA, Benabe JE, Martinez-Maldonad M (1982) Effect of sodium orthovanadate on renal renin secretion in vivo. J Pharmacol Exp Ther 222: 447–451

Lotter EC, McKay LD, Mangiapane ML, Simpson JB, Vogel KE, Porte D Jr, Woods SC (1980) Intraventricular angiotensin elicits drinking in the baboon. Proc Soc Exp Biol Med 163: 48–51

Loudon M, Bing RF, Thurston H, Swales JD (1983) Arterial wall uptake of renal renin and blood pressure control. Hypertension 5: 629–634

Lumbers ER (1971) Activation of renin in human amniotic fluid by low pH. Enzymologia 40: 329–336

211

Lyons HJ, Churchill PC (1975) Renin secretion from rat renal coritcal cell suspension. Am J Physiol 228: 1853–1839

Maack TD, Marion DN, Camargo MJ, Kleinert HD, Laragh JH, Vaughan ED Jr, Atlas SA (1984) Effects of auriculin on blood pressure, renal function, and the renin-aldosterone system in dogs. Am J Med 77: 1069–1075

Maddox DA, Troy JL, Brenner BM (1974) Autoregulation of filtration rate in the absence of macula densa-glomerulus feedback. Am J Physiol 227: 123–131

Maebashi M, Miura Y, Yoshinaga K (1968) Suppressive effect of potassium on renin release. Jpn Circ J 32: 1265–1268

Makra ME, Prior DJ (1985) Angiotensin II can initiate contact-rehydration in terrestrial slugs. J Exp Biol 119: 385–388

Malayan SA, Keil LC, Ramsay DJ, Reid IA (1979) Mechanisms of suppression of plasma renin activity by centrally administered angiotensin II. Endocrinology 104: 672–674

Malling C, Poulsen K (1977) Direct measurement of high molecular weight forms of renin in plasma. Biochim Biophys Acta 491: 542–550

Malvin RL, Vander AJ (1967) Plasma renin activity in marine teleosts and Cetacea. Am J Physiol 213: 1582–1584

Malvin RL, Schiff D, Eiger S (1980) Angiotensin and drinking rates in the euryhaline killifish. Am J Physiol 239: R31–34

Mandich A, Massari A (1994) Presence of angiotensin II imunoreactivity in the ovary of the rainbow trout, Oncorhynchus mykiss. Gen Comp Endocrinol 96: 1–5

Mangiapane ML, Thrascher TN, Keil LC, Simpson JB, Ganong WF (1983) Deficits in drinking and vasopressin secretion after lesions of the nucleus medianus. Neuroendocrinology 37: 73–77

Mann JFE, Phillips I, Dietz R, Haebara H, Ganten D (1978) Effects of central and peripheral angiotensin blockade in hypertensive rats. Am J Physiol 234: H629–H637

Mann JFE, Johnson AK, Ganten D (1980) Plasma angiotensin II: dipsogenic levels and angiotensin generating capacity of renin. Am J Physiol 238: R372–R377

Mann JFE, Schiller PW, Schiffrin EL, Boucher R, Genest J (1981) Brain receptor binding and central action of angiotensin analogs in rats. Am J Physiol 241: R124–R129

Mann JFE, Ganten D, Johnson AK, Rettig R, Ritz E, Unger T (1986) Angiotensin dependent thirst following polyethyleneglycol treatment in the rat. In: de Caro G, Epstein AN, Massi M (eds) The physiology of thirst and salt appetite. NATO ASI Ser, Ser A Life Sci Vol 105. Plenum Press, New York, pp 199–203

Manning PT, Schwartz D, Katsube NC, Holmberg SW, Needleman P (1985) Vasopressin-stimulated release of atriopeptin: endocrine antagonists in fluid homeostasis. Science 229: 395–397

Margolius HS, Horwitz D, Geller RG, Alexander RW, Gill JR Jr, Pisano JJ, Keiser HR (1974) Urinary kallikrein excretion in normal man. Relationships to sodium intake sodium retaining steroids. Circ Res 35: 812–819

Marie J, Seyer R, Lombard C, Desarnaud F, Aumelas A, Jard S, Bonnafous J-C (1990) Affinity chromatography purification of angiotensin II receptor using photoactivatable biotinylated probes. Biochemistry 29: 8943–8950

Marks BH, Garber GB, McCoy FW (1960) The juxtaglomerular apparatus as an extraadrenal site of ACTH action. Proc Soc Exp Biol Med 105: 593–595

Massi M, Epstein AN (1987) The apparent dependence of salt appetite in the pigeon on endogenous angiotensin II. Physiol Behav 41: 155–162

Massi M, Epstein AN (1990) Angiotensin/aldosterone synergy governs the salt appetite of the pigeon. Appetite 14: 181–192

Massi M, de Caro G, Mazzarella L, Epstein AN (1986) The role of the subfornical organ in the drinking behavior of the pigeon Columba livia. Brain Res 381: 289–299

Masson GMC, del Greco F, Corcoran AC, Page IH (1956) Metabolic and pathologic effects of renin in nephrectomized rats. Am J Med Sci 231: 198–204

Mauger J-P, Poggioli J, Guesdon F, Claret M (1984) Noradrenaline, vasopressin and angiotensin increase Ca^{2+} influx by opening a common pool of Ca^{2+} channels in isolated rat liver cells. Biochem J 221: 121–127

Mbassa GK (1989) Peripolar cells from the majority of granulated cells in the kidneys of antelopes and goats. Acta Anat 135: 158–163

McAfee RD, Locke W (1967) Effect of angiotensin amide on sodium isotope flux and short-circuit current of isolated frog skin. Endocrinology 18: 1301–1305

McDonald JK, Zeitman BB, Callahan PX, Ellis S (1974) Angiotensinase activity of dipeptidyl aminopeptidase I (Cathepsin C) of rat liver. J Biol Chem 249: 234–240

McDonald KM, Taher S, Aisenbrey G, deTorrente A, Schrier RW (1975) Effect of angiotensin II and an angiotensin II inhibitor on renin secretion in the dog. Am J Physiol 228: 1562–1567

McGiff JC, Wong PYK (1979) Compartmentalization of prostaglandins and prostacyclin within the kidney: implications for renal function. Fed Proc 38: 89–93

McGiff JC, Crowsaw K, Terragno NA, Lonigro AJ, Strand JC, Williamson MA, Lee JB, Ng KKF (1970) Prostaglandin-like substances appearing in canine renal venous blood during renal ischemia. Circ Res 27: 765–782

McKelvey RW (1963) The presence of a juxtaglomerular apparatus in non-mammalian vertebrates. Anat Rec 145: 259–260

McKenna OC, Angelakos ET (1968a) Adrenergic innervation of the canine kidney. Circ Res 22: 345–354

McKenna OC, Angelakos ET (1968b) Acetylcholinesterase-containing nerve fibers in the canine kiney. Circ Res 23: 645–651

McKinley MJ, Denton DA, Park RG, Weisinger RS (1986) Ablation of subfornical organ does not prevent angiotensin-induced water drinking in sheep. Am J Physiol 250: R1052–R1059

McManus JFA (1948) The periodic acid routine applied to the kidney. Am J Pathol 24: 643–653

McQueen J, Murray GD, Semple PF (1984) Identification of angiotensin II receptor in rat mesenteric artery. Biochem J 223: 659–671

Mendelsohn FAO (1985) Localization and properties of angiotensin receptors. J Hypertens 3: 307–316

Mendelsohn FAO, Quirion R, Saavedra JM, Aguilera G, Catt KJ (1984) Autoradiographic localization of angiotensin II receptors in rat brain. Proc Natl Acad Sci USA 81: 1575–1579

Menzie JW, Hoffman LH, Michelakis AM (1978) Immunofluorescent localization of renin in mouse submaxillary gland and kidney. Am J Physiol 234: E480–E483

Meyer D, Jerusalem C, Walvig F (1967) Untersuchungen zur Feinstruktur der granulierten epitheloiden Zellen präglomerularer Arteriolen in den Nieren von Teleostiern. Z Zellforsch Mikrosk Anat 83: 508–526

Meyer DK, Herrmann M (1978) Inhibitory effect of tyramine-induced release of catecholamine on renin secretion. Naunyn-Schmiedeberg's Arch Pharmacol 280: 191–200

Meyer DK, Abele M, Hertting G (1974) Influence of serotonin on water intake and the renin-angiotensin system in the rat. Arch Int Pharmacodyn Ther 212: 130–140

Meyer P, Menard J, Papanicolaou N, Alexandre JM, Devaux C, Millietz P (1968) Mechanism of renin release following furosemide diuresis in rabbit. Am J Physiol 215: 908–915

Meyers BD, Deen WM, Brenner BM (1975) Effects of norepinephrine and angiotensin II on the determinants of glomerular ultrafiltration and proximal tube fluid reabsorption in the rat. Circ Res 37: 101–110

Miceli MO, Malsbury CW (1983) Feeding and drinking responses in the golden hamster following treatment with cholecystokinin and angiotensin II. Peptides 4: 103–106

Michelakis AM, (1971) The effect of sodium and calcium on renin release in vitro. Proc Soc Exp Biol Med 137: 833–836

Michelakis AM, Caudle J, Liddle GW (1969) In vitro stimulation of renin production by epinephrine, norepinephrine and cyclic AMP. Proc Soc Exp Biol Med 130: 748–753

Migdal SC, Slick GL, Abu-Hamdan D, McDonald FD (1980) Phenytoin, renal function and renin release. J Pharmacol Exp Ther 215: 304–308

Mikami H, Suzuki H, Smeby RR, Ferrario CM (1985) Cerebrospinal fluid angiotensin II immunoreactivity is not derived from the plasma. Hypertension 7: 65–71

Millan MA, Kiss A, Aguilera G (1991) Developmental changes in brain angiotensin II receptors in the rat. Peptides 12: 723–737

Miller RA (1967) Regional responses of interrenal tissue and of chromaffin tissue to hypophysectomy and stress in pigeons. Acta Endocrinol 55: 108–118

Miselis RR (1981) The efferent projections of the subfornical organ of the rat: a circumventricular organ within a neural network subserving water balance. Brain Res 230: 1–23

Miselis RR, Shapiro RE, Hand PJ (1979) Subfornical organ efferents to neural systems for control of body water. Science 205: 1022–1025

Misono K, Whorton RA, Inagami T, Hollifield JW (1976) Renin release (RR) from renal cortical slices: effect of isoproterenol (IP), angiotensin II (AII) and molecular characterization of renin release. Endocrinology 98: A345

Mitsuma T, Nogimori T (1984) Effects of substance P, angiotensin II, oxotremorine and prostaglandin D on thyrotropin secretion in rats. Horm Res 19: 176–184

Miyazaki H, Fukamizu A, Hirose S, Hayashi T, Hori H, Ohkubo H, Nakanishi S, Murakami K (1984) Structure of human renin gene. Proc Natl Acad Sci USA 81: 5999–6003

Mizogami S, Oguri M, Sokabe H, Nishimura H (1968) Presence of renin in the glomerular and aglomerular kidney of marine teleosts. Am J Physiol 215: 991–994

Mizuhira V (1986) Ultrastructure of juxtaglomerular apparatus or juxtaglomerular complex. In: Kokubu T, Yamamoto K (eds) Renin and hypertension. Medical Tribune, Tokyo, pp 9–61 (in Japanese)

Mizuno K, Fukuchi S (1981) A possible role of brain angiotensin II receptor binding in developing hypertension. Jpn Circ J 45: 1111–1115

Mizuno K, Ojima M, Gotoh M, Hashimoto S, Fukuchi S (1985) True renin in human pituitary tissue. J Neurochem 44: 633–636

Moe KE, Weiss ML, Epstein AN (1984) Sodium appetite during captopril blockade of endogenous angiotensin II formation. Am J Physiol 247: R356–R365

Moe O, Tejedor A, Campbell WB, Alpern RJ, Henrich WL (1991) Effects of endothelin on in vitro renin secretion. Am J Physiol 260: E521–E525

Mogenson GJ, Kucharczyk J, Assaf S (1977) Evidence for multiple receptors and neural pathways which subserve water intake initiated by angiotensin II. In: Buckley J, Ferrario CM (eds) Central actions of angiotensin and related hormones. Pergamon Press, New York, pp 493–502

Moore AF, Khairallah PA (1977) Binding of ^{125}I-angiotensin II to rat brain preparation from normotensive and spontaneously hypertensive rats. Physiologist 20: 65

Moore AF, Strong JH, Buckley JP (1981a) Cardiovascular actions of angiotensin in the fowl (Gallus domesticus). I. Analysis. Res Commun Chem Pathol Pharmacol 32: 423–445

Moore AF, Strong JH, Buckley JP (1981b) Cardiovascular actions of angiotensin in the fowl (Gullus domesticus) II. Angiotensin analog agonists and antagonists. Res Commun Chem Pathol Pharmacol 32: 447–457

Moore TJ, Williams GH (1982) Angiotensin II receptors on human platelets. Circ Res 51: 314–320

Moore TJ, Taylor T, Williams GH (1984) Human platelet angiotensin II receptors: regulation by the circulating angiotensin II level. J Clin Endocrinol Metab 58: 778–782

Morgan TA (1971) A microperfusion study of influence of macula densa on glomerular filtration rate. Am J Physiol 220: 186–190

Morild J, Christensen JA, Mikeler E, Bohle A (1988) Peripolar cells in the avian kidney. Virchows Arch [A] 412: 471–477

Morimoto S, Yamamoto K, Horiuchi K, Tanaka H, Ueda J (1970) A release of renin from dog kidney cortex slices. Jpn J Pharmacol 20: 536–545

Morris BJ, Johnson CI (1976) Renin substrate in granules from rat kidney cortex. Biochem J 154: 625–637

Morris BJ, Nixon RL, Johnston CI (1976) Release of renin from glomeruli isolated from rat kidney. Clin Exp Pharmacol Physiol 3: 37–47

Morris JL, Gibbins L (1983) Innervation of the renal vasculature of the toad (Bufo marinus). Cell Tissue Res 231: 357–376

Mujais SK, Kauffman S, Katz AI (1986) Angiotensin II binding sites in individual segments of the rat nephron. J Clin Invest 77: 315–318

Mukai M, Torikata C, Hirose S, Murakami K, Kageyama K (1984) Biochemical and immunohistochemical localization of renin in human pituitary. Lab Invest 51: 425–428

214

Mukherjee A, Kulkarni PV, Haghani Z, Sutko JL (1982a) Identification and characterization of angiotensin II receptors in cardiac sarcolemma. Biochem Biophys Res Commun 105: 575–581

Mukherjee A, Kulkarni PV, McCann SM, Negro-Vilar A (1982b) Evidence for the presence and characterization of angiotensin II receptors in rat anterior pituitary membranes. Endocrinology 110: 665–668

Mukhopadhyay AH, Leavitt L (1978) Evidence for an angiotensin receptor in esophageal smooth muscle of the opossum. Am J Physiol 235: E738–E742

Mukouyama M, Nakajima M, Horiuchi M, Sasamura H, Pratt RE, Dzau VJ (1993) Expression cloning of type 2 angiotensin II receptor reveals a unique class of seven-transmembrane receptors. J Biol Chem 268: 24539–24542

Mullins JJ, Sigmund CD, Kane-Hass C, Gross KW (1989) Expression of the DBA/2J Ren-2 gene in the adrenal gland of transgenic mice. EMBO J 8: 4065–4072

Mulrow PJ (1992) Adrenal renin: regulation and function. Front Neuroendocrinol 13: 47–60

Munday KA, Parsons BJ, Poat JA (1972) Studies on the mechanism of action of angiotensin on ion transport by kidney cortex slices. J Physiol (Lond) 224: 195–206

Munkacsi I (1969) Distribution of the intrarenal monoaminergic nerves in the kidneys of the desert rat (*Dipodomys merriami*) and the white rat (*Rattus norvegicus*). Acta Anat 73: 56–68

Muñoz JM, Braun-Menendez E, Fasciolo JC, Leloir LF (1939) Hypertensin: the substance causing renal hypertension. Nature 144: 980

Murakami K, Ganong WF (1987) Site at which angiotensin II acts to stimulate ACTH secretion in vivo. Neuroendocrinology 46: 231–235

Murakami E, Eggena P, Barret JD, Sambhi MP (1984) Heterogeneity of renin substrate released from hepatocytes and in brain extracts. Life Sci 34: 385–392

Murphy TJ, Alexander RW, Griendling KK, Runge MS, Bernstein K (1991) Isolation of a cDNA encoding the vascular type-1 angiotensin II receptor. Nature 351: 233–236

Murphy TJ, Nakamura Y, Takeuchi K, Alexander RW (1993) A cloned angiotensin receptor isoform from the turkey adrenal gland is pharmacologically distinct from mammalian angiotensin receptors. Mol Pharmacol 44: 1–7

Mutter J, LeMoine J, Tsang B, Kucharczyk J (1984) Central angiotensin-induced water intake and salt appetite in the pig. Brain Res 322: 374–377

Müller J, Barajas L (1972) Electron microscopic and histochemical evidence for a tubular innervation in the renal cortex of the monkey. J Ultrastruct Res 41: 533–549

Myers LS, Steele MK (1989) The brain renin-angiotensin system and the regulation of prolactin secretion in female rats: influence of ovarian hormones. J Neuroendocrinol 1: 299–303

Myers RD, Hall GH, Rudy TA (1973) Drinking in the monkey evoked by nicotine or angiotensin II microinjected in hypothalamic and mesencephalic sites. Pharm Biochem Behav 1: 15–22

Nabika T, Velletri PA, Lovenberg W, Beaven M (1985) Increase in cytosolic calcium and phosphoinositide metabolism induced by angiotensin II and [Arg] vasopressin in vascular smooth muscle cells. J Biol Chem 260: 4661–4670

Naess PA, Christensen G, Kiil F (1993) Inhibitory effect of endothelin on renin release in dog kidneys. Acta Physiol Scand 148: 131–136

Naftilan AJ, Oparil S (1978) Inhibition of renin release from rat kidney slices. Am J Physiol 235: F62–F68

Nagatsu I, Nagatsu T, Yamamoto T, Glenner GG, Mehl JW (1970) Purification of aminopeptidase A in human serum and degradation of angiotensin II by the purified enzyme. Biochim Biophys Acta 198: 255–270

Nairn RC, Masson GMC, Corcoran AC (1956) The production of serous effusions in nephrectomized animals by the administration of renal extracts and renin. J Pathol Bacteriol 71: 155–163

Nairn RC, Fraser KB, Chadwick CS (1959) The histological localization of renin with fluorescent antibody. Br J Exp Pathol 40: 155–163

Nakajima M, Mathews DC, Hewitson T, Kincaid-Smith P (1989) Modified immunogold labelling applied to the study of protein droplets of glomerular disease. Virchows Arch [A] 415: 489–499

Nakajima T, Khosla MC, Sakakibara S (1978) Comparative biochemistry of renins and angiotensins in the vertebrates. Jpn Heart J 19: 799–805

Nakamura Y, Nishimura H, Khosla MC (1982) Vasodepressor action of angiotensin in conscious chickens. Am J Physiol 243: H456–H462

Nakamura Y, Madey MA, Nishimura H, Quach D, Barajas L (1992) Lack of control of renin release by adrenergic nervous system in the aglomerular toadfish. Gen Comp Endocrinol 88: 62–75

Nakayama T, Nakajima T, Sokabe H (1973) Comparative studies on angiotensins. III. Structure of fowl angiotensin and its identification by DNS-method. Chem Pharm Bull 21: 2085–2087

Nakayama T, Nakajima T, Sokabe H (1977) Comparative studies on angiotensins. IV. Structure of snake (*Elaphe climocophora*) angiotensin. Chem Pharm Bull 25: 3255–3260

Naruse K, Takii Y, Inagami T (1981) Immunohistochemical localization of renin in luteinizing hormone-producing cells of rat pituitary. Proc Natl Acad Sci USA 78: 7579–7583

Naruse K, Inagami T, Celio MR, Warkman RJ, Takii T (1982) Immunocytochemical evidence that angiotensin I and II are formed by intracellular mechanisms in juxtaglomerular cells. Hypertension 4 (Suppl II): 70–74

Naruse K, Murakoshi M, Osamura RY, Naruse M, Toma H, Watanabe K, Demura H, Inagami T, Shizume K (1985) Immunohistological evidence for renin in human endocrine tissues. J Clin Endocrinol Metab 61: 172–177

Naruse K, Naruse M, Obana K, Demura R, Demura H, Inagami T, Shizume K (1986) Renin in the rat pituitary coexists with angiotensin II and depends on testosterone. Endocrinology 118: 2470–2476

Naruse M, Inagami T (1982) Markedly elevated specific renin levels in the adrenal in genetically hypertensive rats. Proc Natl Acad Sci USA 79: 3295–3299

Naruse M, Naruse K, Inagaki T, Inagami T (1984a) Immunreactive renin in mouse adrenal gland. Localization in the inner cortical region. Hypertension. 6: 275–280

Naruse M, Naruse K, Shizume K, Inagami T (1984b) Gonadotropin-dependent renin in the rat testes. Proc Soc Exp Biol Med 177: 337–342

Nash FD, Rostorfer HH, Bailie MD, Wathen RL, Schneider EG (1968) Renin release. Relation to renal sodium load and dissociation from hemodynamic changes. Circ Res 22: 473–487

Nasjletti A, Masson GMC (1969) Effect of corticosteroids on plasma angiotensinogen and renin activity. Am J Physiol 217: 1396–1400

Nasjletti A, Matsunaga M, Masson GMC (1971) Effects of sex hormones on the renal pressor system. Can J Physiol Pharmacol 49: 292–301

Natarajan R, Nadler J (1992) Platelet-derived growth factor is a potent inhibitor of angiotensin II-induced aldosterone synthesis. Mol Cell Endocrinol 83: 57–63

Navar LG, Burke TJ, Robinson RR, Clapp JR (1974) Distal tubular feedback in the autoregulation of single nephron glomerular filtration rate. J Clin Invest 53: 516–525

Needleman P, Douglas JR Jr, Jakschik B, Stoecklein PB, Johnson EM Jr (1974) Release of renal prostaglandin by catecholamines: relationship to renal endocrine function. J Pharmacol Exp Ther 188: 453–460

Newton MA, Laragh JH (1968) Effects of glucocorticoid administration on aldosterone excretion and plasma renin in normal subjects, in essential hypertension and in primary aldosteronism. J Clin Endocrinol Metab 28: 1014–1022

Neyses L, Locher R, Wehling M, Pech H, Tenschert W, Vetter W (1984) Angiotensin II binding to human mononuclear cells: receptor or free fluid endocytosis? Clin Sci 66: 605–612

Ng KKF, Vane JR (1967) Conversion of angiotensin I to angiotensin II. Nature 216: 762–766

Nicolaïdis S, El Ghissassi M (1991a) Organum cavum pre-lamina terminalis, an undescribed brain organ in rats – morphology and physiology. Am J Physiol 260: R396–R406

Nicolaïdis S, El Ghissassi M (1991b) Angiotensin and sodium interaction in the organum cavum pre-lamina terminalis: electrophysiological and drinking responses. Brain Res Bull 27: 469–473

Nicolaïdis S, Fitzsimons JT (1975) La dépendance de la prise d'eau induite par l'angiotensine II envers la fonction vasomotrice cérébrale locale chez le rat. C R Acad Sc Paris 281: 1417–1420

216

Nicolaïdis S, El Ghissassi M, Thornton SN (1991) Rostro-sagittal brain: site of integration of hydrational signals in body fluid regulation and drinking. In: Ramsay DJ, Booth DA (eds) Thirst-physiological and psychological aspects. Springer, Berlin Heidelberg New York, pp 207–218

Nielsen AH, Poulsen K (1988) Is prorenin of physiological and clinical importance? J Hypertens 6: 949–958

Nikcols GA, Nikcols MA, Helwig JJ (1990) Binding of parathyroid hormone and parathyroid hormone-related protein to vascular smooth muscle of rabbit renal miscovessels. Endocrinology 126: 721–727

Nilsson O (1965) The adrenergic innervation of the kidney. Lab Invest 14: 1392–1395

Nishimura H (1980a) Evolution of the renin-angiotensin system. In: Pang PKT, Epple A (eds) Evolution of vertebrate endocrine systems. Graduate Studies, Texas Tech Univ No 21, pp 373–404

Nishimura H (1980b) Comparative endocrinology of renin and angiotensin. In: Johnson AJ, Anderson RR (eds) The renin-angiotensin system. Plenum Press, New York, pp 29–77

Nishimura H (1985) Evolution of the renin-angiotensin system and its role in control of cardiovascular function in fishes. In: Foreman RE, Gorbman A, Dodd JM, Olsson R (eds) Evolutionary biology of primitive fishes. Plenum Press, New York, pp 275–293

Nishimura H (1987) Role of the renin-angiotensin system in osmoregulation. In: Pang PKT, Schreibman MP (eds) Vertebrate endocrinology: fundamentals and biomedical implications, vol 2. Academic Press, San Diego, pp 157–187

Nishimura H, Bailey JR (1982) Intrarenal renin-angiotensin system in primitive vertebrates. Kidney Int 22: S185–192

Nishimura H, Madey AM (1989) Signals controlling renin release in aglomerular toadfish. Fish Biochem Physiol 7: 323–329

Nishimura H, Ogawa M (1973) The renin-angiotensin system in fishes. Am Zool 13: 823–838

Nishimura H, Oguri M, Ogawa M, Sokabe H, Imai M (1970) Absence of renin in kidneys of elasmobranchs and cyclostomes. Am J Physiol 218: 911–915

Nishimura H, Ogawa M, Sawyer WH (1973) Renin-angiotensin system in primitive bony fishes and a holocephalian. Am J Physiol 224: 950–956

Nishimura H, Sawyer WH, Nigrelli RF (1976) Renin, cortisol and plasma volume in marine teleost fishes adapted to dilute media. J Endocrinol 70: 47–59

Nishimura H, Crofton JT, Norton VM, Share L (1977) Angiotensin generation in teleost fish determined by radioimmunoassay and bioassay. Gen Comp Endocrinol 32: 236–247

Nishimura H, Norton VM, Bumpus FM (1978) Lack of specific inhibition of angiotensin II in eels by angiotensin antagonists. Am J Physiol 235: H95–103

Nishimura H, Lunde LG, Zucker A (1979) Renin Response to hemorrhage in the aglomerular toadfish, *Opsanus tau*. Am J Physiol 237: H105–H111

Nishimura H, Madey MA, Mugaas JN, Khosla MC, Crofton JT (1981a) Radioimmunoassay of fowl angiotensin I. Gen Comp Endocrinol 45: 262–272

Nishimura H, Nakamura Y, Taylor AA, Madey MA (1981b) Renin-angiotensin and adrenergic mechanisms in control of blood pressure in the fowl. Hypertension 3 (Suppl I): 141–149

Nishimura H, Nakamura Y, Sumner RP, Khosla MC (1982) Vasopressor and depressor actions of angiotensin in the anesthetized fowl. Am J Physiol 242: H314—H324

Nishimura H, Walker OE, Patton CM, Madison AN, Chiu AT, Keisler J (1994) Novel angiotensin receptor subtypes in fowl. Am J Physiol 267: R1174–R1181

Nishizuka Y (1984) The role of protein kinase C in cell surface signal transduction and tumour production. Nature 308: 693–698

Nolly H, Fasciolo JC (1971) Renin-angiotensin system and homestasis in *Bufo arenarum*. Comp Biochem Physiol 39A: 833–841

Nolly HL, Fasciolo JC (1972) The renin-angiotensin system throughout the phylogenetic scale. Comp Biochem Physiol 41A: 249–254

Nolly HL, Fasciolo JC (1973) The specificity of the renin-angiotensinogen reaction through the phylogenetic scale. Comp Biochem Physiol 44A: 639–645

Nolly HL, Reid IA, Ganong WF (1974) Effect of theophylline and adrenergic blocking drugs on the renin response to norepinephrine in vitro. Circ Res 35: 575–579

Norman JT (1991) The role of angiotensin II in renal growth. Renal Physiol Biochem 14: 175-185

Nothstine SA, Davis JO, de Roos RM (1971) Kidney extracts and ACTH on adrenal steroid secretion in a turtle and a crocodilian. Am J Physiol 221: 726-732

Nussberger J, Brunner DB, Waeber B, Brunner HR (1985) True versus immunoreactive angiotensin II in human plasma. Hypertension 7 (Suppl I): 1-7

Oates JA, Whorton AR, Gerkens JF, Brach RA, Hollifield JW, Frolich JC (1979) The participation of prostaglandins in the control of renin release. Fed Proc 38: 72-74

Oberling Ch (1927) L'existence d'une housse neuromusculaire au niveau des artéres glomérulaires de l'Homme. Acad Sci Paris 184: 1200-1202

Oberling Ch (1944) Further studies on the preglomerular cellular apparatus. Am J Pathol 230: 155-171

Oberling Ch, Hatt PY (1960a) Ultrastructure de l'appareil juxtaglomérulaire du rat. Comp Rend Acad Sci Paris 250: 929-932

Oberling Ch, Hatt PY (1960b) Étude de l'appareil juxtaglomérulaire du rat au microscope electronique. Ann Anat Pathol Paris 5: 441-474

Ochiai H, Nakai Y (1984) Ultrastructural and morphometric studies on the neurohypophysial nerve terminals of the rat following adminstration of angiotensin II. Neuroendocrinology 39: 496-502

Ogawa M (1977) Evolution of juxtaglomerular apparatus. Gunma Symp Endocrinol 14: 75-81

Ogawa M, Hirano T (1982) Studies on the nephron of freshwater stingray, *Potamotrygon magdalenae*. Zool Magazine 91: 101-105

Ogawa M, Oguri M (1978) Occurrence of the renin-angiotensin system in the vertebrates. Jpn Heart J 19: 791-798

Ogawa M, Sokabe H (1971) The macula densa site of avian kidneys. Z Zellforsch 120: 29-36

Ogawa M, Oguri M, Sokabe H, Nishimura H (1972) Juxtaglomerular apparatus in the vertebrates. Gen Comp Endocrinol (Suppl 3): 374-381

Oguri M (1978) Presence of juxtaglomerular cells in the holocephalian kidney. Gen Comp Endocrinol 36: 170-173

Oguri M (1980) A histological investigation on the juxtaglomerular cell granules in fish kidneys. Bull Jpn Soc Sci Fish 46: 797-800

Oguri M, Sokabe H (1968) Juxtaglomerular cells in the teleost kidneys. Bull Jpn Soc Sci Fish 34: 882-888

Oguri M, Sokabe H (1974) Comparative histology of the corpuscles of Stannius and the juxtaglomerular cells in the kidneys of teleosts. Bull Jpn Soc Sci Fish 40: 545-549

Oguri M, Kamiya K, Sokabe H (1969) A histological study on the juxtaglomerular cells in the kidney of Japanese mackerel. Bull Jpn Soc Sci Fish 35: 737-742

Oguri M, Ogawa M, Sokabe H (1970) Absence of juxtaglomerular cells in the kidneys of Chondrichthyes and Cyclostomi. Bull Jpn Soc Sci Fish 36: 881-884

Oguri M, Ogawa M, Sokabe H (1972) Juxtaglomerular cells in aglomerular teleosts. Bull Jpn Soc Sci Fish 38: 195-200

Ohkubo H, Kageyama R, Ujihara M, Hirose T, Inayama S, Nakanishi S (1983) Cloning and sequence analysis of cDNA for rat angiotensinogen. Proc Natl Acad Sci USA 80: 2196-2200

Ohkubo H, Nakayama K, Tanaka T, Nakanishi S (1986) Tissue distribution of rat angiotensinogen mRNA and structural analysis of its heterogeneity. J Biol Chem 261: 319-323

Okahara T, Abe Y, Yamamoto K (1977) Effects of dibutyryl cyclic AMP and propranolol on renin secretion in dogs. Proc Soc Exp Biol Med 156: 213-218

Okamura T, Inagami T (1984) Release of active and inactive renin from hog renal cortical slices in vitro. Am J Physiol 246: F765-F771

Okamura T, Clements DL, Inagami T (1981) Renin, angiotensins, and angiotensin converting enzyme in neuroblastoma cells: evidence for intracellular formation of angiotensins. Proc Natl Acad Sci USA 78: 6940-6943

Okawara Y, Kobayashi H (1988) Enhancement of water intake by captopril (SQ14225), an angiotensin I-converting enzyme inhibitor, in the goldfish, *Carassius auratus*. Gen Comp Endocrinol 69: 114-118

Okawara Y, Karakida T, Yamaguchi K, Kobayashi H (1985) Diurnal rhythm of water intake and plasma angiotensin II in the Japanese quail (*Coturnix coturnix japonica*). Gen Comp Endocrinol 57: 89–92

Okawara Y, Karakida T, Aihara M, Yamaguchi K, Kobayashi H (1987) Involvement of angiotensin II in water intake in the Japanese eel, *Anguilla japonica*. Zool Sci 4: 523–528

Okkels MH (1929) Sur l'existence d'une spécialisation morphologique au niveau du pôle vasculaire du glomerule rénal chez la grenouille. C R Acad Sci 188: 193–195

Oldfield BJ (1991) Neurochemistry of the circuitry subserving thirst. In: Ramsay DJ, Booth DA (eds) Thirst-physiological and psychological aspects. Springer, Berlin Heidelberg New York, pp 176–193

Oliver G (1897) The action of renal extracts on the peripheral vessels. J Physiol 21: Proc Physiol Soc pxii

Oliver WJ, Gross F (1967) Effect of testosterone and duct ligation on submaxillary renin principle. Am J Physiol 213: 341–346

Olivereau M, Lemonine AM (1969) Identification de "cellules juxtaglomérulaires" chez deux téléostéens: le muge et l'anguille. Bull Assoc Anat 142: 1260–1269

Olson Al, Robillard JE, Kisker CT, Smith BA, Perlman S (1991) Negative regulation of angiotensinogen gene expression by glucocorticoids in fetal sheep liver. Pediatr Res 30: 256–260

Olson KR (1984) Distribution of flow and plasma skimming in isolated perfused gills of three teleosts. J Exp Biol 109: 97–108

Olson KR, Kullman D, Narkates AJ, Oparil S (1986) Angiotensin extraction by trout tissues in vivo and metabolism by the perfused gill. Am J Physiol 250: R532–R538

Olson KR, Lipke D, Datta Munsi JS, Morita A, Ghosh TK, Kunwar G, Ahmad M, Roy PK, Sigh ON, Nasar SST, Pandey A, Oduleye SO, Kullman D (1987) Angiotensin-converting enzyme in organs of air-breathing fish. Gen Comp Endocrinol 68: 486–491

Oparil S, Ehrlich EN, Lindheimer MD (1975) Effect of progesterone on renal sodium handling in man: relation to aldosterone excretion and plasma renin activity. Clin Sci Mol Med 49: 139–147

Opdyke DF, Holcombe R (1976) Response to angiotensin I and II and to the AI-converting enzyme inhibitor in a shark. Am J Physiol 231: 1750–1753

Opgenorth TA, Burnett Jr JC, Granger JP, Scriven TA (1986) Effects of atrial natriuretic peptide on renin secretion in nonfiltering kidney. Am J Physiol 250: F798–F801

Osborn JL, Kopp UC, Thames MC, Dibona GF (1984) Interactions among renal nerves, prostaglandins, and renal arterial pressure in the regulation of renin release. Am J Physiol 247: F706–F713

Osborne MJ, Droz B, Meyer P, Morel F (1975) Angiotensin II: renal localization in glomerular mesangial cells by autoradiography. Kidney Int 8: 245–254

Osswald H, Schmitz HJ, Kemper R (1978) Renal action of adenosine: Effect on renin secretion in the rat. Naunyn-Schmiedeberg's Arch Pharmacol 303: 95–99

Otsuka K, Assaykeen TA, Goldfein A, Ganong WF (1970) Effect of hypoglycemia on plasma renin activity in dogs. Endocrinology 87: 1306–1317

Ouali R, Poulette S, Penhoat A, Saez JM (1992) Characterization and coupling of angiotensin-II receptor subtypes in cultured bovine adrenal fasciculata cells. J Steroid Biochem Mol Biol 43: 271–280

Ouali R, LeBrethon MC, Saez Jm (1993) Identification and characterization of angiotensin-II receptor subtypes in cultured bovine and human adrenal fasciculata cells and PC12W cells. Endocrinology 133: 2766–2772

O'Toole LB, Armour KJ, Decourt C, Hazon N, Lahlou B, Henderson IW, (1990) Secretory patterns of 1α-hydroxycorticosterone in the isolated perifused internal gland of dogfish, *Scyrhiolinus canicula*. J Mol Endocrinol 5: 55–60

Page IH, Helmer OM (1940) A crystalline pressor substance (angiotonin) resulting from the reaction between renin and renin-activator. J Exp Med 71: 29–42

Paglin S, Stukenbrok H, Jamieson JD (1984) Interaction of angiotensin II with dispersed cells from the anterior pituitary of the male rat. Endocrinology 114: 2284–2292

219

Pagnan A, Pessina AC, Thiene G, DalPalu C (1978) The natural history of hypertension in turkeys. Clin Sci Mol Med 55: 213S-215S

Palaic D, Le Morvan P (1971) Angiotensin tachyphylaxis in guinea pig aortic strips. J Pharmacol Exp Ther 179: 522-531

Palkovits M, Zaborsky L (1977) Neuroanatomy of central cardiovascular control. Nucleus tractus solitarii: afferent and efferent neuronal connections in relation to the baroreceptor reflex arch. Prog Brain Res 47: 9-34

Palkovits M, deJong W, van der Wal B, deWied D (1970) Effect of adrenocorticotropic and growth hormones on aldosterone production and plasma renin activity in chronically hypophysectomized sodium-deficient rats. J Endocrinol 47: 243-250

Pals DT, Couch SJ (1993) Renin release induced by losartan (Dup753), an angiotensin II receptor antagonist. Clin Exp Hypertens 15: 1-13

Pandey KN, Maki M, Inagami T (1984a) Detection of renin mRNA in mouse testis by hybridization with renin cDNA probe. Biochem Biophys Res Commun 125: 662-667

Pandey KN, Melner MH, Parmentier M, Inagami T (1984b) Demonstration of renin activity in purified rat Leydig cells: evidence for the existence of an endogenous inactive (latent) form of enzyme. Endocrinology 115: 1753-1759

Pandey KN, Misono KS, Inagami T (1984c) Evidence for intracellular formation of angiotensins: coexistence of renin and angiotensin-converting enzyme in Leidig cells of rat testis. Biochem Biophys Res Commun 122: 1337-1343

Pang PKT, Wang R, Shan J, Karpinski E, Benishin CG (1990) Specific inhibition of long-lasting, L-type calcium channels by synthetic parathyroid hormone. Proc Natl Acad Sci USA 87: 623-627

Panthier J-J, Foote S, Chambraud B, Strosberg AD, Corvol P, Rougeon F (1982) Complete amino acid sequence and maturation of the mouse submaxillary gland renin precursor. Nature 298: 90-92

Park CS, Malvin RL (1978) Calcium in the control of renin release. Am J Physiol 235: F22-H25

Park CS, Han DS, Fray JCS (1981) Calcium in the control of renin secretion: Ca^{2+} influx as an inhibitory signal. Am J Physiol 240: F70-F74

Parmentier M, Inagami T, Pochet R, Descline JC (1983) Pituitary-dependent renin-like immunoreactivity in the rat testis. Endocrinology 112: 1318-1323

Passo SS, Assaykeen TA, Goldfien A, Ganong WF (1971) Effect of alpha and beta-adrenergic blocking agents on the increase in renin secretion produced by stimulation of the medula oblongata in dogs. Neuroendocrinology 7: 97-104

Paul M, Printz MP, Harms E, Unger T, Lang RE, Ganten D (1985) Localization of renin (EC 3.4.23) and converting enzyme (EC 3.4.15.1) in nerve endings of rat brain. Brain Res 334: 315-324

Peach MJ (1977) Renin-angiotensin system: biochemistry and mechanisms of action. Physiol Rev 57: 313-370

Peach MJ (1981) Molecular actions of angiotensin. Biochem Pharmacol 30: 2745-2751

Peach MJ, Bumpus FM, Khairallah PA (1971) Release of adrenal catecholamines by angiotensin I. J Pharmacol Exp Ther 176: 366-376

Pearce RM (1909) An experimental study of the influence of kidney extracts and of the serum of animals with renal lesions upon the blood pressure. J Exp Med 11: 430-442

Pedersen EB, Poulsen K (1983) Agression-provoked huge release of submaxillary mouse renin to saliva. Acta Endocrinol 104: 510-512

Penit J, Faure M, Jard S (1983) Vasopressin and angiotensin II receptors in rat aortic smooth muscle cells in culture. Am J Physiol 244: E72-E82

Perroteau I, Netchitailo P, Homo-Delarche F, Delarue C, Lihrmann I, Leboulenger F, Vaudry H (1984) Role of exogenous and endogenous prostaglandins in steroidogenesis by isolated frog interrenal gland: evidence for dissociation in acrenocorticotropin and angiotensin action. Endocrinology 115: 1765-1773

Perrott MN, Balment RJ (1985) Drinking behavior and the renin angiotensin system in euryhaline and stenohaline fish. J Endocrinol 107 Abstr 193

Perrott MN, Grierson CE, Hazon N, Balment RJ (1992) Drinking behavior in sea water and fresh water teleosts, the role of the renin-angiotensin system. Fish Physiol Biochem 10: 161–168

Persson AEG, Boberg U (1988) The juxtaglomerular apparatus. Elsevier, Amsterdam

Peter K (1970) Über die Nierenkanälchen des Menschen und einiger Säugetiere. Anat Anz 30: 114–124

Peter S (1976) Ultrastructural studies on the secretory process in the epitheloid cells of the juxtaglomerular apparatus. Cell Tissue Res 168: 45–53

Peters G, Fitzsimons JT, Peters-Haefeli L (1975) Control mechanisms of drinking. Springer, Berlin Heidelberg New York

Petersen EP, Camara CG, Abhold RH, Wright JW, Harding JW (1984) Characterization of angiotensin binding to gerbil brain membranes using [^{125}I] angiotensin III as the radioligand. Brain Res 321: 225–235

Petersen EP, Camara CG, Abhold RH, Wright JW, Harding JW (1985a) Characterization of angiotensin binding in the African green monkey. Brain Res 341: 139–146

Petersen EP, Wright JW, Harding JW (1985b) Characterization of angiotensin II binding sites in African green monkey uterus. Life Sci 36: 177–182

Pettinger WA, Keeton K (1975) Altered renin release and propranolol potentiation of vasodilatory drug hypotension. J Clin Invest 55: 236–243

Pettinger WA, Marchelle M, Augusto L (1971) Renin suppression by DOC and Nacl in the rat. Am J Physiol 221: 1071–1074

Pettinger WA, Augusto L, Leon AS (1972) Alteration of renin release by stress and adrenergic receptor and related drugs in unanesthetized rats. In: Bloor CM (ed) Comparative pathophysiology of circulatory disturbances. Plenum Press, New York, pp 105–117

Phillips MI (1978) Angiotensin in the brain. Neuroendocrinology 25: 354–377

Phillips MI (1987a) Brain angiotensin. In: Gross PM (ed) Circumventricular organs and body fluids, vol III. CRC Press, Boca Raton, pp 163–182

Phillips MI (1987b) Functions of angiotensin in the central nervous system. Annu Rev Physiol 49: 413–435

Phillips MI, Felix D (1976) Specific angiotensin II receptive neurons in the cat subfornical organ. Brain Res 109: 531–540

Phillips MI, Hoffman WE (1977) Sensitive sites in the brain for the blood pressure and drinking responses to angiotensin II. In: Buckley JP, Ferrario CM (eds) Central actions of angiotensin and related hormons. Pergamon Press, New York, pp 325–356

Phillips MI, Stenstrom B (1985) Angiotensin II in the rat brain comigrates with authentic angiotensin II in high pressure liquid chromatography. Circ Res 56: 212–219

Phillips MI, Weyhenmeyer J, Felix D, Ganten D, Hoffman WE (1979) Evidence for an endogenous brain renin angiotensin system. Fed Proc 38: 2260–2266

Phillips MI, Speakman EA, Kimura B (1993) Tissue renin-angiotensin system. In: Raizada MK, Phillips MI, Summers C (eds) Cellular and molecular biology of the renin-angiotensin system. CRC Press, Boca Raton, pp 97–130

Phillis JW, Wu PH (1981) Catecholamines and the sodium pump in excitable cells. Prog Neurobiol (Oxf) 17: 141–184

Pickering GW, Prinzmetal M (1938) Some observations on renin, a pressor substance contained in normal kidney, together with a method for its biological assay. Clin Sci 3: 211–227

Pitcock JA, Hartroft PM, Newmark LN (1959) Increased renal pressor activity (renin) in sodium deficient rats and correlation with juxtaglomerular cell granulation. Proc Soc Exp Biol Med 100: 868–869

Plotsky PM, Sutton SW, Bruhn TO, Ferguson AV (1988) Analysis of the role of angiotensin II in mediation of adrenocorticotropin secretion. Endocrinology 122: 538–545

Plunkett LM, Saavedra JM (1985) Increased angiotensin II binding affinity in the nucleus tratus solitalius of spontaneously hypertensive rats. Proc Natl Acad Sci USA 82: 7721–7724

Polanco MJ, Mata MI, Agapito MT, Recio JM (1990) Angiotensin-converting enzyme distribution and hypoxia response in mammal, bird and fish. Gen Comp Endocrinol 79: 240–245

Polanco MJ, Miguel JL, Agapito MT, Recio JM (1992) Characterization of chicken lung angiotensin I-converting enzyme. J Endocrinol 132: 261–268

Porchet M, Dhainaut-Courtois (1988) Neuropeptides and monoamines in annelids. In: Thorndyke MC, Goldsworthy GJ (eds) Neurohormones in invertebrates. Cambridge Univ Press, Cambridge, New York, p 225

Poth MM, Heath RG, Ward M (1975) Angiotesin-converting enzyme in human brain. J Neurochem 25: 83–85

Potkay S, Gilmore JP (1973) Autoregulation of glomerular filtration in renin-depleted dogs. Proc Soc Exp Biol Med 143: 508–513

Pratt RE, Ouellette AJ, Dzau VJ (1983) Biosynthesis of renin: multiplicity of active and intermediate forms. Proc Natl Acad Sci USA 80: 6809–6813

Printz MP, Printz JM, Gregory TJ (1978) Identification of angiotensinogen in animal brain homogenates. Circ Res 43 (Suppl I): 21–27

Printz MP, Ganten D, Unger T, Phillips MI (1982) Minireview: the brain renin angiotensin system 1982. In: Ganten D, Phillips MI, Scholkens BA (eds) The renin angiotensin system in the brain. Springer, Berlin Heidelberg New York, pp 3–52

Prinzmetal M, Friedman B (1936) Pressor effects of kidney extracts from patients and dogs with hypertension. Proc Soc Exp Biol Med 35: 122–124

Pritchard PCH (1967) Living turtles of the world. TFH Publ, Hong Kong, pp 141, 144

Propper CR, Johnson WE (1994) Angiotensin II induces water absorption behavior in two species of desert anurans. Horm Behar 28: 41–52

Quali R, Poulette S, Penhoat A, Saez JM (1992) Characterization and coupling of angiotensin II receptor subtypes in cultured bovine adrenal fasciculata cells. J Steroid Biochem Mol Biol 43: 271–280

Quinlan JT, Phillips MI (1981) Immunoreactivity for an angiotensin II-like peptide in the human brain. Brain Res 205: 212–218

Quinn SJ, Williams GH (1992) Regulation of aldosterone secretion. In: James VHT (ed) Adrenal gland. Raven Press, New York, pp 159–189

Rainey WE, Byrd EW, Sinnokrot RA, Carr BR (1991) Angiotensin-II activation of cAMP and corticosterone production in bovine adrenocortical cells: effects of nonpeptide angiotensin-II antagonists. Mol Cell Endocrinol 81: 33–41

Raizada MK, Yang JW, Phillips MI, Fellows RE (1981) Rat brain cells in primary culture characterization of angiotensin II binding sites. Brain Res 207: 343–356

Raizada MK, Muther TF, Sumners C (1984a) Increased angiotensin II receptors in neuronal cultures from hypertensive rat brain. Am J Physiol 247: C364–C372

Raizada MK, Stenstrom B, Phillips MI, Sumners C (1984b) Angiotensin II in neuronal cultures from brains of normotensive and hypertensive rats. Am J Physiol 247: C115–C119

Ramirez VD, Feder HH, Sawyer CH (1984) The role of brain catecholamines in the regulation of LH secretion: a critical inquiry. Front Neuroendocrinol 8: 27–84

Ramsay DJ, Booth DA (1991) Thirst. Springer, Berlin Heidelberg New York

Ramsay DJ, Reid IA (1975) Some central mechanisms of thirst in the dog. J Physiol (London) 253: 517–525

Ramsay DJ, Keil LC, Sharpe MC, Shinsako J (1978) Angiotensin II infusion increases vasopressin, ACTH and 11-hydroxycorticoid secretion. Am J Physiol 324: R66–R71

Ramsay DJ, Reid IA, Brown C (1979) Mechanism of the dipsogenic action of tetradecapeptide renin substrate in dogs. Endocrinology 105: 947–951

Re RN (1984) Cellular biology of the renin-angiotensin systems. Arch Intern Med 144: 2037–2041

Re R, Parab M (1984) Effects of angiotensin II on RNA synthesis by isolated nuclei. Life Sci 34: 647–651

Re RN, Fallon JT, Dzau VJ, Quay SC, Haber E (1982) Renin synthesis by canine aortic smooth muscle cells in culture. Life Sci 30: 99–106

Reboreda JC, Segura T (1989) Water balance effects of systemic and intracerebroventricular administration of angiotensin II in the toad *Bufo arenarum*. Comp Biochem Physiol 93: 505–509

Regoli D, Vane JR (1964) A sensitive method for the assay of angiotensin. Br J Pharmacol 23: 351–359

Regoli D, Riniker B, Brunner HR (1963) The enzymatic degradation of various angiotensin II derivatives by serum, plasma or kidney homogenates. Biochem Pharmacol 12: 637–646

Regoli D, Park WK, Rioux F (1974) Pharmacology of angiotensin. Pharmacol Rev 26: 69–123

Reid IA (1977) Is there a brain renin-angiotensin system? Circ Res 41: 147–153

Reid IA (1984) Actions of angiotensin II on the brain: mechanisms and physiologic role Am J Physiol 246: F533–F543

Reid IA, Morris BJ, Ganong WF (1978) The renin angiotensin system. Annu Rev Physiol 40: 377–410

Rice KK, Richter CP (1943) Increased sodium chloride and water intake of normal rats treated with desoxycorticosterone acetate. Endocrinology 33: 106–115

Richards AM, Nicholls MG (1993) Interactions between renin and atrial natriuretic factor. In: Robertson JIS, Nicholls MG (eds) The renin-angiotensin system, vol 1. Gower Medical Publ, London, pp 36.1–36.17

Richardson D, Stella A, Leonetti G, Bartorelli A, Zanchetti A (1974) Mechanisms of renal release of renin by electrical stimulation of the brain stem in the cat. Circ Res 34: 425–434

Richoux JP, Cordonnier JL, Bouhnik J, Clauser E, Corvol P, Menard J, Grignon G (1983) Immunocytochemical localization of angiotensinogen in rat liver and kidney. Cell Tissue Res 233: 439–451

Richter CP (1936) Increased salt appetite in adrenalectomized rats. Am J Physiol 115: 155–161

Richter CP (1956) Salt appetite of mammals: its dependence on instinct and metabolism. In: Autuoli M (ed) L'Instinct dans le Comportement des Animaux et de I'Homme. Masson, Paris, pp 577–629

Riegger AJG, Lever AF, Miller AJ, Morton JJ, Slack B (1977) Correction of renal hypertension in the rat by prolonged infusion of saralasin inhibitors. Lancet 2: 1317–1319

Rittel W, Iselin B, Kappeler H, Riniker B, Schwyzer R (1957) Synthese eines hochwirksamen Hypertensin II Amides. Helv Chim Acta 40: 614–624

Ritthaler T, Scholz H, Ackermann M, Riegger G, Kurtz A, Kramer BK (1995) Effects of endothelins on renin secretion from isolated mouse renal juxtaglomerular cells. Am J Physiol 268: F39–F45

Rivier C, Vale W (1983) Effect of angiotensin II on ACTH release in vivo: role of corticotropin-releasing factor. Regul Pept 7: 253–258

Rix E, Ganten D, Schüll B, Ungerth T, Taugner R (1981) Converting enzyme in the choroid plexus brain and kidney; immunocytochemical and biochemical studies in rats. Neurosci Lett 22: 125–130

Rix E, Ganten D, Stock G, Taugner R (1982) Immunocytochemical demonstration of renin and converting enzyme in rat and mouse brain. In: Ganten D, Printz M, Phillips MI, Schörkens BA (eds) The renin angiotensin system in the brain. Springer, Berlin Heidelberg New York, pp 126–136

Robb CA, Davis JO, Johnston CI, Hartroft PM (1969) Effects of deoxycorticosterone on plasma renin activity in conscious dogs. Am J Physiol 216: 884–889

Robberecht W, Denef C (1987) Evidence for paracrine interaction between gonadotrophs and somatotrophs through an opioid peptide. J Endocrinol Invest 10 (Suppl 3): 69

Robberecht W, Denef C (1988) Stimulation and inhibition of pituitary growth hormone release by angiotensin II in vitro. Endocrinology 122: 1496–1504

Robberecht W, Andries M, Denef C (1992) Stimulation of prolactin secretion from rat pituitary by luteinizing hormone-releasing hormone: evidence against mediation by angiotensin II acting through a (Sar^1-Ala^8)-angiotensin II-sensitive receptor. Neuroendocrinology 56: 185–194

Robertson AL Jr, Khairallah PA (1971) Angiotensin II: rapid localization in nuclei of smooth and cardiac muscle. Science 172: 1138–1139

Robertson JIS, Nicholls MG (1993) The renin-angiotensin system, vols 1 and 2. Gower Medical Publ, London

Robinson DG Jr (1975) Gerbil classification and nomenclature. Gerbil Digest 2: 3

Rocchini AP, Barger AC (1979) Renin release with carotid occulsion in the conscious dog: role of renal arterial pressure. Am J Physiol 236: H108–H111

Rogers TB (1984) High affinity angiotensin II receptors in myocardial sarcolemmal membranes. J Biol Chem 259: 8106–8114

Rogers TB, Gaa ST, Allen IS (1986) Identification and characterization of angiotensin II receptors on cultured heart myocytes. J Pharmacol Exp Ther 236: 438–444

Rohr YU (1966) Zum Feinbau des Subfornikal-Organs der Katze. I. Der Gefäss-Apparat. Z Zellforsch 73: 246–271

Rojo-Ortega JM, Chretien M, Genest J (1972) Renin-angiotensin system in chronically hypophysectomized rats. Horm Metab Res 4: 500–505

Rolls BJ, Ramsay DJ (1975) The elevation of endogenous angiotensin and thirst in the dog. In: Peters G, Fitzsimons JT, Peters-Haefeli L (eds) Control mechanisms of drinking. Springer, Berlin Heidelberg New York, pp 74–78

Rolls BJ, Rolls ET (1982) Thirst. Cambridge Univ Press, Cambridge, pp 46, 52–55

Rolls BJ, Phillips PA, Ledingham JGG, Forsling ML, Morton JJ, Crowe MJ (1986) Human thirst: the controls of water intake in health men. In: de Caro AN, Epstein AN, Massi M (eds) The physiology of thirst and sodium appetite. Plenum Press, New York, pp 521–526

Rondeau JJ, McNicoll N, Escher E, Melonche S, Ong H, de Lean A (1990) Hydrodynamic properties of the angiotensin II receptor from bovine adrenal zona glomerulosa. Biochem J 268: 443–448

Rosengerger L, Triggle DJ (1978) Calcium, calcium translocation, and specific calcium antagonists. In: Weiss GB (ed) Calcium in drug action. Plenum Press, New York, pp 3–31

Rosenthal J, Boucher R, Rojo-Ortega JM, Genest J (1969) Renin activity in aortic tissue of rats. Can J Physiol Pharmacol 47: 53–56

Rosenthal JH, Pfeifle B, Michailov ML, Pschorr J, Jacob ICM, Dahlheim H (1984) Investigations of components of the renin-angiotensin system in rat vascular tissue. Hypertension 6: 383–390

Rosivall L, Carmines PK, Navar LG (1984) Effects of renal arterial angiotensin I infusion on glomerular dynamics in sodium replete dogs. Kidney Int 26: 263–268

Rosset E, Veyrat R (1971) Release of renin by human kidney slices, in vitro effect of angiotensin II, norepenephrine and aldosterone. Rev Enrop Etudes Clin Biol 16: 792–794

Rougeon F, Chambraud B, Foote S, Panthier J-J, Nageotte R, Corvol P (1981) Molecular cloning of a mouse submaxillary gland renin cDNA fragment. Proc Natl Acad Sci USA 78: 6367–6371

Rouzaire-Dubois B, Devynck MA, Chevillotte E, Meyer P (1975) Angiotensin receptors in rat uterine membranes. FEBS Lett 55: 168–172

Rowland NE, Fregly MJ (1988) Characteristics of thirst and sodium appetite in mice (Mus musculus). Behav Neurosci 102: 969–974

Rowland NE, Rozelle A, Riley PJ, Fregly MJ (1992) Effect of nonpeptide angiotensin receptor antagonists on water intake and salt appetite in rats. Brain Res Bull 29: 389–393

Ruyter JNC (1925) Über einen merkwürdigen Abschnitt der Vasa Afferentia in der Mäuseniere. Z Zellforsch 2: 242–248

Ryan GB, Coghlan JP, Scoggins BA (1979) The granulated peripolar epithelial cell: a potential secretory component of the renal juxtaglomerular complex. Nature 277: 655–656

Ryan GB, Alcorn D, Coghlan JP, Hill PA, Jacobs R (1982) Ultrastructural morphology of granule release from juxtaglomerular myoepitheloid and peripolar cells. Kidney Int 22 (Suppl 12): S3-S8

Ryan JW (1967) Renin-like enzyme in the adrenal gland. Science 158: 1589–1590

Ryan JW (1970) Specificity of the renin-like enzyme of rabbit uterus. Biochem J 116: 159–160

Ryan JW, Ferris TF (1967) Release of renin-like enzyme from the pregnant uterus of the rabbit. Biochem J 105: 16c-17c

Ryan JW, Ryan US, Schultz DR, Whitker C, Chung A (1975) Subcellular localization of pulmonary angiotensin-converting enzyme (kininase II). Biochem J 146: 497–499

Saavedra JM (1992) Brain and pituitary angiotensin. Endocr Rev 13: 329–380

Saavedra JM, Chevillard C (1982) Vasopressin reversibly increases angiotensin converting enzyme in specific hypothalamic nuclei of Bratteboro rats. Brain Res 246: 157–160

Saavedra JM, Correa FMA, Plunkett LM, Israel A, Kurihara M, Shigematsu K (1986a) Binding of angiotensin and atrial natriuretic peptide in brain of hypertensive rats. Nature 320: 758–760

Saavedra JM, Israel A, Plunkett LM, Kurihara M, Shigematsu K, Correa FMA (1986b) Quantitative distribution of angiotensin II binding sites in rat brain by autoradiography. Peptide 7: 679–687

Sakai RR, Nicolaidis S, Epstein AN (1986) Salt appetite is suppressed by interference with angiotensin II and aldosterone. Am J Physiol 251: R762-R768

Sakakibara S, Sokabe H, Nakajima T, Khosla MC, Kumagaye S, Watanabe TX (1985) Evolution of angiotensins and some problems in their synthetic preparation. In: Lofts B, Holms WN (eds) Current trends in comparative endocrinology. Hong Kong Univ Press, Hong Kong, pp 1157–1159

Sakuta H, Sekiguchi M, Okamoto K, Sakai Y (1991) Endogenous angiotensin II receptors in *Xenopus* oocytes and eggs. Eur J Pharmacol 208: 31–39

Salazar FJ, Fiksen-Olsen MJ, Opgenorth TJ, Granger JP, Burnett JC Jr, Romero JC (1986) Renal effects of ANP without changes in glomerular filtration rate and blood pressure. Am J Physiol 251: F532-F536

Saltman S, Baukal A, Waters S, Bumpus FM, Catt KJ (1975) Competitive binding activity of angiotensin II analogues in an adrenal cortex radioligand-receptor assay. Endocrinology 97: 275–282

Salzet M, Verger-Bocquet M, Wattez C, Malecha J (1992) Evidence for angiotensin-like molecules in the central nervous system of the leech *Theromyzon tessulatum* (O.F.M.). A possible diuretic effect. Comp Biochem Physiol 101A: 83–90

Salzet M, Wattez C, Baert J-L, Malecha J (1993) Biochemical evidence of angiotensin II-like peptide and proteins in the brain of the rhynchobdellid leech *Theromyzon tessulatum*. Brain Res 631: 247–255

Salzet M, Bulet P, Wattez C, Verger-Bocquet M, Malecha J (1995) Structural characterization of a diuretic peptide from the central nervous system of the leech *Erpobdella octoculata*. J Biol Chem 270: 1575–1582

Sandberg K, Bor M, Ji H, Markwick A, Millan MA, Catt KJ (1990) angiotensin II-induced calcium mobilization in oocytes by signal transfer through gap junctions. Science 249: 298–301

Sandberg K, Ji H, Millan MA (1991) Amphibian myocardial angiotensin II receptors are distinct from mammalian AT_1 and AT_2 receptor subtypes. FEBS Lett 284: 281–283

Sanford B, Stephens GA (1988) The effects of adrenocorticotropin hormone and angiotensin II on adrenal corticosteroid secretions in the freshwater turtle, *Pseudemys scripta*. Gen Comp Endocrinol 72: 107–114

Santos RAS, Baracho NCV (1992) Angiotensin-(1–7) is a potent antidiuretic peptide in rats. Brazilian J Med Biol Res 25: 651–654

Santos RAS, Campagnole-Santos MJ, Baracho NCV, Fontes MAP, Silva LCS, Neves LAA, Oliveira DR, Caligiorne SM, Rodrigues ARV, Gropen Jr C, Carvalho WS, Simoes e Silva AC, Khosla MC (1994) Characterization of a new angiotensin antagonist selective for angiotensin-(1–7): Evidence that the actions of angiotensin-(1–7) are mediated by specific angiotensin receptors. Brain Res Bull 35: 293–298

Saruta T, Matsuki S (1975) The effects of cyclic AMP, theophyline, angiotensin II and electrolytes upon renin release from rat kidney slices. Endocrinol Jpn 22: 137–140

Saruta T, Cook R, Kaplan NM (1972) Adrenocortical steroidogenesis: studies on the mechanism of action of angiotensin and electrolytes. J Clin Invest 51: 2239–2245

Sasaki K, Yamamoto Y, Bardhan S, Iwai N, Murray JJ, Hasegawa M, Matsuda Y, Inagami T (1991) Cloning and expression of a complementary DNA encoding a bovine adrenal angiotensin II type-1 receptor. Nature 351: 230–233

Sasanuma H, Hein L, Krieger JE, Pratt RE, Kobilka BK, Dzau VJ (1992) Cloning, characterization, and expression of two angiotensin receptor (AT-1) isoforms from the mouse genome. Biochem Biophys Res Commun 185: 253–259

225

Saye JA, Singer HA, Peach MJ (1984) Role of endothelium in conversion of angiotensin I to angiotensin II in rabbit aorta. Hypertension 6: 216–221

Saylor DL, Perez RA, Absher DR, Baisden RH, Woodruff ML, Joyner WL, Rowe BP (1992) Angiotensin II binding sites in the hamster brain: Localization and subtype distribution. Brain Res 595: 98–106

Schaffenburg CA, Haas E, Goldblatt H (1960) Concentration of renin in kidneys and angiotensinogen in serum of various species. Am J Physiol 199: 788–792

Schalekamp MADH, Derkx FHM (1993) Renal actions of angiotensin II in man: normal and abnormal. Gower Medical Publ, London pp 27.1–27.9

Schelling P, Ganten D, Heckl R, Hayduk K, Hutchinson JS, Sponer G, Ganten U (1977) On the origin of angiotensin-like peptides in cerebrospinal fluid. In: Buckley JP, Ferrario CM (eds) Central actions of angiotensin and related hormones. Pergamon press, New York, pp 519–526

Schelling P, Sponer G, Ganten U (1978) Components of the renin-angiotensin system in cerebrospinal fluid (CSF) and in periventricular brain tissue. Naunyn-Schmiedebergs Arch Pharmacol 302 (Suppl): R42

Schelling P, Ganten U, Sponer G, Unger T, Ganten D (1980) Components of the renin-angiotensin system in the cerebrospinal fluid of rats and dogs with special consideration of the origin and the fate of angiotensin II. Neuroendocrinology 31: 297–308

Schelling P, Fischer H, Ganten D (1991) Angiotensin and cell growth: a link to cardiovascular hypertrophy? J Hypertens 9: 3–15

Scheuer DB, Thrasher TN, Keil LC, Ramsay DJ (1989) Mechanism of inhibition of renin response to hypotension by atrial natriuretic factor. Am J Physiol 257: R194–R203

Schiavone MT, Santos RAS, Brosnihan KB, Khosla MC, Ferrario CM (1988) Release of vasopressin from the rat hypothalamo-neurohypophysial system by angiotensin-(1–7) heptapeptide. Proc Natl Acad Sci USA 85: 4095–4098

Schiavone MT, Khosla MC, Ferrario CM (1990) Angiotensin-(1–7): evidence for novel actions in the brain. J Cardiovasc Pharmacol 16 (Suppl4): S19–S24

Schiffrin EL, Genest J (1982) Mechanism of captopril-induced drinking. Am J Physiol 242: R136–R140

Schiffrin EL, Genest J (1983) Tonin-angiotensin II system. In: Genest J, Kuchel O, Hamet P, Cantin M (eds) Hypertension, 2nd edn: Tonin-angiotensin II system, Sec 4, Chap 20. McGraw-Hill, New York, pp 309–319

Schiffrin EL, Gutkawska J, Genest J (1981) Mechanism of captopril-induced renin release in conscious rats. Proc Soc Exp Biol Med 167: 327–332

Schiffrin EL, Gutkowska J, Genest J (1984) Effect of angiotensin II and deoxycorticosterone on vascular angiotensin II receptors in rats. Am J Physiol 246: H608–H614

Schiffrin EL, Garcia R, Konrad EM (1993) Interactions of angiotensin II and atrial natriuretic peptide. In: Raizada MK, Phillips MI, Sumners C (eds) Cellular and molecular biology of the renin-angiotensin system. CRC Press, Boca Raton, pp 433–468

Schirar A, Capponi AM, Catt KJ (1980) Elevation of uterine angiotensin II receptors during early pregnancy in the rat. Endocrinology 106: 1521–1527

Schlatter E, Salomonsson M, Persson AEG, Greger R (1989) Macula densa cells sense luminal NaCl concentration via furosemide sensitive Na^+ $2Cl^-$ K^+ cotransport. Pflügers Arch Eur J Physiol 414: 286–290

Schleiffer R, Gairard A (1985) Angiotensin II enhances calcium exchanges in isolated rat aorta. Eur J Pharmacol 111: 129–132

Schnermann J, Persson AEG, Agerup B (1973) Tubuloglomerular feedback. Nonlinear relation between glomerular hydrostatic pressure and loop of Henle perfusion rate. J Clin Invest 52: 862–869

Schoenenberg P, Kehrer P, Müller AF, Gaillard RC (1987) Angiotensin II potentiates corticotropin releasing activity of CRH_{41} in rat anterior pituitary cells: mechanism of action. Neuroendocrinology 45: 86–90

Scholkens BA (1978) Influence of somatostatin on blood pressure and plasma renin activity in the rat. Arzneim Forsch 28: 802–803

Schrager EE, Osborne MJ, Johnson AK, Epstein AN (1975) Entry of angiotensin into cerebral ventricles and circumventricular structures. In: Davies DS, Reid JL (eds) Central actions of drugs in blood pressure regulation. University Park Press, Baltimore, pp 65–67

Schramme C, Denef C (1983) Stimulation of prolactin release by angiotensin II in superfused rat anterior pituitary cell aggregates. Neuroendocrinology 36: 483–485

Schultzberg M (1984) Overviews of colocalization of peptides in the peripheral nervous system. In: Chan-Palay V, Palay SL (eds) Coexistence of neuroactive substances in neurons. John-Wiley & Sons, New York, pp 225–244

Schwalts AG, Lindenmayer GE, Allen JC (1975) The sodium-potassium adenosine triphosphatease: pharmacological, physiological and biochemical aspects. Pharmacol Rev 27: 3–134

Schwartz J, Cherny R (1992) Intercellular communication within the anterior pituitary influencing the secretion of hypophysial hormones. Endocr Rev 13: 453–475

Schwarz E, Schwarz HK (1943) The wild and commercial stocks of the house mouse, *Mus musculus* Linnaeus. J Mammal 24: 59–72

Schwob JE, Johnson AK (1977) Angiotensin-induced dipsogenesis in domestic fowl (*Gallus gallus*). J Comp Physiol Psychol 91: 182–188

Sealey JE, Laragh JH (1975) 'Prorenin' in human plasma? Methodological and physiological implications. Circ Res 36/37: I10–I16

Sealey JE, Clark I, Bull MB, Laragh JH (1970) Potassium balance and the control of renin secretion. J Clin Invest 49: 2119–2127

Sechi LA, Valentin J-P, Griffin CA, Schambelan M (1993) Autoradiographic characterization of angiotensin II receptor subtypes in rat intestine. Am J Physiol 265: G21–G27

Seltzer A, Tsutsumi K, Shigematsu K, Saavedra JM (1993) Reproductive hormones modulate agniotensin II AT receptors in the dorsomedial arcuate nucleus of the female rat. Endocrinology 133: 939–941

Semple PF (1980) The effects of haemorrhage and sodium depletion on plasma concentrations of angiotensin II and [des-Asp^1] angiotensin II in the rat. Endocrinology 107: 771–773

Semple PF, Norton JJ (1976) Angiotensin II and angiotensin III in rat blood. Circ Res 38 (Suppl II): 122–126

Semple PF, Macrae WA, Norton JJ (1980) Angiotensin II in human cerebrospinal fluid may be an immunoassay artifact. Clin Sci 59: 61s–64s

Sen I (1985) Evidence of essential disulfide bonds in angiotensin II binding sites of the rabbit hepatic membranes. Inactivation by dithiothreitol. Biochim Biophys Acta 813: 103–110

Sernia C, Sinton L, Thomas WG, Pascoe W (1985) Liver angiotensin II receptors in the rat: binding properties and regulation by dietary Na$^+$ and angiotensin II. J Endocrinol 106: 103–111

Sernia C, Lello P, Thomas WG (1990) Angiotensin receptors in an Australian marsupial, the brushtail possum, *Trichosurus vulpecula*. Gen Comp Endocrinol 77: 116–126

Setler PE (1971) Drinking induced by injection of angiotensin II into the hypothalamus of the rhesus monkey. J Physiol (Lond)217: 59P–60P

Severs WB, Summy-Long J, Taylor JS, Connor JD (1970) A central effect of angiotensin: release of pituitary pressor material. J Pharm Exp Ther 174: 27–34

Severs WB, Summy-Long J, Daniels-Severs AE (1973) Effect of a converting enzyme inhibitor (SQ20881) on angiotensin-induced drinking. Proc Soc Exp Biol Med 142: 203–204

Sexton JM, Britton SL, Beirwaltes WH, Fiksen-Olsen MJ, Romero JC (1979) Formation of angiotensin III from [des-Asp^1] angiotensin I in the mesenteric vasculature. Am J Physiol 237: H218–H223

Seyama Y, Ishikawa M, Yamashita S (1979) Renin-like activity in the plasma and kidney of a snake, *Elaphe quadrivirgata*. Chem Pharm Bull 27: 2008–2020

Seymour A, Zehr JE (1979) Influence of renal prostaglandin synthesis on renin control mechanisms in the dog. Circ Res 45: 13–25

Sgro S, Ferguson AV, Renaud LP (1984) Subfornical organ-supraoptic nucleus connections: an electrophysiologic study in the rat. Brain Res 303: 7–13

Shade RE, Grim CE (1975) Suppression of renin and aldosterone by small amounts of DOCA in normal man. J Clin Endocrinol Metab 40: 652–658

Shade RE, Davis JO, Johnson JA, Witty RT (1972) Effects of renal arterial infusion of sodium and potassium on renin secretion in the dog. Circ Res 31: 719–727

Shade RE, Davis JO, Johnson JA, Gotshall RW, Spielman WS (1973) Mechanism of action of angiotensin II and antidiuretic hormone on renin sectretion. Am J Physiol 224: 926–929

Shai SY, Fishel RS, Martin BM, Berk BC, Bernstein KE (1992) Bovine angiotensin-converting enzyme cDNA cloning and regulation. Increased expression during endothelial cell growth arrest. Circ Res 70: 1274–1281

Sharpe LG, Swanson LW (1974) Drinking induced by injections of angiotensin into forebrain and mid-brain sites of the monkey. J Physiol 239: 595–622

Shaw HB, Lond MD, Lond FRCP (1906) Auto-intoxication: its relation to certain disturbances of blood pressure. Lancet 1: 1295–1306

Shenker Y, Bates ER, Egan BH, Hammond J, Grekin RJ (1988) Effect of vasopressors on atrial natriuretic factor and hemodynamic function in humans. Hypertension 12: 20–25

Shibata S, Briggs AH (1966) The relationships between electrical and mechanical events in rabbit aortic strips. J Pharmacol Exp Ther 153: 466–470

Shimada K, Yazaki Y (1978) Binding sites for angiotensin II in human mononuclear leukocytes. J Biochem 84: 1013–1015

Shimizu K, Share L, Claybaugh JR (1973) Potentiation by angiotensin II of the vasopressin response to an increasing plasma osmolality. Endocrinology 93: 42–50

Shoji M, Share L, Crofton JT (1989) Effect on vasopressin release of microinjection of angiotensin II into the paraventricular nucleus of conscious rats. Neuroendocrinology 50: 327–333

Shopsin B, Sathananthan G, Gershon S (1973) Plasma renin response to lithium in psychiatric patients. Clin Pharmacol Ther 14: 561–564

Sigmund CD, Gross KW (1991) Structure, expression, and regulation of the murine renin genes. Hypertension 18: 446–457

Silldorff EP, Stephens GA (1992) The pressor response to exogenous angiotensin I and its blockade by angiotensin II analogues in the American alligator. Gen Comp Endocrinol 87: 141–148

Siller WG (1971) Structure of the kidney. In: Bell DG, Freeman BM (eds) Physiology and biochemistry of domestic fowl. Academic Press, New York, pp 197–231

Silverman AJ, Barajas L (1974) Effect of reserpine on the juxtaglomerular cells and renal nerves. Lab Invest 30: 723–731

Silvermann AJ, Hoffman DL, Zimmerman EA (1981) The descending afferent connections of the paraventricular nucleus of the hypothalamus (PVN) Brain Res Bull 6: 47–61

Silverstein E, Friedland J (1990) Angiotensin converting enzyme inhibitor captopril suppresses a genetic polydipsic behavior. Pharm Biochem Behav 37: 831–833

Simerly RB, Swanson LW, Gorski RA (1984) Demonstration of a sexual dimorphism in the distribution of serotonin-immunoreactive fibers in the medial preoptic nucleus of the rat. J Comp Neurol 225: 151–166

Simon E, Gerstberger R, Gray DA (1992) Central nervous angiotensin II responsiveness in birds. Progr Neurobiol 39: 179–207

Simonnet G, Rodriguez F, Fumoux F, Czernichow P, Vincent JD (1979) Vasopressin release and drinking induced by intracranial injection of angiotensin II in monkey. Am J Physiol 237: R20–R25

Simonnet G, Carayon A, Alard M, Cesselin F, Lagoguey A (1984) Evidence for an angiotensin II-like material and for a rapid metabolism of angiotensin II in the rat brain. Brain Res 304: 93–103

Simon-Opperman C, Gray D, Szczepanska-Sadowska E, Simon E (1984) Arterial hypertension at elevated plasma level of Na$^+$, ADH and AII in ducks with high chronic salt intake. Clin Exp Hypertens 6A: 2117–2121

Simon-Oppermann C, Gray DA, Simon E (1986) Independent osmoregulatory control of central and systemic angiotensin II concentration in dogs. Am J Physiol 250: R918–R925

228

Simpson FO, Devine CE (1964) Adrenergic nerve terminals in arterioles of sheep kidney. Proc Univ Otago Med Sch 42: 26–27

Simpson FO, Devine CE (1996) Fine Structure of autonomic neuromuscular contacts in arterioles of sheep renal cortex. J Anat 100: 127–137

Simpson JB, Routtenberg A (1972) The subfornical organ and carbachol-induced drinking. Brain Res 45: 135–152

Simpson JB, Routtenberg A (1973) Subfornical organ: site of drinking elicitation by angiotensin II. Science 181: 1172–1175

Simpson JB, Routtenberg A (1975) Subfornical lesions reduce intravenous angiotensin-induced drinking. Brain Res 88: 154–161

Simpson JB, Epstein AN, Camardo JS (1975) Ablation or competitive blockade of subfornical organ (SFO) prevents thirst of intravenous angiotensin. Fed Proc 34: 374

Singh R, Husain A, Ferrario C, Speth RC (1984) Rat brain angiotensin II receptors: effects of intracerebroventricular angiotensin II infusion. Brain Res 303: 133–139

Singh R, Harding JW, Speth RC (1986) Effect of intraventricular infusion of an angiotensin II antagonist on ^{125}I-angiotensin II binding in the rat. Eur J Pharmacol 120: 319–327

Sirett NE, McLean AS, Bray JJ, Hubbard JI (1977) Distribution of angiotensin II receptors in rat brain. Brain Res 122: 299–312

Sirett NE, Thornton SN, Hubbard JI (1979a) Angiotensin binding and pressor activity in the rat ventricular system and midbrain. Brain Res 166: 139–148

Sirett NE, Thornton SN, Hubbard JI (1979b) Brain angiotensin binding and central sarcosine 8 angiotensin responses in normal rats and the New Zealand strain of genetically hypertensive rats. Clin Sci 56: 607–611

Sirett NE, Bray JJ, Hubbard JI (1981) Localization of immunoreactive angiotensin II in the hippocampus and striatum of rat brain. Brain Res 217: 405–411

Skadhauge E (1981) Osmoregulation in Birds. Springer, Berlin Heidelberg New York, p 22

Skeggs LT, Marsh WH, Kahn JR, Shumway NP (1954) The existence of two forms of hypertensin. J Exp Med 99: 275–282

Skeggs LT, Lentz KE, Kahn JR, Dorer FE, Levine M (1969) Pseudorenin. A new renin-forming enzyme. Circ Res 25: 451–462

Skeggs LT, Dorer FE, Levine M, Lentz E, Kahn JR (1980) The biochemistry of the renin-angiotensin system. In: Johnson JA, Anderson RR (eds) The renin-angiotensin system. Plenum Press, New York pp 1–27

Skinner SL, McCubbin JW, Page IH (1963) Renal baroreceptor control of renin secretion. Science 141: 814–816

Skorecki KL, Ballermann BJ, Rennke HG, Brenner BM (1983) Angiotensin II receptor regulation in isolated renal glomeruli. Feb Proc 42: 3064–3070

Skøtt O (1988) Do osmotic forces play a role in renin secretion? Am J Physiol 255: F1–F10

Skøtt O, Baumbach L (1985) Effects of adenosine on renin release from isolated rat glomerili and kidney slices. Pfluegers Arch Eur J Physiol 404: 232–237

Skøtt O, Taugner R (1987) Effects of extracellular osmolarity on renin release and on the ultrastructure of the juxtaglomerular epithelioid cell granules. Cell Tissue Res 249: 325–329

Smith CL (1966) Rapid demonstration of juxtaglomerular granules in mammals and birds. Stain Technol 41: 291–294

Smith JB (1986) Angiotensin-receptor signaling in cultured vascular smooth muscle cells. Am J Physiol 250: F759–F769

Smith JB, Brock TA (1983) Analysis of angiotensin-stimulated sodium transport in cultured smooth muscle cells from rat aorta. J Cell Physiol 114: 284–290

Smith JB, Smith L, Brown ER, Barnes D, Sabir MA, Davis JS, Farese RV (1984) Angiotensin II rapidly increases phosphatidate-phosphoinositide synthesis and phosphoinositide hydrolysis and mobilizes intracellular calcium in cultured arterial muscle cells. Proc Natl Acad Sci USA 81: 7812–7816

Smith JB, Smith L, Higgins BL (1985) Temperature and nucleotide dependence of calcium release by myo-inositol 1,4,5-triphosphate in cultured vascular smooth muscle cells. J Biol Chem 260: 14413–14416

Smith JM, Mouw DR, Vander AJ (1979) Effect of parathyroid hormone on plasma renin activity and sodium excretion. Am J Physiol 236: F311–F319

Smith RD, Chiu AT, Wong PC, Herblin WF, Timmermans PBMWM (1992) Pharmacology of nonpeptide angiotensin II receptor antagonists. Annu Rev Pharmacol Toxicol 32: 135–165

Smith WL, Bell TG (1978) Immunohistochemical localization of the prostaglandin-forming cyclooxygenase in renal cortex. Am J Physiol 235: F451–F457

Snadberg K, Ji H, Iida T, Catt KJ (1992) Intercellular communication between follicular angiotensin receptors and *Xenopus laevis* oocytes: mediation by an inositol 1,4,5-triphosphate-dependent mechanism. J Cell Biol 117: 157–167

Snapir N, Robinson B, Godschalk M (1976) The drinking response of the chicken to peripheral and central administration of ANG II. Pharm Biochem Behav 5: 5–10

Snyder RA, Watt KWK, Wintroub BV (1985) A human platelet angiotensin I-processing system. Identification of components and inhibition of angiotensin-converting enzyme by product. J Biol Chem 260: 7857–7860

Sobel D, Vagnucci A (1982) Angiotensin II mediated ACTH release in rat pituitary cell culture. Life Sci 30: 1281–1286

Sofroniew MV, Eckstein F, Schrell U, Cuello AC (1984) Evidence for colocalization of neuroactive substances in hypothalamic neurons. In: Chan-Palay V, Palay SL (eds) Coexistence of neuroactive substances in neurons. John Wiley & Sons, New York, pp 73–90

Sokabe H (1974) Phylogeny of the renal effects of angiotensin. Kidney Int 6:263–271

Sokabe H, Nakajima T (1972) Chemical structure and role of angiotensins in the vertebrates. Gen Comp Endocrinol (Suppl 3): 382–392

Sokabe H, Ogawa M (1974) Comparative studies of the juxtaglomerular apparatus. Int Rev Cytol 37: 270–327

Sokabe H, Mizogami S, Murase T, Sakai F (1966) Renin and euryhalinity in the Japanese eel, *Anguilla japonica*. Nature 212: 952–953

Sokabe H, Ogawa M, Oguri M, Nishimura H (1969) Evolution of the juxtaglomerular apparatus in the vertebrate kidneys. Tex Rep Biol Med 27: 867–885

Sokabe H, Nishimura H, Ogawa M, Oguri M (1970) Determination of renin in the corpuscles of Stannius of the teleost. Gen Comp Endocrinol 14: 510–516

Sokabe H, Nishimura H, Kawabe K, Tenmoku S, Arai T (1972) Plasma renin activity in varing hydrated states in bullfrog. Am J. Physiol 222: 142–146

Sokabe H, Oide H, Ogawa M, Utida S (1973) Plasma renin activity in Japanese eels (*Anguilla japonica*) adapted to seawater or in dehydration. Gen Comp Endocrinol 21: 160–167

Somlyo AV, Somlyo AP (1968) Electromechanical and pharmacomechanical coupling in vascular smooth muscle. J Pharmacol Exp Ther 159: 129–145

Song K, Allen AM, Paxinos G, Mendelsohn FAO (1992) Mapping of angiotensin II receptor subtype heterogeneity in rat brain. J Comp Neurol 316: 467–484

Soubrier F, Alhenc-Gelas F, Hubert C, Allegrini J, John M, Tregear G, Corvol P (1988) Two putative active centers in human angiotensin I-converting enzyme revealed by molecular cloning. Proc Natl Acad Sci USA 85: 9386–9390

Speth RC, Harik SI (1985) Angiotensin II receptor binding sites in brain microvessels. Proc Natl Acad Sci USA 82: 6340–6343

Speth RC, Valloton MB, Chernicky C, Khosla MC, Ferrario CM (1983) Angiotensin II receptors in dog brain. Fed Proc 42: 494

Speth RC, Singh R, Smeby RR, Ferrario CM, Husain A (1984) Restricted dietary sodium intake alters peripheral but not central angiotensin II receptors. Neuroendocrinology 38: 387–392

Speth RC, Wamsley JK, Gehlert DR, Chernicky CL, Barnes KL, Ferrario CM (1985) Angiotensin II receptor localization in the canine CNS. Brain Res 326: 137–143

Spinedi E, Negro-Vilar A (1983) Angiotensin II and ACTH release: site of action and potency relative to corticotropin releasing factor and vasopressin. Neuroendocrinology 37: 446–453

Spinedi E, Rodriguez G (1986) Angiotensin II and adrenocorticotropin release: mediation by endogenous corticotropin-releasing factor. Endocrinology 119: 1397–1402

Sraer J, Baud L, Cosyns J-P, Verroust P, Nivez M-P, Ardaillou R (1977) High affinity binding of ^{125}I-angiotensin II to rat glomerular basement membranes. J Clin Invest 59: 69–81

Sraer JD, Sraer J, Ardaillou R, Mimoune O (1974) Evidence for renal glomerular receptors for angiotensin II. Kidney Int 6: 241–246

Stallone JN, Nishimura H, Khosla MC (1988) Angiotensin II vascular receptors in fowl aorta: binding specificity and modulation by divalent cations and guanine nucleotides. J Pharm Exp Ther 251: 1076–1082

Stallone JN, Nishimura H, Nasjletti A (1990) Angiotensin II binding sites in aortic endothelium of the domestic fowl. Am J Physiol 258: R777–R782

Stamler JF, Raizada MK, Fellows RE, Phillips MI (1980) Increased specific binding of angiotensin II in the organum vasculosum of the lamina terminalis area of the spontaneously hypertensive rat brain. Neurosci Lett 17: 173–178

Steele MK (1987) Effects of angiotensins injected into various brain areas on luteinizing hormone release in female rats. Neuroendocrinology 40: 401–405

Steele MK, Ganong WF (1986a) Effects of catecholamine-depleting agents and receptor blockers on basal and angiotensin II- or norepinephrine-stimulated luteinizing hormone release in female rats. Endocrinology 119: 2728–2736

Steele MK, Ganong WF (1986b) Brain-angiotensin system and secretion of luteinizing hormone and prolactin. Front Neuroendocrinol 9: 99–113

Steele MK, Myers LS (1990) In vivo studies on paracrine actions of pituitary angiotensin II in stimulating prolactin release in rats. Am J Physiol 258: E619–E624

Steele MK, Negro-Vilar A, McCaan SM (1981) Effect of angiotensin II on in vivo and in vitro release of anterior pituitary hormones in the female rat. Endocrinology 109: 893–899

Steele MK, McCann SM, Negro-Vilar A (1982) Modulation by dopamine and estradiol of the central effects of angiotensin II on anterior pituitary hormone release. Endocrinology 111: 722–729

Steele MK, Gallo RV, Ganong WF (1983) A possible role for the brain renin-angiotensin system in the regulation of LH secretion. Am J Physiol 245: R805–R810

Steelle MK, Gallo RV, Ganong WF (1985) Stimulatory or inhibitory effects of angiotensin II upon LH secretion in overiectomized rats: a function of gonadal steroids. Neuroendocrinology 40: 210–216

Steele MK, Stephenson KN, Meredith JM, Levine JE (1992) Effects of angiotensin II on LHRH release, as measured by in vitro microdialysis of the anterior pituitary gland of conscious female rats. Neuroendocrinology 55: 276–281

Stephens GA (1981) Blockade of angiotensin pressor activity in the freshwater turtle. Gen Comp Endocrinol 45: 364–371

Stephens GA (1984) Angiotensin and norepinephrine effects on isolated vascular strips from a reptile. Gen Comp Endocrinol 54: 175–180

Stephens GA, Creekmore JS (1984) Response of plasma renin activity to hypotension and angiotensin converting enzyme inhibitor in the turtle. J Comp Physiol 154: 287–294

Stephens GA, Robertson FM (1985) Renal responses to diuretics in the turtle. J Comp Physiol 155: 387–393

Stephens GA, Davis JO, Freeman RH, Watkins BE (1978) Effects of sodium and potassium salts with anions other than cloride on renin secretion in the dog. Am J Physiol 234: F10–F15

Strittmater SM, Lo MM, Javitch JA, Snyder SH (1984) Autoradiographic visualization of angiotensin-converting enzyme in rat brain with [³H] captopril: localization to a striato-nigral pathway. Proc Natl Acad Sci USA 81: 1599–1603

Sturgeon RD, Brophy PD, Levitt RA (1973) Drinking elicited by intracranial microinjection of angiotensin in the cat. Pharm Biochem Behav 1: 353–355

Sugiura M, Cantor EH, Spector S (1982) Angiotensin II ligand binding in brain microvessels of normotensive and hypertensive rats. Fed Proc 41: 1653

Sugiura N, Hagiwara H, Hirose S (1992) Molecular cloning of porcine soluble angiotensin-binding protein. J Biol Chem 267: 18067–18072

Sullivan MJ, Beltz TG, Johnson AK (1990) Amastatin potentiation of drinking induced by blood-borne angiotensin: evidence for mediation by endogenous brain angiotensin. Brain Res 510: 237–241

231

Sumitomo T, Suda T, Nakano Y, Tozawa F, Yamada M, Demura H (1991) Angiotensin II increases the corticotropin-releasing factor messenger ribonucleic acid level in the rat hypothalamus. Endocrinology 128: 2248–2252

Summy-Long J, Severs WB (1974) Angiotensin and thirst: studies with a converting enzyme inhibitor and a receptor antagonist. Life Sci 15: 569–582

Sumners C, Tang W, Zelezna B, Gault T, Raizada MK (1991) Functional angiotensin II receptor subtypes in neuronal and astrocyte glial cultures from one-day-old rat brain. FASEB J 5: A871

Sutherland LE (1966) Immunological and functional aspects of juxtaglomerular cells. Thesis, University of Toronto, Toronto

Suzuki H, Saruta T, Brosnihan KB, Ferrario CM (1987) Angiotensin II-induced drinking in water-deprived and nephrectomized dogs. Jpn Heart J 28: 211–219

Suzuki S, Franco-Saenz R, Mulrow PJ (1981) The role of renal prostaglandins in the renin response to isoproterenol in the rat in vitro. Endocrinology 108: 1654–1657

Swanson GN, Hanesworth JM, Sardinia MF, Coleman JKM, Wright JW, Hall KL, Miller-Wing AV, Stobb JW, Cook VI, Harding EC, Harding JW (1992) Discovery of a distinct binding site for angiotensin II (3–8), a putative angiotensin IV receptor. Regul Pept 40: 409–419

Swanson LW, Sawchenko PE (1983) Hypothalamic integration: organization of the paraventricular and supraoptic nuclei. Annu Rev Neurosci 6: 269–324

Swanson LW, Marshall GR, Needleman P, Sharpe LG (1973) Characterization of central angiotensin II receptors involved in elicitation of drinking in the rat. Brain Res 49: 441–446

Symonds EM, Stanley MA, Skinner SL (1968) Production of renin by in vitro cultures of human chorion and uterine muscle. Nature 217: 1152–1153

Szczepanska-Sadowska E (1991) Hormonal inputs to thirst. In: Ramsay DJ, Booth DA (eds) Thirst-physiological and psychological aspects. Springer Berlin, Heidelberg New York, pp 110–130

Tagawa H, Vander AJ (1970) Effects of adenosine compounds on renal function and renin secretion in dogs. Circ Res 26: 327–338

Tagawa H, Vander AJ, Bonjour JP, Malvin RL (1971) Inhibition of renin secretion by vasopressin in unanesthetized sodium-deprived dogs. Am J Physiol 220: 949–951

Taher MS, McLain KF, McDonald KM, Schrier RW (1976) Effect of beta-adrenergic blockade on renin response to renal nerve stimulation. J Clin Invest 57: 459–465

Takahashi K, Hisa H, Satoh S (1984) Effects of alpha-agonist on renin and prostaglandin E2 release in anesthetized dogs. Am J Physiol 247: E604–E608

Takahashi K, Hisa H, Satoh S (1991) Serotonin-induced renin release in the dog kidney. Eur J Pharmacol 193: 315–320

Takahashi S, Fukamizu A, Hasegawa T, Yokoyama M, Nomura T, Katsuki M, Murakami K (1991) Expression of the human angiotensinogen gene in transgenic mice and transfected cells. Biochem Biophys Res Commun 180: 1103–1106

Takei Y (1977a) Angiotensin and water intake in the Japanese quail (Coturnix coturnix japonica) Gen Comp Endocrinol 31: 364–372

Takei Y (1977b) The role of the subfornical organ in drinking induced by angiotensin in the Japanese quail, Coturnix coturnix japonica. Cell Tissue Res 185: 175–181

Takei Y (1987) In vivo characterization of angiotensin II receptors and converting enzymes in the quail and eel. Zool Sci 4: 1088

Takei Y (1988) Changes in blood volume after alteration of hydromineral balance in conscious eels, Anguilla japonica. Comp Biochem Physiol 91A: 293–297

Takei Y (1992a) Vasoactive hormones and body fluid homeostasis: a phylogenetic consideration. In: Hill RB, Kuwasawa K, McMahon BR, Kuramoto T (eds) Comparative physiology. S Karger, Basel, pp 239–249

Takei Y (1992b) Role of peptide hormones in fish osmoregulation. In: Rankin JC, Jensen FB (eds) Fish ecophysiology. Chapman & Hall, London, pp 136–160

Takei Y, Hasegawa Y (1990) Vasopressor and depressor effects of native angiotensins and inhibition of these effects in the Japanese quail. Gen Comp Endocrinol 79: 12–22

Takei Y, Hatakeyama I (1987) Changes in blood volume after hemorrhage and injection of hypertonic saline in the conscious quail, Coturnix coturnix japonica. Zool Sci 4: 803–811

232

Takei Y, Kobayashi H (1990) Hormonal regulation of water and sodium intake in birds. In: Wada M, Ishii S, Scanes CG (eds) Endocrinology of birds: molecular to behavioral. Japan Sci Soc Press, Tokoyo/Springer, Berlin Heidelberg New York, pp 171-183

Takei Y, Kobayashi H, Yanagisawa M, Bando T (1979a) Involvement of catecholaminergic nerve fibers in angiotensin II-induced drinking in the Japanese quail, *Coturnix coturnix japonica*. Brain Res 174: 229-244

Takei Y, Hirano T, Kobayashi H (1979b) Angiotensin and water intake in the Japanese eel, *Anguilla japonica*. Gen Comp Endocrinol 38: 466-475

Takei Y, Uemura H, Kobayashi H (1985) Angiotensin and hydromineral balance: with special reference tō induction of drinking behavior. In: Lofts B, Holmes WN (eds) Current trends in comparative endocrinology. Hong Kong Univ Press, Hong Kong, pp 933-936

Takei Y, Okawara Y, Kobayashi H (1988a) Water intake induced by water deprivation in the quail, *Coturnix coturnix japonica*. J Comp Physiol B 158: 519-525

Takei Y, Okawara Y, Kobayashi H (1988b) Drinking induced by cellular dehydration in the quail, *Coturnix coturnix japonica*. Comp Biochem Physiol 90A: 291-296

Takei Y, Okubo J, Yamaguchi K (1988c) Effects of cellular dehydration on drinking and plasma angiotensin II level in the eel, *Anguilla japonica*. Zool Sci 5: 43-51

Takei Y, Stallone JN, Nishimura H, Campanile CP (1988d) Angiotensin II receptors in the fowl aorta. Gen Comp Endocrinol 69: 205-216

Takei Y, Okawara Y, Kobayashi H (1989) Control of drinking in birds. In: Hughes MR, Chadwick AC (eds) Progress in avian osmoregulation. Leeds Philos and Literary Soc Ltd, Leeds, pp 1-12

Takei Y, Silldorff EP, Hasegawa Y, Watanabe TX, Nakajima K, Stephens GA, Sakakibara S (1993a) New angiotensin I isolated from a reptile, *Alligator mississippiensis*. Gen Comp Endocrinol 90: 214-219

Takei Y, Hasegawa Y, Watanabe TX, Nakajima K, Hazon N (1993b) A novel angiotensin I isolated from an elasmobranch fish. J Endocrinol 139: 281-285

Takemoto Y, Nakajima T, Hasegawa Y, Watanabe TX, Sokabe H, Kumagae S, Sakakibara S (1983) Chemical structures of angiotensins formed by incubating plasma with the kidney and the corpuscles of stannius in the chum salmon, *Oncorhynchus keta*. Gen Comp Endocrinol 51: 219-227

Tallant EA, Jaiswal N, Diz DI, Ferrario CM (1991) Human astrocytes contain two distinct angiotensin receptor subtypes. Hypertension 18: 32-39

Tanaka J, Kaba H, Saito H, Seto K (1985a) Electro-physiological evidence that circulating angiotensin II sensitive neurons in the subfornical organ alter the activity of hypothalamic paraventricular neurohypophyseal neurons in the rat. Brain Res 342: 361-365

Tanaka J, Kaba H, Saito H, Seto K (1985b) Subfornical organ neurons with efferent projections to the hypothalamic paraventricular nucleus: an electrophysiological study in the rat. Brain Res 346: 151-154

Tanaka T, Gresik E, Michelakis A, Barka T (1980) Immunocytochemical localization of renin in kidneys and submandibular glands of SWR/J and C57B1/6J mice. J Histochem Cytochem 28: 1113-1118

Tang S-S, Rogg H, Schumacher R, Dzau VJ (1992) Characterization of nuclear angiotensin-II-binding sites in rat liver and comparison with plasma membrane receptors. Endocrinology 131: 374-380

Tanigawa H, Allison DJ, Assaykeen TA (1972) Comparison of the effects of various catecholamines on plasma renin activity alone and in the presence of adrenergic blocking agents. In: Genest J, Kiow E (eds) Hypertension 172. Springer, Berlin Heidelberg New York, pp 37-44

Tannenbaum J, Splawinski JA, Oates JA, Nies AS (1975) Enhanced renal prostaglandin production in the dog. I. Effects of renal function. Circ Res 36: 197-203

Tarjan E, Denton DA, McBurnie MI, Weisinger RS (1988) Water and sodium intake of wild and New Zealand rabbits following angiotensin. Peptides 9: 677-679

Taugner C, Poulsen K, Hackenthal E, Taugner R (1979) Immunocytochemical localization of renin in mouse kidney. Histochemistry 62: 19-27

Taugner R, Hackenthal E (1989) The juxtaglomerular apparatus. Springer, Berlin Heidelberg New York, pp 61–67

Taugner R, Hackenthal E, Inagami T, Nobiling R, Poulsen K (1982a) Vascular and tubular renin in the kidneys of mice. Histochemistry 75: 473–484

Taugner R, Hackenthal E, Rix E, Nobiling R, Poulsen K (1982b) Immunocytochemistry of the renin-angiotensin system: renin, angiotensinogen angiotensin I, angiotensin II and converting enzyme in the kidneys of mice, rats, and tree shrews. Kidney Int 22 (Suppl 12): S33–S43

Taugner R, Buhrle CP, Nobiling R (1984a) Ultrastructural changes associated with renin secretion from the juxtaglomerular apparatus of mice. Cell Tissue Res 237: 459–472

Taugner R, Mannek, E, Nobiling R, Bührle CP, Hackenthal E, Ganten D, Inagami T, Schröder H (1984b) Coexistence of renin and angiotensin II in epitheloid cell secretory granules of rat kidney. Histochemistry 81: 39–45

Taugner R, Buhrle CP, Nobiling R, Kirschke H (1985a) Coexistence of renin and catepsin B in epithelioid cell secretory granules. Histochemistry 83: 103–108

Taugner R, Whalley A, Angermüller S, Bührle CP, Hackenthal E (1985b) Are the renin-containing granules of juxtaglomerular epitheloid cells modified lysosomes? Cell Tissue Res 239: 575–587

Taugner R, Yokota S, Buhrle CP, Hackenthal E (1986) Cathepsin D coexists with renin in the secretory granules of juxtaglomerular epithelioid cells. Histochemistry 84: 19–22

Taugner R, Metz R, Rosivall L (1988) Macrophagic phenomena in renin granules. Cell Tissue Res 251: 229–231

Taylor AA, Davis JO, Breitenbach RP, Hartroft PM (1970) Adrenal steroid secretion and a renal-pressor system in the chicken (Gallus domesticus). Gen Comp Endocrinol 14: 321–333

Tewksbury DA, Dart RA, Travis J (1981) The amino terminal amino acid sequence of human angiotensinogen. Biochem Biophys Res Commun 99: 1311–1315

Thames MD, Jarecki M, Donald DE (1978) Neural control of renin secretion in anesthetized dogs. Interaction of cardiopulmonary and carotid baroreceptor. Circ Res 42: 237–245

Thomas WG, Sernia C (1985) Regulation of rat brain angiotensin II (AII) receptors by intravenous AII and low dietary Na^+. Brain Res 345: 54–61

Thomas WG, Greenland KJ, Shinkel TA, Sernia C (1992) Angiotensinogen is secreted by pure rat neuronal cell cultures. Endocrinology 588: 191–200

Thompson CI, Epstein AN (1991) Salt appetite in rat pups: Ontogeny of angiotensin II-aldosterone synergy. Am J Physiol 260: R421–R429

Thornton SN, Baldwin BA (1985) Drinking in the goat in response to simultaneous peripheral and central infusions of angiotensin II. Physiol Behav 35: 753–755

Thrasher TN, Ramsay DJ (1986) The organum vasculosum laminae terminalis. In: de Caro G, Epstein AN, Massi M (eds) The physiology of thirst and sodium appetite. Plenum Press, New York, pp 327–332

Thrasher TN, Simpson JB, Ramsay DJ (1982) Lesions of the subfornical organ block angiotensin induced drinking in the dog. Neuroendocrinology 35: 68–72

Thurau K (1964) Renal hemodynamics. Am J Med 36: 698–719

Thurau K (1975) Modification of angiotensin-mediated tubuloglomerular feedback by extracellular volume. Kidney Int 8: S202–S207

Thurau K, Kramer K (1959) Weitere Untersuchungen zur myogenen Natur der Autoregulation des Nierenkreislaufes Aufhebung der verhalten des Glomerulus-Filtrates. Pflüegers Arch Eur J Physiol 269: 77–93

Tian Y, Balla T, Baukal AJ, Catt KJ (1995) Growth responses to angiotensin II in bovine adrenal glomerulosa cells. Am J Physiol 268: E135–E144

Tigerstedt R, Bergman PG (1898) Niere und Kleislauf. Scand Arch Physiol 8: 223–271

Timmermans PBMWM, Carini DJ, Chiu AT, Duncia JV, Price WA Jr, Wells GJ, Wong PC, Johnson AL, Wexler RR (1991) The discovery of a new class of highly specific nonpeptide angiotensin II receptor antagonists. Am J Hypertens 4: 275S–281S

Timmermans PBMWM, Chiu AT, Herblin WF, Smith RD, Wong PC (1992) Multiple angiotensin receptors: Selective ligands and functional correlates. Jpn J Pharmacol 58 (Suppl 2): 135–141

234

Tobian L, Janecek J, Tomboulian A (1959a) Correlation between granulation of juxtaglo-merular cells and extractable renin in rats with experimental hypertension. Proc Soc Exp Biol Med 100: 94–96

Tobian L, Tomboulian A, Janecek J (1959b) Effect of high perfusion pressure on the granu-lation of juxtaglomerular cells in an isolated kidney. J Clin Invest 38: 605–610

Tobian L, Braden M, Maney J (1964) The effect of unilateral renal denervation on the secretion of renin. J Lab Clin Med 64: 1011

Tokuda C, Kimura K, Kamishima Y (1995) Angiotensin II suppresses water absorption through the ventral skin of Japanese tree-frogs in vitro. Zool Sci 12: 203–206

Trahair JF, Ryan GB (1988) Immunohistochemical identification of plasma proteins in cyto-plasmic granules of peripolar cells of the sheep. J Anat 160: 109–115

Trahair JF, Gall JAM, Alcorn D, Coghlan JP, Fernley R, Perschow J, Grattage LP, Johnston CI, Ryan CB (1989) Immunohistochemical study of peripolar cells of the sheep. J Anat 162: 125–132

Tran D-Y, Hoff KS, Hillyard SD (1992) Effects of angiotensin II and bladder condition on hydration behavior and water uptake in the toad, *Bufo woodhousei*. Comp Biochem Physiol 103A: 127–130

Trippodo NC, McCaa RE, Guyton AC (1976) Effect of prolonged angiotensin II infusion on thirst. Am J Physiol 230: 1063–1066

Tronik D, Dreyfus M, Babinet C, Rougeon F (1987) Regulated expression of the Ren-2 gene in transgenic mice derived from parental strains carrying only the Ren-1 gene. EMBO J 6: 983–987

Tsien R, Pozzan T, Rink TJ (1982) Calcium homeostasis in intact lymphocytes: cytoplasmic free calcium monitored with a new, intracellularly trapped fluorescent indicator. J Cell Biol 94: 325–334

Tsuneki K, Takei Y, Kobayashi H (1978) Parenchymal fine structure of the subfornical organ in the Japanese quail (*Coturnix coturnix japonica*). Cell Tissue Res 191: 405–419

Tsutsumi K, Saavedra JM (1991) Angiotensin-II receptor subtypes in median eminence and basal forebrain areas involved in regulation of pituitary function. Endocrinology 129: 3001–3008

Tuominen RK, Hudson PM, McMillan MK, Ye H, Stachowiak MK, Hong J-S (1991) Long-term activation of protein kinase C by angiotensin II in cultured bovine adrenal medullary cells. J Neurochem 56: 1292–1298

Ueda H (1976) Renin and nervous system. Jpn Heart J 17; 521–526

Ueda H, Yasuda H, Takabatake Y, Iizuka M, Iizuka T, Ihori M, Yamamoto M, Sabamoto Y (1967) Increased renin release evoked by mesencephalic stimulation in the dog. Jpn Heart J 8: 498–506

Ueda H, Nakanishi H, Abe Y (1978) Effect of glucagon on renin secretion in the dog. Eur J Pharmacol 52: 85–92

Uemura H (1964) Effects of water deprivation on the hypothalamo-hypophysial neurosecre-tory system of the grass parakeet, *Melopsittacus undulatus*. Gen Comp Endocrinol 4: 193–198

Uemura H, Kobayashi H, Okawara Y, Yamaguchi K (1983) Neuropeptides and drinking in birds. In: Mikami S, Honma K, Wada M (eds) Avian endocrinology. Japan Sci Soc Press, Tokyo/Springer, Berlin Heidelberg New York, pp 255–262

Uemura H, Tezuka Y, Hasegawa C, Kobayashi H (1994) Immunohistochemical investigation of neuropeptides in the central nervous system of the amphioxus, *Branchiostoma belcheri*. Cell Tissue Res 277: 279–287

Umezawa H, Aoyagi T, Suda H, Hamada M, Takeuchi T (1976) Bestatin, an inhibitor of aminopeptidase B, produced by actinomycetes. J Antibiotics 29: 97–99

Unger T, Ganten D, Ludwig G, Lang RE (1986) The brain-renin–angiotensin system: update. In: de Caro G, Epstein AN, Massi M (eds) The physiology of thirst and sodium appetite. NATO ASI Ser, Ser A Life Sci, vol 105. Plenum Press, New York, pp 123–133

Uno S, Guo D-F, Nakajima M, Ohi H, Imada T, Haramatsu R, Nakakubo H, Nakamura N, Inagami T (1994) Glucocorticoid induction of rat angiotensin II type 1_A receptor gene promoter. Biochem Biophys Res Commun 204: 210–215

Unsicker K, Axelsson S, Owman Ch, Svensson K-G (1975) Innervation of the male genital tract and kidney in the amphibia, *Xenopus laevis* Daudin, *Rana temporaria* L., and *Bufo bufo* L. Cell Tissue Res 160: 453–484

Urata H, Kinoshita A, Misono KS, Bumpus FM, Husain A (1990) Identification of a highly specific chymase as the major angiotensin II-forming enzyme in the human heart. J Biol Chem 265: 22348–22357

Uva B, Vallarino M (1982) Renin-angiotensin system and osmoregulation in the terrestrial chelonian *Testudo hermanni* Gamelin. Comp Biochem Physiol 71A: 449–451

Uva B, Masini MA, Hazon N, O'Toole LB, Henderson IW, Ghiani P (1992) Renin and angiotensin converting enzyme in elasmobranchs. Gen Comp Endocrinol 86: 407–412

Vale W, Vaughan J, Smith M, Yamamoto G, Rivier J, Rivier C (1983) Effects of synthetic ovine corticotropin-releasing factor, glucocorticoids, catecholamines neurohypophyseal peptides and other substances on cultured corticotropin cells. Endocrinology 113: 1121–1131

Vallarino M (1984) Seasonal kidney and plasma renin concentration in *Testudo hermanni* Gmelin. Comp Biochem Physiol 79A: 529–531

Vallotton MB (1987) The renin-angiotensin system. Trends Pharmacol Sci 8: 69–74

Van Breemen CP, Aaronson CP, Loutzenhiser R (1979) Sodium-calcium interactions in mammalian smooth muscle. Pharmacol Rev 30: 167–208

Van Dongen WJ, Van der Heijden CA (1969) Demonstration of renal juxtaglomerular granules and evaluation of index of granulation in toad, *Bufo bufo*. Z Zellforsch 94: 40–45

Van Eekelen J, Anke M, Phillips MI (1988) Plasma angiotensin II levels at moment of drinking during angiotensin II intravenous infusion. Am J Physiol 255: R500-R506

Van Houten M, Schiffrin EL, Mann JFE, Posner BI, Boucher R (1980) Radioautographic localization of specific binding sites for blood-borne angiotensin II in the rat brain. Brain Res 186: 480–485

Vander AJ (1965) Effect of catecholamines and the renal nerves on renin secretion in anesthetized dogs. Am J Physiol 209: 659–662

Vander AJ (1967) Control of renin release. Physiol Rev 47: 359–382

Vander AJ (1968) Inhibition of renin release in the dog by vasopressin and vasotocin. Circ Res 23: 605–609

Vander AJ (1970) Direct effects of potassium on renin secretion and renal function. Am J Physiol 219: 445–459

Vander AJ, Carlson J (1969) Mechanism of the effects of frusemide on renin secretion in anesthetized dogs. Circ Res 25: 145–152

Vander AJ, Geelhoed GW (1965) Inhibition of renin secretion by angiotensin II. Proc Soc Exp Biol Med 120: 393–403

Vander AJ, Miller R (1964) Control of renin secretion in the dog. Am J Physiol 207: 537–545

Vandongen R (1975) Inhibition of renin secretion in the isolated rat kidney by antidiuretic hormone. Clin Sci Mol Med 49: 73–76

Vandongen R, Greenwood DM (1975) The stimulation of renin secretion by non-vasoconstrictor infusion of adrenaline and noradrenaline in the isolated rat kidney. Clin Sci Mol Med 49: 609–612

Vandongen R, Peart WS (1974) The inhibition of renin secretion by alpha-adrenergic stimulation in the isolated rat kidney. Clin Sci Mol Med 47: 471–479

Vandongen R, Peart WS, Boyd GW (1973) Adrenergic stimulation of renin secretion in the isolated perfused rat kidney. Circ Res 32: 290–296

Vandongen R, Peart WS, Boyd GW (1974) Effect of angiotensin II and its nonpressor derivatives on renin secretion. Am J Physiol 226: 277–282

Vandongen R, Tunney A, Mahoney D, Barden A (1981) Dissociation of beta-adrenergic stimulation of renin secretion and prostacyclin synthesis in the rabbit kidney. Prostaglandins 21: 1007–1014

Vaughan ED Jr, Gavras H, Laragh JH, Koss MN (1973) Vascular permeability factor dissociation from the angiotensin II induced pressor and drinking responses. Nature 242: 334–335

Velletri PA, Aquilano DR, Bruckwick E, Tsai-Morris CH, Dufau ML, Lovenberg W (1985) Endocrinological control and cellular localization of rat testicular angiotensin-converting enzyme (EC 3.4.15.1). Endocrinology 116: 2516–2522

Verger-Bocquet M, Wattez C, Salzet M, Malecha J (1992) Immunocytochemical identification of peptidergic neurons in compartment 4 of the supraesophageal ganglion of the leech *Theromyzon tessulatum* (O.F.M.). Can J Zool 70: 856–865

Vesely DL (1981) Angiotensin II stimulates guanylate cyclase activity in aorta, heart, and kidney. Am J Physiol 240: E391–E393

Villarreal D, Freeman RH, Davis JO, Diets JR, Echtenkamp SF (1982) Effects of sodium chloride on prostacyclin-stimulated renin release in dogs with filtering and nonfiltering kidneys. Proc Soc Exp Biol Med 171: 34–40

Vincent S, Sheen W (1903) The effect of intravascular injections of extracts of animal tissues. J Physiol (Lond) 29: 242–265

Viskoper RJ, Maxwell MH, Lupu AN, Rossenfield S (1977) Renin stimulation by isoproterenol and theophylline in the isolated perfused kidney. Am J Physiol 232: F248–F253

Volicer L, Loew C (1971) Penetration of angiotensin II into the brain. Neuropharmacology 10: 631–636

Volmert RF, Firman JD (1991) Water and NaCl intake of chicks as mediated by angiotensin II, renin, or salt deficiency. Physiol Behav 50: 921–927

Volpe M, Pepino P, Lambo G, Pignalosa S, Mele AF, Rubattu S, Condorelli G, Covino E, Trimarco B (1991) Modulatory role of angiotensin-II in the secretion of atrial natriuretic factor in rabbits. Endocrinology 128: 2427–2431

Voth D, Agusten M, Schipp R, Lübcke H (1971) Untersuchungen zur Wirkung von Angiotensin II auf einen autorhythmischen glatten Gefässmuskel (Pfortader der Ratte). Arch Kreislaufforsch 65: 41–57

Wada M, Kobayashi H, Farner DS (1975) Induction of drinking in the white-crowned sparrow, *Zonotrichia leucophrys gambelii*, by intracranial injection of angiotensin II. Gen Comp Endocrinol 26: 192–197

Waeber B, Nussberger J, Brunner HR (1990) Angiotensin-converting enzyme inhibitors in hypertension. In: Laragh JH, Brenner BM (eds) Hypertension: pathophysiology, diagnosis, and management. Raven Press, New York, pp 2209–2232

Wågermark J, Ungerstedt U, Ljungqvist A (1968) Sympathetic innervation of the juxtaglomerular cells of the kidney. Circ Res 22: 149–153

Wallis CJ, Printz MP (1980) Adrenal regulation of regional brain angiotensinogen content. Endocrinology 106: 337–342

Warren DJ, Ferris TE (1970) Renin secretion in renal hypertension. Lancet 1: 159–162

Watkins BE, Davis JO, Lohmeier TE, Freeman RH (1976) Intrarenal site of action of calcium on renin secretion in dogs. Circ Res 39: 847–853

Weber MA, Stokes GS, Gain JM (1974) Comparison of the effects on renin release of beta adrenergic antagonist with differing properties. J Clin Invest 54: 1413–1419

Weekley LB (1984) Central neuroendocrine regulation of renin secretion rate: a hypothesis. J Theor Biol 111: 609–613

Wei L, Alhenc-Gelas F, Corvol P, Clauser E (1991) The two homologous domains of human angiotensin I-converting enzyme are both catalytically active. J Biol Chem 266: 9002–9008

Weichert G (1965) Über die Wirkung fractionierter Nieren extrakte von Vertebraten verschiedener Evolutionsstadien auf den Blutdruck der nephrektomierten Ratte. Pflügers Arch 284: 147–159

Weidmann P, Massry SG, Coburn JW, Maxwell MH, Atleson J, Kleeman CR (1972) Blood pressure effects of acute hypercalcemia. Studies in patients with chronic renal failure. Ann Intern Med 76: 741–745

Weinberger MH, Aoi W, Henry DP (1975) Direct effect of beta-adrenergic stimulation on renin release by the rat kidney slice in vitro. Circ Res 37: 318–324

Weindl A (1973) Neuroendocrine aspects of circumventricular organs. In: Ganong WF, Martini L (eds) Frontiers in neuroendocrinology. Oxford Univ Press, New York, pp 3–32

Weindl A, Schweisfurth H, Sofroniew MV, Dahlheim H (1977) Distribution of converting enzyme in the rat brain: high activities in the subfornical organ, area postrema, and choroid plexus. Acta Endocrinol 85 (Suppl 212): 158

Weisinger RS, Denton DA, McKinley MJ (1977) Inhibition of water intake by ouabain administration in sheep. Pharmacol Biochem Behav 7: 121–128

Weisinger RS, Considine P, Denton DA, Leksell LG, McKinley MJ, Mouw D, Müller A, Tarjan E (1982) Sodium concentration of the cerebrospeinal fluid in the salt appetite of sheep. Am J Physiol 242: R51–R63

Weisinger RS, Denton DA, McKinley MJ, Müller AF, Targan E (1986) Angiotensin and sodium appetite of sheep. Am J Physiol 251: R690–R699

Weisinger RS, Denton DA, DiNicolantonio R, McKinley MJ, Müller AF, Tarjan E (1987) Role of angiotensin in sodium appetite of sodium-deplete sheep. Am J Physiol 253: R482–R488

Weisinger RS, Denton DA, DiNicolantonio R, McKinley MJ (1988) The effect of captopril or enalaprilic acid on the Na appetite of Na deplete rats. Clin Exp Pharmacol Physiol 15: 55–65

Welches WR, Santos RAS, Chappell MC, Brosnihan KB, Greene LJ, Ferrario CM (1991) Evidence that prolyl endopeptidase participates in the processing of brain angiotensin. J Hypertens 9: 631–638

Weld MM, Fryer JN (1987) Stimulation by angiotensin I and II of ACTH release from goldfish pituitary cell columns. Gen Comp Endocrinol 68: 19–27

White CS, Levitt RA, Boyer S (1972) Drinking elicited by CNS injection of angiotensin in the rat. Psychon Sci 26: 283–284

Wideman RF Jr, Braun EJ, Anderson GL (1981) Microanatomy of the renal cortex in the domestic fowl. J Morphol 168: 249–267

Wideman RF Jr, Nishimura H, Bottje WG, Glahn RP (1993) Reduced renal arterial perfusion pressure stimulates renin release from domestic fowl kidneys. Gen Comp Endocrinol 89: 405–414

Wiegmann TB, MacDougall ML, Savin VJ (1990) Glomerular effects of angiotensin II require intrarenal factors. Am J Physiol 258: F717–F721

Wilcox CS (1978) Effect of increasing the plasma magnesium concentration on renin release from the dog kidney: Interactions with calcium and sodium. J Physiol (Lond) 284: 203–217

Wilcox CS, Aminoff MJ, Kurtz AB, Slater JDH (1974) Comparison of the renin response to dopamine and noradrenaline in normal subjects and patients with automonic insufficiency. Clin Sci Mol Med 46: 481–488

Williams TA, Hooper NM, Turner AJ (1991) Characterization of neuronal and endothelial forms of angiotensin converting enzyme in pig brain. J Neurochem 57: 193–199

Williams GH, McDonnell LM, Raux MC, Hollenberg NK (1974) Evidence for different angiotensin II receptors in rat adrenal glomerulosa and rabbit vascular smooth muscle cells. Circ Res 34: 384–390

Willoughby EJ, Peaker M (1979) Birds. In: Maloiy GMO (ed) Comparative physiology of osmoregulation in animals, vol 2. Academic Press, New York, pp 1–55

Wilson CM, Cherry M, Taylor BA, Wilson JD (1981) Genetic and endocrine control of renin activity in the submaxillary gland of the mouse. Biochem Genet 19: 509–523

Wilson JX (1984a) The renin-angiotensin system in nonmammalian vertebrates. Endocr Rev 5: 45–61

Wilson JX (1984b) Coevolution of the renin-angiotensin system and the nervous control of blood circulation. Can J Zool 62: 137–147

Wilson KM, Sumners C, Hathaway S, Fregly MJ (1986) Mineralocorticoids modulate central angiotensin II receptors in rats. Brain Res 382: 87–96

Wilson W (1952) A new staining method for demonstrating the granules of the granules of the juxtaglomerular complex. Anat Rec 112: 497–507

Winer N, Choksi DS, Walkenhorst WG (1971) Effects of cyclic AMP, sympathomimetic amines and adrenergic receptor antagonists on renin secretion. Circ Res 29: 239–248

Wintroub BU, Klickstein LB, Kaempfer CE, Austin KF (1981) A human neutrophil-dependent pathway for generation of angiotensin II; purification and physiochemical characterization of the plasma protein substrate. Proc Natl Acad Sci USA 78: 1204–1208

Witty RT, Davis JO, Shade RE, Johnson JA, Prewitt RL (1972) Mechanisms regulating renin release in the dog with thoracic caval constriction. Circ Res 32: 339–347

Wolf G, Handel PJ (1966) Aldosterone-induced sodium appetite: Dose-response and specificity. Endocrinology 78: 1120–1124

Woodward RM, Miledi R (1991) Angiotensin II receptors in *Xenopus* oocytes. Proc Soc Lond B 244: 11–19

Worley RTS, Rich GT, Pryor JS (1978) Effect of calcium ionophore Br-X537a on renin synthesis and release in *Amphiuma means* kidney culture. Nature 271: 174–176

Wright GB, Alexander RW, Ekstein LS, Gimbrone MA Jr (1982) Sodium, divalent cations, and guanine nucleotides regulated the affinity of the rat mesenteric artery angiotensin II receptor. Circ Res 50: 462–469

Wright JW, Harding JW (1988) A reevaluation of angiotensin III's potency as a pressor and dipsogenic agent in normotensive and hypertensive animal models. In: Harding JW, Wright JW, Speth RC, Barnes CD (eds) Angiotensin and blood pressure regulation. Academic Press, New York, pp 209–230

Wright JW, Harding JW (1992) Regulatory role of brain angiotensins in the control of physiological and behavioral responses. Brain Res Rev 17: 227–262

Wright JW, Morseth S, Mana MJ, LaCrosse E, Petersen EP, Harding JW (1984) Central angiotensin III induced dipsogenicity in rats and gerbils. Brain Res 295: 121–126

Wright JW, Sullivan MJ, Petersen EP, Harding JW (1985) Brain angiotensin II and III binding and dipsogenicity in the rabbit. Brain Res 358: 376–379

Wright JW, Morseth SL, Sallivan MJ, Harding JW (1986) Comparison of angiotensin II and III induced dipsogenicity and pressor action. In: de Caro G, Epstein AN, Massi M (eds) The physiology of thirst and sodium appetite. NATO ASI Ser, Ser A Life Sci, Vol 105. Plenum Press, New York, pp 187–192

Yamada C, Kobayashi H (1987) Immunoreactive angiotensin II in the corpuscles of Stannius of the rainbow trout, *Salmo gairdneri.* Zool Sci 4: 387–390

Yamada C, Noji S, Shioda S, Nakai Y, Kobayashi H (1990) Intragranular colocalization of arginine vasopressin- and angiotensin II-like immunoreactivity in the hypothalamo-neurohypophysial system of the goldfish, *Carassius auratus.* Zool Sci 7: 257–263

Yamaguchi K (1981) Effects of water deprivation on immunoreactive angiotensin II levels in plasma, cerebroventricular perfusate and hypothalamus of the rat. Acta Endocrinol 97: 137–144

Yamaguchi K, Nishimura H (1988) Angiotensin II-induced relaxation of fowl aorta. Am J Physiol 255: R591–R599

Yamaguchi K, Negoro H, Hama H, Kamoi K, Sakaguchi T (1979) A comparison between the effects of angiotensin II and angiotensin III injected into the third ventricle on vasopressin secretion in conscious rats. Folia Endocrinol Jpn 55: 1286–1295

Yamaguchi K, Hama H, Sakaguchi T, Negoro H, Kamoi K (1980) Effects of intraventricular injection of Sar^1-Ala^8-angiotensin II on plasma vasopressin level increased by angiotensin II and by water deprivation in conscious rats. Acta Endocrinol 93: 407–412

Yamaguchi K, Koike M, Hama H (1985) Plasma vasopressin response to peripheral administration of angiotensin in conscious rats. Am J Physiol 248: R249–R256

Yamaguchi K, Karakida T, Koike M, Hama H (1988) Central effects of catecholamine antagonists on angiotensin-induced vasopressin secretion in conscious rats. Acta Endocrinol (Copenh) 118: 82–86

Yamamoto K, Hasegawa T, Miyazaki M, Ueda J (1969) Control of renin secretion in the anesthetized dog. II. Relationships between renin secretion, plasma sodium concentration and GFR in the perfused kidney. Jpn Circ J 33: 593–600

Yamamoto K, Okahara T, Abe Y, Ueda J, Kishimoto T, Morimoto S (1973) Effects of cyclic AMP and dibutyryl cyclic AMP on renin release in vivo and in vitro. Jpn Circ J 37: 1271–1276

Yamamoto K, Iwao H, Abe Y, Morimoto S (1984) Effect of Ca on renin release in vitro and in vivo. Jpn Circ J 38: 1127–1131

Yamashita H, Osaka T, Kannan H (1984) Effects of electrical and chemical stimulation of the paraventricular nucleus on neurons in the subfornical organ of cats. Brain Res 323: 176–180

Yanagawa N, Capparelli AW, Jo OD, Freidal A, Barrett JD, Eggena P (1991) Production of angiotensiongen and renin-like activity by rabbit proximal tubular cells in culture. Kidney Int 39: 938–941

Yanagisawa M, Kurihara H, Kimura S, Tomobe Y, Kobayashi M, Mitsui Y, Yazaki Y, Goto K, Masaki T (1988) A novel potent vasoconstrictor peptide produced by vascular endothelial cells. Nature 332: 411–415

Yang H-YT, Neff NH (1972) Distribution and properties of angiotensin converting enzyme of rat brain. J Neurochem 19: 2443–2450

Yang Z-F, Epstein AN (1991) Blood-borne and cerebral angiotensin and the genesis of salt intake. Horm Behav 25: 461–476

Ye MQ, Healy DP (1992) Characterization of an angiotensin type-1 receptor partial cDNA from rat kidney: evidence for a novel AT_{1B} receptor subtype. Biochem Biophys Res Commun 185: 204–210

Yokosawa H, Inagami T, Haas E (1978) Purification of human renin. Biochem Biopys Res Commun 83: 306–312

Young CE, McDonald IR (1978) The effect of intravenous infusion of angiotensin II in drinking in the Australian marsupial *Trichosurus vulpecula*. J Physiol 280: 77–85

Young DB, Guyton DB (1977) Steady state aldosterone dose-response relationships. Circ Res 40: 138–142

Yung WH, Chiu KW (1985) Contractile response of the isolated dorsal aorta of the snake to angiotensin II and norepinephrine. Gen Comp Endocrinol 60: 259–265

Zanchetti A, Stella A (1975) Neural control of renin release. Clin Sci Mol Med 48: 215S–225S

Zawada Jr ET, Johnson M (1984) The effects of changes in serum calcium and parathormone on plasma renin activity in intact mongrel dogs. Adv Exp Med Biol 178: 311–318

Zehr JE, Feigl EO (1973) Suppression of renin activity by hypothalamic stimulation. Circ Res 32/33: I17–I27

Zelcer E, Sperelakis N (1981) Angiotensin induction of active responses in cultured reaggregates of rat aortic smooth muscle cells. Blood vessels 18: 263–279

Zhuo J, Mendelsohn FAO (1993) Intrarenal angiotensin II receptors. In: Robertson JIS, Nicholls MG (eds) The renin–angiotensin system. Gower Medical Publ, London, pp 25.1–25.14

Zimmerman KW (1933) Über den Bau des Glomerulus der Säugetiere. Z Mikrosk Anat Forsch 32: 176–278

Subject Index

241

244

Springer-Verlag
and the Environment

We at Springer-Verlag firmly believe that an international science publisher has a special obligation to the environment, and our corporate policies consistently reflect this conviction.

We also expect our business partners – paper mills, printers, packaging manufacturers, etc. – to commit themselves to using environmentally friendly materials and production processes.

The paper in this book is made from low- or no-chlorine pulp and is acid free, in conformance with international standards for paper permanency.